Cambridge Studies in Historical Geography 15

THE UNDERDRAINING OF FARMLAND IN ENGLAND DURING THE NINETEENTH CENTURY

Cambridge Studies in Historical Geography

Series editors:
ALAN R. H. BAKER   J. B. HARLEY   DAVID WARD

Cambridge Studies in Historical Geography encourages exploration of the philosophies, methodologies and techniques of historical geography and publishes the results of new research within all branches of the subject. It endeavours to secure the marriage of traditional scholarship with innovative approaches to problems and to sources, aiming in this way to provide a focus for the discipline and to contribute towards its development. The series is an international forum for publication in historical geography which also promotes contact with workers in cognate disciplines.

THE UNDERDRAINING OF FARMLAND IN ENGLAND DURING THE NINETEENTH CENTURY

A. D. M. PHILLIPS

Lecturer in Historical Geography, University of Keele

The right of the
University of Cambridge
to print and sell
all manner of books
was granted by
Henry VIII in 1534.
The University has printed
and published continuously
since 1584.

CAMBRIDGE UNIVERSITY PRESS

CAMBRIDGE
NEW YORK   PORT CHESTER
MELBOURNE   SYDNEY

CAMBRIDGE UNIVERSITY PRESS
Cambridge, New York, Melbourne, Madrid, Cape Town, Singapore, São Paulo, Delhi

Cambridge University Press
The Edinburgh Building, Cambridge CB2 8RU, UK

Published in the United States of America by Cambridge University Press, New York

www.cambridge.org
Information on this title: www.cambridge.org/9780521105804

First published 1989
This digitally printed version 2009

A catalogue record for this publication is available from the British Library

Library of Congress Cataloguing in Publication data
Phillips, A. D. M.
The underdraining of farmland in England during the nineteenth
century / A. D. M. Phillips.
    p.   cm. – (Cambridge studies in historical geography: 15)
Bibliography.
Includes index.
ISBN 0-521-36444-2
1. Drainage – Economic aspects – England – History – 19th century.
2. Agriculture – Economic aspects – England – History – 19th century.
3. Farms – England – History – 19th century. I. Title. II. Series.
HD1683.G62E56 1989
338.1′62–dc19   88-34698 CIP

ISBN 978-0-521-36444-7 hardback
ISBN 978-0-521-10580-4 paperback

Contents

Figures

Tables

Preface

The theme of land drainage has long attracted the attention of historians and historical geographers alike. However, the compass of the theme is broad and, from the nineteenth century, recognition has been made of distinct, component parts of the process. J. Bailey Denton, writing in 1861, suggested a fourfold schema of land drainage that still holds currency: the drainage and reclamation of fenland and marshland; the control and management of rivers in their valleys; the provision and maintenance of ditches and minor watercourses; and the underdraining of farmland. The last stage of this classification forms the subject-matter of this book. The study considers the need for and spread of underdraining in England in the nineteenth century; examines the technical and economic factors in the adoption of the improvement; and assesses the impact of underdraining on agricultural practice in that period.

In writing this book, great reliance has been placed on data derived from records of estates and of the loans for underdraining made by the government and various land-improvement companies in the nineteenth century. The gathering of such material was facilitated by the detailed knowledge and kindness of the archivists and staff of the county record offices at Exeter, Newcastle upon Tyne and Northampton, of the Department of Palaeography and Diplomatic of Durham University, and of the Public Record Office. I owe special thanks to P. I. King, R. M. Gard and J. Fewster for directing me to relevant and often uncatalogued manuscript collections. However, this account could not have been produced without the generosity and co-operation of owners of private collections of manuscripts. I am deeply indebted to the following for access to material in their possession: Lord Barnard (the Raby papers); the Duke of Buccleuch (the Buccleuch Barnwell and Boughton papers); H.R.H. the Duke of Cornwall (the Duchy of Cornwall papers); Mr B. Fisher (the Lands Improvement Company papers); Lord Howick (the Grey papers at Howick); the late Duke of Northumberland (the Northumberland papers); the late seventh Earl Spencer (the Spencer papers); the Marquess of Tavistock and the trustees of the Bedford estates (the Bedford papers); and Sir Richard Baker Wilbraham (the Wilbraham papers at Rode Hall).

My interest in land drainage was first aroused while undertaking an undergraduate study of mossland reclamation in nineteenth-century Cheshire. It was channelled into underdraining by H. C. Darby, who suggested the topic for postgraduate research. Over the years, friends and colleagues in the departments of Geography first at University College London and subsequently at Keele University have aided in the unblocking of many draining problems. Discussions with Michael Thompson on the nature of landlord investment in nineteenth-century agriculture have proved invaluable, while at the beginning of the project Ted Collins and Eric Jones offered stimulating advice on the role of underdraining as an agricultural improvement. At various stages, Hugh Clout, Michael Jahn and John Walton not only helped in the collection of data but also provided much sound counsel. The maps and diagrams were drawn by Muriel Patrick of Keele University with her customary skill and care. My greatest debt, however, is to Hugh Prince, who supervised my postgraduate work and encouraged its transformation to book form.

Abbreviations

AH	*Agricultural History*
AHR	*Agricultural History Review*
BPP	British Parliamentary Papers
DPD	Department of Palaeography and Diplomatic, Durham University
DRO	Devon Record Office
EcHR	*Economic History Review*
JRASE	*Journal of the Royal Agricultural Society of England*
KU	Keele University
NRO	Northamptonshire Record Office
NdRO	Northumberland Record Office
PRO	Public Record Office

1

Debates about underdraining

In the analysis of eighteenth- and nineteenth-century agriculture, considerable use has been made of the dichotomy between heavy lands and light lands to explain changes in farming systems and agricultural productivity.[1] The free-draining light lands had experienced marked agricultural progress both economically and technically from at least the seventeenth and eighteenth centuries, founded on the adoption of a system of grain and livestock farming being integrated by the use of rotations which incorporated the growth of cereals and fodder crops, especially the turnip. It has generally been argued that the light-land mixed-farming systems based on turnip husbandry and high feeding made that sector of agriculture more dynamic, productive and prosperous than any other in the eighteenth century and for the greater part of the following century.[2]

Agricultural systems on the clay-based heavy lands were much less advanced. Both the heaviness and moisture-retentiveness of such soils made them difficult to work, compressed the working season and rendered them unsuitable for the growth of fodder crops, especially turnips, for feeding stock through the winter. As a result, farming practices on heavy lands lacked the flexibility of those on light lands. On arable, rotations were dominated by wheat, oats or beans, and a bare fallow. Wheat was recognized as the main cash product and fallows persisted as a means of cleansing land after grain crops, being accepted as the penalty for the wheat crop. In grassland areas, meadow and pasture were strictly delimited and immune from the plough. Winter fodder came from meadow land, not fodder crops, and both the area of profitable summer grazing and the number of stock were restricted. The inability to grow fodder crops prevented farmers on heavy lands adopting the mixed-farming systems of the light lands and limited their development of more profitable enterprises. The technical solution to the problems of the heavy lands was identified as the adoption of underdraining, which aimed at remedying the physical difficulties inherent in such soils and represented a major advance both in efficiency and the conservation of cultivated land over existing methods of surface draining.[3]

Underdraining is an agricultural technique to improve the physical condition of soils for crop production. Its aim is to remove surplus water from soil.[4] Generally, water can enter the soil from above by rain or from below through rising groundwater. Both sources of water, if not removed or disposed only slowly, cause the water table within the soil to rise and eventually produce surface waterlogging, so inhibiting crop growth. The provision of underdrains acts as an outlet to this excessive soil water, lowering the level of saturation and reducing waterlogging in the upper layers of the soil so as to encourage plant development. The depth and spacing of drains act to control the water level in the soil and the deeper and denser the drains the lower the water table becomes, producing by the time drains stop flowing a drier soil for agricultural activity. The ultimate intention of underdraining is to reproduce as far as possible the condition of free-draining land, workable all year round save during and immediately after rainfall and, where after-rain excess soil water is removed quickly, leaving an optimum soil moisture content for both plant growth and cultivation.[5]

Besides being a technique to improve the soil water regime, the adoption of efficient underdraining schemes has the potential to lead to economic changes in the agricultural systems practised, providing increases both in the intensity of cultivation and in productivity. Studies of present-day underdraining systems have revealed that better drainage has allowed the soil to be cultivated with greater ease. As a result, cultivation costs are reduced, as are the number of machine-use days and the amount of labour time. In addition it has been shown that improved drainage renders fertilizers and manures more effective and that the maintenance of a water table well below the surface significantly reduces livestock poaching of land, both factors aiding more intensive cultivation.[6]

Present-day accounts have further demonstrated the increased productivity associated with the improvement's adoption. Better drainage enables soils to be treated for spring cultivations between one and five days earlier. Earlier cultivation facilitates earlier sowing, resulting in a longer growing season and in itself productive of higher yields. A deeper, drier soil permits the greatest development of plant root systems, again promoting increased yields, and also provides the opportunity to switch to new crops. The adoption of underdraining can improve productivity in at least four ways: by assuring the existing yields of crops in the same farming system; by increasing crop yields in the same farming system; by the use of new crops and more flexible rotations in the same farming system; and by a complete change in land use to a new, more productive farming system.[7]

Clearly the adoption of underdraining offered considerable potential for agricultural progress on heavy soils. Agriculturalists had long realized the need for the improvement, and descriptions of the process date from the seventeenth century.[8] However, it was not until the nineteenth century that both the materials for and the systems of underdraining were perfected, rendering the practice an effective technique and in the view of contemporary agriculturalists one of the fundamental bases of increased agricultural productivity on heavy lands.[9]

Present-day studies have acknowledged the potential of underdraining to revalue heavy-land agriculture in the nineteenth century, J. T. Coppock for example considering it 'by far the most important improvement to the land itself' during the period'.[10] Yet the significance of the improvement in agricultural change in that century has engendered extensive debate among agricultural historians and others. As one of a select number of technical advances in the period intended to improve productivity, the extent of adoption and the effectiveness of underdraining have been challenged in the light of differing concepts of agricultural development in the nineteenth century. At a temporal level, those who have argued for marked progress in agricultural output in the nineteenth century have seen underdraining as a vital component in that process. Thus, J. D. Chambers and G. E. Mingay regarded it as a major technical element in their view of the agricultural revolution, contributing from the 1840s to 'the fulfilment of the promise of plenty for all'. Indeed, for Mingay underdraining was 'the most important and most capital-absorbing of the productive improvements of the nineteenth century'. Again, underdraining was identified by F. M. L. Thompson as one of the economic, physical and technical changes that allowed farming to move 'from being an extractive industry... into being a manufacturing industry' that was the essence of his second agricultural revolution from 1815 to 1880. On the other hand, those who place the major changes in agricultural improvement in the seventeenth century and earlier minimize the importance of underdraining, E. Kerridge reporting that the technique in the nineteenth century had but limited application.[11] At a spatial level, underdraining looms large in the conflicting views of the relative balance in the productivity and prosperity of farming systems on the two main soil types of the country, the light and heavy lands, over the nineteenth century. Thus, H. C. Darby suggested that underdraining made possible a revaluation of agriculture in general on the claylands from the 1840s, comparable to that experienced on the light soils in the eighteenth century. A stronger claim came from R. W. Sturgess, who argued that the improvement prefaced a revolution on the clays of the country by facilitating the extension of livestock enterprises. Yet E. J. T. Collins and E. L. Jones, intent on preserving the light lands as the leading sector in English agriculture, disputed the extent of agricultural progress on the claylands in the third quarter of the nineteenth century, maintaining that investment in underdraining, the necessary basis for such change, was poor.[12]

Such divergent assessments of the value of underdraining are a product not simply of the partiality of agricultural historians and others in ranking technical innovations in agriculture but more importantly of the lack of reliable data on the improvement. All too often, underdraining has been invoked or rejected as a technical factor in agricultural change without recourse to quantitative data on the innovation, its spread and its effects on output and farming systems. Without such material, the significance of the improvement in nineteenth-century agriculture will always be open to question, as indeed will all technical innovations of the period. O. R. McGregor has rightly perceived that 'the all-important subject of [under] drainage...has not yet received the attention its importance warrants as a means to increased productivity and as a source of capital expenditure'.[13] The purpose of the present study is to remedy this neglect by examining the extent and agricultural effect of underdraining and the factors involved in its spread in England during the nineteenth century. For such an enterprise, a detailed examination of the variety of opinion in the existing literature on the major aspects of the improvement's adoption and contribution to agriculture provides an essential starting point.

Amount and distribution of underdraining

Amount of land drained

A fundamental measure of the importance of underdraining as a nineteenth-century agricultural improvement is the amount of land so treated. However, there is a lack of statistical evidence of the area drained and few quantitative attempts have been made to estimate the extent to which the improvement was adopted.[14] Most accounts rely on qualitative assessments of the importance of underdraining in the nineteenth century. They also present contrasting views: J. H. Clapham could note that less than a possible maximum amount had been drained, but nearly all that was absolutely essential had been done by 1870 and G. E. Fussell recorded that millions were spent and thousands of acres drained, while Kerridge claimed that 'large-scale pipe drainage...did not affect, far less revolutionize, more than a part of English agriculture'.[15] In general, there is agreement that underdraining was a widespread improvement, a concordance resulting from the common use of the same secondary, nineteenth-century sources, especially the county prize essays of the Royal Agricultural Society and James Caird's account of English agriculture in 1850–1.[16] However, Fussell acknowledges that these sources provide no real measure of the extent of the improvement, to which L. Hoelscher adds that most of the estimates emanating from such sources were vague.[17]

To remedy this lack of statistical evidence, a number of attempts have been made to quantify the area drained in the nineteenth century, predominantly at a national level. Five detailed estimates exist which provide acreage or some other statistical measure of land drained, the bases for which have varied from contemporary calculations of the area drained and the area needing draining, and applications for nineteenth-century government draining loans, to present-day assessments of nineteenth-century underdraining survival. Collins and Jones rejected the view of underdraining as a widespread improvement and considered that the amount drained in England and Wales between 1850 and 1880 was small: in all, some 3 million acres or 16 per cent of the area requiring draining had been treated in this period.[18] In opposition to Collins and Jones, Sturgess argued that the lack of adequate underdraining applied only to the claylands of East Anglia, the east midlands and the Weald. He established a regional distinction in draining activity, claiming that in the area north and west of Leicestershire underdraining was widely adopted between 1850 and 1880. In this area and period, he calculated that some £16 million was spent on the improvement, a sum sufficient to drain 2 million acres, representing a high proportion of the land that needed to be treated.[19] B. D. Trafford rejected both views and maintained that underdraining was an extensive and popular improvement between 1840 and 1890. Using a contemporary estimate of drain-pipe production, he calculated that between 10 and 12 million acres of England and Wales were drained in this period, virtually all the land that required treatment.[20] The most recent estimates by F. H. W. Green and by M. Robinson support Trafford's conclusion. Both have used the reports of the Ministry of Agriculture's draining advisers for approving grants for underdraining in the decade 1971–80. From these Green deduced that 11.6 million acres had been drained in England and Wales by 1880, while Robinson determined that by 1890 some 12–14 million acres had been improved, representing just over 50 per cent of the agricultural area of England and Wales.[21]

That these estimates should vary so widely not only demonstrates that the amount of land drained in the nineteenth century is unknown, but also indicates the equivocal nature of the evidence used. However, as they represent the most serious attempts to gauge the area drained in the nineteenth century, a detailed examination of them is warranted.

Collins and Jones based their assumption that too little land was drained on three sources: the 1870 Agricultural Returns which recorded that in 11 out of 42 counties draining activity had been negligible and in only 6 was it completed or not required; the estimate of J. Bailey Denton, a leading nineteenth-century draining engineer, that by 1880 only 3 million acres out of 20 million requiring draining had been treated, about 16 per cent of the total; and the 1873 estimate of Caird, an Inclosure Commissioner and agriculturalist, of land drained, which was claimed to correspond closely to that of Denton.

Each of these sources may be regarded as unreliable. The 1870 returns contain no information on acreage drained. They are qualitative statements, to which as was noted in the Agricultural Returns themselves 'agriculturalists who are well acquainted with the manner in which our farming is carried on may not probably attach much importance'.[22]

Denton's figures are only estimates and will bear more than one interpretation. Collins and Jones only use Denton's 1880 figures, but, as he produced a similar set in 1873, a comparison of the two is valuable, as is a comparison of Denton's and Caird's 1873 estimates. In 1873 Denton considered that 3 million acres had been drained in England and Wales.[23] This figure comprised two components, 1.5 million acres drained with money borrowed under the various government draining loans and land improvement acts – the public component; and 1.5 million acres drained with money derived from the private funds of landowners – the private component. Denton calculated that 20 million acres required draining in England and Wales. The acreage drained expressed as a percentage of the total requiring draining gives a figure of 15 per cent, an amount corresponding to the proportion suggested by Collins and Jones to have been drained. Seventeen million acres remained undrained, but Denton admitted that of this amount only 8 million acres would be capable of profitable draining, that is draining that would yield in increased rent 5 per cent on the outlay.[24] He did not expand this statement but, assuming the 3 million acres he had stated as drained was 'profitable draining', then the total area which had required draining and which would provide 'profitable draining' would amount to no more than 11 million acres. That landlords and tenants would drain land which would not pay for the cost of the improvement is unlikely, and Denton's 3 million acres drained is more realistically expressed as a percentage of the total of land that would pay for the improvement. In this case, the proportion increases to 27 per cent.

Denton was inconsistent in his estimates and by 1880 he had revised several of his figures. He still considered that 3 million acres had been drained in England and Wales, but altered the ratio of public to private money.[25] With the money borrowed under the government draining loans and land-improvement acts, Denton now suggested that 1 million acres had been drained in England and Wales, and that the remaining 2 million acres had been drained with private funds. He reduced the total area requiring draining to 18.455 million acres. The amount drained expressed as a percentage of the total requiring the improvement gives, as in 1873, a low value, 16 per cent, a figure that was adopted by Collins and Jones. But, as in 1873, Denton admitted that only half the undrained land would pay for draining, that is yielding 5 per cent return on the outlay.[26] Assuming again that the 3 million acres drained by 1880 had been profitable draining, the total area which required draining and would provide 'profitable draining' was 10.7 million acres. The 3 million acres drained represented 28 per cent of the total area that would pay for draining, a value that is much higher than the 16 per cent accepted by Collins and Jones.

Collins and Jones were also of the opinion that Caird's estimate of land drained in 1873 agreed with that of Denton in 1880. However, Caird thought that only 10 million acres required draining in England, Wales and Scotland, as against Denton's 1880 figure of 18.455 million acres in England and Wales alone.[27] Caird also claimed that, with the money borrowed under the government draining loans and land improvement acts, 2 million acres had been drained in England, Wales and Scotland. He did not include the acreage drained from private funds. If draining financed privately were included, there would be little correspondence between Caird's estimate and that of Denton which incorporated the private capital element.

Not only do the estimates and their interpretation vary, but as estimates their reliability may be questioned. In 1873, Denton claimed that 20 million acres required draining. As the total cultivated area of England and Wales at the time was 26.5 million acres, the idea that 20 million acres required draining was, as the *Edinburgh Review* noted, exaggerated.[28] Denton's figure of 3 million acres drained by either 1873 or 1880 is also questionable. The 1873 estimate according to the *Edinburgh Review* was 'purely speculative and probably inaccurate'.[29] Caird's figures differed distinctly from those of Denton. In estimating the total area drained with money borrowed under the government draining loans and the land improvement acts in England, Wales and Scotland, Caird suggested 2 million acres in 1873, while for the same three countries Denton in 1880 calculated the acreage as 1.333 million. Yet the estimated acreage drained under the government draining loans and land improvement acts should have been fairly uniform, as the Inclosure Commissioners, the body that administered the loans, published the amount lent under such acts and this sum only required division by the average cost of draining per acre.

No such calculation could be made for privately funded draining. To account for this sector, Denton put forward ratios attempting to define the relationship of underdraining financed by public money to that privately financed. Such ratios could be little more than guesswork, and the variations between those proposed reveal this fact. In 1873, Denton recorded the ratio between public and private expenditure on underdraining at 1:1; in 1880 his ratio rose to 1:2; in the *Edinburgh Review*, Caird was reported as estimating the ratio at 1:3; while the *Edinburgh Review* in 1880 considered that Caird had underestimated the private sector.[30] That Denton doubled the estimate of private draining in seven years and that Caird, who as an Inclosure Commissioner had as much opportunity as Denton of gauging the amount of land drained, put the ratio even higher are indications of the uncertainty existing in measuring the acreage drained privately.

Collins and Jones believed that too little land was drained between 1850 and 1880, but what has emerged is the unreliability of the figures advanced by Denton and Caird on which they depend. These are crude estimates, which when inspected closely give contrasting results, and their calculation makes their approximation to reality difficult, if not impossible, to gauge. Despite these drawbacks, both the proportion – some 15 to 20 per cent – and the area – some 2 to 3 million acres – suggested by Collins and Jones have received wide currency in the subsequent literature.[31]

Sturgess in his estimate claimed that a considerable amount of draining, over 2 million acres, took place in the north and west of England between 1850 and 1880. This figure was derived from the fact that 'of the first £2⅓ million borrowed from the government and improvement companies [for draining purposes] between 1846 and 1855, 83 per cent was spent on estates in counties to the north and west of Leicestershire'.[32] Sturgess regarded this percentage as a constant and applied it to the total amount lent under the government draining loans and land improvement acts by 1880. He further assumed that, using Denton's 1880 ratio of public to private draining of 1:2, the sum spent on draining, both publicly and privately funded, in the north and west of England would be some £16 million, which would represent just over 2 million acres drained.

The important element in Sturgess' estimate is the percentage constant of draining expenditure in the north and west. This figure was derived from notices of applications in the *London Gazette* for draining loans. But inconsistencies occur in the material. The loans recorded in the *London Gazette* referred only to the government draining loans of 1846 and 1850 and those made under the Private Money Draining Act, 1849. The improvement companies were not required to publish notices of applications there.[33] The pattern that Sturgess' figures are supposed to indicate may not bear any relationship to that of draining loans issued by the improvement companies. The applicability of a percentage derived from the government draining loans and the Private Money Draining Act to the total amount lent for draining under all the land improvement acts is questionable.

Furthermore, the sums recorded in the *London Gazette* notices were not necessarily the sums borrowed by landowners. A notice of application in the *London Gazette* indicated that a loan was intended to be made on the estate named: the loan itself may not have been taken up or, if it was, not to the amount stated in the notice. This fact is demonstrated in the following example. In a study of agrarian change in Lancashire, T. W. Fletcher was the first to use the notices of applications for draining loans in the *London Gazette* in an attempt to indicate the widespread adoption of draining in that county.[34] He noted that the six largest loans applied for in February 1847 under the Public Money Draining Act, 1846, and recorded in the *London Gazette*, came from landowners with estates in the north and northwest, three in fact from Lancashire: the earls of Carlisle and of Lonsdale applied for £47,134 and £30,060 respectively; the duke of Sutherland for £38,000 mainly for his Shropshire and Staffordshire properties; the earl of Derby for £34,000; T. Clifton for £25,000; and the earl of Ellesmere for £20,000, a total of £194,194 for draining purposes.[35] However, of this amount applied for, only £36,000, some 19 per cent, was actually borrowed. No evidence exists of the earls of Derby and of Ellesmere taking up their proposed loans; the duke of Sutherland borrowed only £6,000; and the earls of Carlisle and of Lonsdale and T. Clifton contracted loans of £10,000 each.[36] The applications for draining loans in the *London Gazette* cannot be regarded as a precise measure of draining carried out.

The pattern that Sturgess suggested would seem inaccurate and cannot be applied to other and later draining loans. However, the total amount borrowed under the government draining loans and the land improvement acts is at least known. In calculating the total area drained in the north and west at over 2 million acres, Sturgess had to make an estimate of the amount drained with private funds. To do this, he adopted Denton's 1880 ratio of public to private draining of 1:2. The use of this ratio at once places Sturgess' figures in the same realm of unreliability as those advanced by Denton.

The estimate of Trafford of 12 million acres drained between 1840 and 1890 is also based on data calculated by Denton. Rejecting Denton's 1880 figure of 3 million acres drained as too low, Trafford relied on drain-pipe makers' evidence provided by Denton in 1855, published in 1863 and republished in 1883.[37] Denton assumed in 1855 that there were some 2,800 brickyards making drain-pipes, producing on average for the year 150,000 pipes, making a total of 420 million pipes. As 1,250 pipes were needed on average to drain an acre, Denton calculated that this number of pipes would drain 336,000 acres a year. Deducting a quarter of this acreage as temporary or substandard draining, Denton concluded that a quarter of a million acres were drained annually in Great Britain. Of this amount, Trafford found it 'hard to imagine fewer than 200,000 acres per year being drained in a satisfactory manner in England and Wales'.[38] As Denton's pipe-production figures referred to 1855, Trafford took them as being 'broadly representative of the period 1840–90'.[39] On that basis he considered that 12 million acres were drained in England and Wales between 1840 and 1890.

Objections can be readily raised to the calculation of this figure. Even if we assume Denton's numbers of brickyards, of pipes made and of acres drained to be correct, they refer solely to the year 1855. Although they may have some relevance to that date, Trafford provides no evidence to show that what applied to 1855 was typical of the period from 1840 to 1890. The belief in a constant pipe production and acreage drained over these fifty years reveals a lack of awareness of economic movements in agriculture in the second half of the nineteenth century. Denton's estimates in 1855 came at the height of the mid-century draining boom: the government draining loans had just been established and the improvement companies were beginning to provide additional funds. They may not be applicable to the period after the late 1870s when English agriculture was beset with depression. A brief examination of drain-pipe makers specified as such in Post Office and Kelly's directories for groups of counties revealed a decline in numbers after 1870. The number of drain-pipe makers recorded in the directories for Bedfordshire, Berkshire, Buckinghamshire, Huntingdonshire, Northamptonshire and Oxfordshire fell from 27 in 1869 to 19 in 1885–7 and to 15 in 1890–1; the number in the directories for Cambridgeshire, Norfolk and Suffolk declined from 20 in 1869 to 14 in 1888; and the number for Staffordshire, Warwickshire and Worcestershire from 40 in 1880, to 24 in 1888 and to 22 in 1896.[40] Finally, it should be added that, whereas Trafford used Denton's 1855 drain-pipe figures to estimate an area drained between 1840 and 1890 of 12 million acres, Denton was of the opinion in 1880 that the information he gathered in 1855 corroborated his view that only 3 million acres had been drained satisfactorily in England and Wales by 1880.[41]

The estimates of both Green and Robinson are not based on any nineteenth-century calculations but on the 1971–80 reports of Ministry of Agriculture draining advisers on applications for present-day underdraining government grants. In these reports, amongst other information, advisers detail, where known, the drainage history of the area to benefit. In this history, they attempt to identify three elements: old drains present but not working; some old drains still working; and no evidence of old drains. According to both Green and Robinson, old drains may be taken as dating from before 1939. From these data, Green calculated on a county basis the percentage of the area submitted for underdraining grants in 1972–3 in which old drains were reported as partly functioning, and the percentage of the area underdrained in 1976–7 for which failure of old drains was given as the reason for grant application. Finding a similarity in the resulting distributions, he assumed that these two percentages could be regarded as representing the proportion of cultivated land with pre-1939 underdraining. As he believed little underdraining was carried out between 1880 and 1939, the resulting area, 11.6 million acres, formed the amount of land drained in England and Wales before 1880.[42]

Robinson made more extensive use of these reports. In the decade 1971–80, there were nearly 125,000 grant applications for underdraining involving about 2.38 million acres. This material had been collated on a parish basis by the Ministry of Agriculture. For parishes with grant applications, Robinson identified the proportion of the area intended to be drained that was reported to possess old drains. The resulting proportion was assumed to be representative of all old draining activity in the whole parish. From such proportions, the total area in each parish presumed to have old drains was calculated and summed to produce a figure of 14 million acres in England and Wales. To allow for possible overestimation, Robinson reduced this total to 12 million acres, an area representing the extent of underdraining between 1850 and 1890. The parish areas of old drains were then mapped as a percentage of agricultural land to indicate the spatial distribution of the improvement.[43]

The methods used by Green and Robinson to produce estimates of land drained in England and Wales in the nineteenth century from these data warrant scrutiny. To obtain his figure, Green equated the area in a county submitted for underdraining grants in 1972–3 and 1976–7 with its total cultivated area. Such an equation supposes that all cultivated land requires draining. However, much agricultural land is free-draining, not in need of the improvement, and the incorporation of such land into Green's calculations would produce a significant exaggeration of the area drained by 1880.

The assumption made by Robinson, that the proportion that the area with old drains formed of the area submitted to benefit from underdraining grants within a parish was applicable to the whole parish, must be questioned. The reliability of such an approach is dependent on the number of grant applications in a parish and the area that they represent of that parish; the greater the area covered the more reliable the resulting proportion was likely to be. However, no indication is given of the proportion that grant areas formed of each parish, and the distribution map of draining intensity provides considerable grounds for doubting the reliability of the approach. Thus, virtually all the Pennine area in the northern counties is shown as possessing over 50 per cent and in many cases over 75 per cent of its agricultural land drained. These were areas of poor-quality agricultural land, largely given over to rough pasture, where the capital investment in underdraining would never have paid and would rarely have been contemplated, a view supported by H. M. E. Holt's study of upland farming between 1840 and 1880 on a number of estates in Cumberland, Westmorland and Northumberland.[44] The conclusion cannot be avoided that the inspected areas in a parish were not always a representative sample, a failing that debases the accuracy of Robinson's estimate.

More serious problems are presented by the source used by both Green and Robinson to estimate their areas of nineteenth-century underdraining. The data on old drains in the Ministry of Agriculture's draining advisers' reports are statistically imprecise. Drainage history constitutes but a minor part of the standardized reports on underdraining applications, forming one section out of thirty requiring completion by the advisers.[45] Discussions with these advisers reveal that they and farmers are largely unaware of the pre-1939 drainage history of most fields, and as no detailed survey is undertaken their reports on old drains involve considerable guesswork.[46] The section on drainage history provides the opportunity only to record the presence or absence of old drains in the area for which a grant application for underdraining has been made. No attempt is made to determine the proportion of the area proposed to be drained that contains old drains. If old drains are discovered in some part of the benefit area, it is assumed that the whole of that area had been drained before 1939, a situation not established by field observation and providing a questionable base on which to calculate total areas of old drains. No information on the date of old drains found in the benefit area is recorded, and the efforts of both Green and Robinson to relate their areas with old drains to specific years in the nineteenth century – 1880 and 1890 respectively – should be regarded as conjectural, going beyond what is available in the draining advisers' reports. Indeed, the nature of the data used by Green and by Robinson renders uncertain their estimates of land drained in the nineteenth century.

These five estimates of the amount of land drained in England in the nineteenth century exhibit wide differences. The variation is largely a consequence of the ambiguous nature of the evidence used: none of the evidence submitted is sufficient to support any claim. As a result, the accuracy of all the estimates may be questioned and little confidence can be placed in estimates which are so variable and unreliable. The fact that these estimates vary so much warrants a re-examination of the area underdrained in the nineteenth century.

Distribution of underdraining

In the search for national estimates of amount of land drained, the spatial distribution of the improvement has received less thorough attention. Most accounts have related the regional incidence of underdraining to the occurrence of claylands, and, by producing a map of the distribution of claylands in England and Wales, Darby has attempted to locate those areas of potential draining activity.[47] However, within the broad expanses of English claylands, there is disagreement on the regional impact of the improvement. Based on the notices of application for draining loans in the *London Gazette*, Sturgess drew a distinction between the claylands to the north and west of Leicestershire, which were well drained, and those of the east midlands, East Anglia and the Weald, where the improvement was not widely adopted.[48] This pattern has received support from Robinson in his analysis of the Ministry of Agriculture's draining advisers' reports.[49] Against this, Thompson located the major impact of underdraining in the midland counties, as he considered that these contained most of the heavy, wet lands.[50] However, Mingay could express the opinion that on the arable clay areas of the south and east there had been 'a heavy expenditure on drainage', a view endorsed by H. C. Prince.[51]

Examination of county studies of nineteenth-century agriculture adds little clarification of the problem. The closed-system approach that marks most of these accounts tends to preclude extra-county comparisons. In his analysis of Northumbrian agriculture, S. Macdonald described the extensive draining of claylands after 1840, and C. S. Davies, R. E. Porter and Fletcher recorded that large portions of the counties of Cheshire and Lancashire had been drained since the 1850s. However, D. B. Grigg and T. W. Beastall noted the importance of underdraining to clayland agriculture in Lincolnshire from the 1820s, while J. Thirsk and J. Imray claimed that draining in Suffolk had done much to establish the agricultural prosperity of that county in the third quarter of the nineteenth century. The widespread adoption of the improvement from the middle of the century was reported by A. G. Parton and B. M. Short on the clays of Surrey, Kent and Sussex. B. R. Dittmer and Thompson brought attention to the value of underdraining in Wiltshire, while in the midlands J. R. Walton and R. C. Gaut detailed extensive use of the improvement in Oxfordshire and Worcestershire.[52] The conclusion to be drawn from these accounts is the lack of spatial variation in the incidence of underdraining in the nineteenth century, the improvement being as important on the clays of the north and west as those of the south and east.

Much of the confusion over the regional impact of the improvement is a product of the lack of reliable, comparative data on nineteenth-century draining activity. But equally important has been the failure to provide measures of the amount of land in need of draining throughout the country so that the relative intensity of the improvement may be determined. To achieve this, more precise assessments than at present available are needed of the types of land requiring underdraining. For example, the Soil Survey's maps of England and Wales at 1:250,000 and 1:1,000,000 reveal that, although there is a general need for draining claylands, clayland soils vary greatly in texture, some taking on the character of loams and requiring little draining, while many soil types other than clays would benefit from the improvement.[53] Further, the role of physical factors other than soil type in influencing the spatial distribution of underdraining has not been fully examined. As draining was a technique for removing excess soil water, a strong relationship may have existed between the regional intensity of the improvement and the pattern of rainfall in the nineteenth century, as some have suggested.[54] Indeed, as the influence of weather on agricultural prosperity has been frequently stressed in the period, particularly in the wet seasons of the late 1870s and early 1880s, there may be a temporal connection between rainfall trends and the adoption of underdraining.[55] In the absence of data on these factors, the spatial distribution of the improvement in the nineteenth century must remain unclear.

The chronology of the spread of underdraining

Despite the general discussions of the spread of underdraining there is no clear picture of the timing, rate and spatial pattern of its adoption in the nineteenth century. Few studies have provided a precise chronology of either area drained or draining expenditure during the period.[56] In the absence of such data, two surrogate measures of the temporal adoption of underdraining have been proposed, related to technical change in the improvement and to trends in capital investment in nineteenth-century agriculture.

The time series derived from technical developments is based on the introduction of new draining systems and materials and of government financial assistance for the improvement: the supersedence of Joseph Elkington's method of spring draining and the East Anglian system of hollow drains that existed in 1800 by James Smith's scheme of shallow, close, parallel drains popularized in the 1830s and 1840s and by Josiah Parkes' system of deep, parallel drains publicized in the 1840s; the development of more permanent alternatives to stones, bushes and straw as drain fill with the use of tiles from the 1820s, aided by their exemption from tax, and with the dominance of pipes from the 1840s; the invention of machines from 1835 onwards to produce tiles and then pipes cheaply and in great numbers; and the establishment of government loans in 1846 and 1850 and of a group of government-sponsored improvement companies in the 1840s and 1850s to provide capital for the improvement.[57]

From these developments, it has been suggested that from 1800 to the 1820s little underdraining was carried out, as both systems and materials were unreliable.[58] With the greater availability of tiles from the 1820s, underdraining was introduced in many parts of the country, for example in Cumberland, Lancashire and Lincolnshire.[59] The appearance in the 1830s of Smith's draining system and of tile-making machines encouraged a slight growth of activity,[60] but it was the combined effect of drain-pipe-making machines, Parkes' deep-draining system and the government draining loans that produced a rapid expansion of draining in the country. Lord Ernle wrote of the large-scale adoption of the improvement following from the 1846 government loan, a view supported by many.[61] However, with the techniques well established, scholars have differed in their identification of the main period of draining in the nineteenth century: Jones suggested the 1840s and 1850s; Green and Hoelscher preferred the 1850s; Sturgess extended the period to the 1850s and 1860s; Clapham, B. W. Adkin and Mingay proposed a high level of activity from the 1840s to the 1870s; while Trafford and Robinson maintained a uniform intensity of draining until 1890.[62]

The temporal pattern of underdraining derived from trends in nineteenth-century agricultural investment differs from that based on technical change. Underdraining, along with farm buildings, represented the major form of capital investment in nineteenth-century farming.[63] Capital investment was related to agricultural prosperity, which was measured usually in terms of landlords' rent and depended much on price fluctuations in the market, high prices producing expansion, low prices contraction.[64] Thompson has argued that capital investment in agriculture tended to be proportionally higher in periods of agricultural distress, when landlords laid out money on improvements to maintain existing rents or to prevent drastic reductions. He pointed to the 1820s and 1830s, which witnessed a fall in agricultural prices and rents from the levels of the Napoleonic Wars, as a period when landlords began to invest large sums in agricultural improvement.[65] Indeed, D. C. Moore specifically regarded the adoption of draining and associated improvements as a response by some landlords to the low prices of the period 1815–35, attempting to offset these by increased production. This investment trend has been distinguished on certain estates in Lincolnshire and East Anglia.[66] Again, Thompson suggested that the sharp depression in prices between 1848 and 1852 coincided with another period of heavy landlord investment. Only the extensive depression in agricultural prices from the late 1870s, according to Thompson, did not produce increased capital investment: after a few years of trying to maintain rents by a growth in outlay, landlords from the middle of the 1880s allowed estate improvements to decline.[67] Correspondingly, in periods of agricultural prosperity – the late 1830s to the early 1840s and the early 1850s to the late 1870s – it can be assumed that capital expenditure was proportionally not as great.[68]

This inverse relationship between the level of capital investment and agricultural prosperity has not been uniformly endorsed. In an analysis of the Holkham estate, S. Wade Martins found that high levels of capital outlay coincided with times of farming prosperity and that investment followed the pattern of boom and slump rather than being a foil to it.[69] B. A. Holderness' chronology of investment on various East Anglian estates differed from Thompson's set pattern between 1831 and 1870.[70] Amongst others, P. J. Perry has suggested a correspondence between extensive landlord investment in agricultural improvement and the period of agricultural prosperity and rising rent from 1851 to 1878, in many ways a view of nineteenth-century agricultural investment reflecting Ernle's picture of a lack of outlay in the period 1815–36 contrasting with great expenditure in his period of high farming from 1837 to 1874.[71]

The time series of draining activity derived from developments in draining techniques and from capital investment trends can be seen as mutually contradictory and internally variable. Both technical change and investment trends in agriculture influenced the spread of the improvement but in themselves they do not reveal the extent and pattern of its adoption through time. Without specific data on the time and rate of adoption of underdraining, it is difficult to evaluate its role as a nineteenth-century agricultural improvement and to assess the factors involved in that adoption.

Capital provision of underdraining

As an agricultural improvement, the adoption of underdraining necessitated capital outlay. With increasing technical efficiency, its cost rose, estimated to reach by the middle of the century with the use of machine-made pipes and regular systems of drain layout between £4 and £8 per acre, an outlay that was as large, if not greater, than that incurred for parliamentary enclosure.[72] The level of availability and the provision of capital were therefore critical factors in the spread of the improvement. Three possible sources of such investment – landlords, tenants and the state – have been recognized but little is known about their precise role in financing underdraining during the nineteenth century.[73]

Because of its cost by the middle of the century, underdraining has been identified as part of the fixed equipment of farming provided by landlords. Thompson and D. Spring have described a general relationship between landlord and tenant capital in financing permanent agricultural improvements in the century.[74] They argued that the distinction between the two was unclear up to 1815, with landowners shifting the burden of repairs and improvement on to tenants in a period of prosperity. Between 1815 and 1835, the responsibility for outlay on permanent improvements was clarified so that by the 1830s the fixed capital of farming became in theory a landlord charge. The development of landlord control of underdraining, a major element of the fixed capital of agriculture, has been invoked to account for its spread. The potential capital base for underdraining that the landlord possessed was likely to be much greater than that of a tenant, and Mingay related an expansion of draining activity in the 1850s to landlords' financing of the improvement.[75]

At present, insufficient analyses of estate expenditure on draining have been undertaken to substantiate this pattern of landlord responsibility. However, a range of factors has been suggested as influencing the level of landlord investment in the improvement.[76] Of these, estate size and rate of return on investment have received most attention. The capital pool for draining that could be provided by landlords was not uniform throughout the country. Much depended on the economic resources of individual estates and these varied, being largely a product of size.[77] Thompson's analysis of John Bateman's digest of the 'Return of owners of land in England and Wales 1873', revealed that the range of estate size throughout the country and within each county was considerable.[78] A direct relationship between estate size and capital investment in fixed agricultural improvement has been proposed, the larger estates witnessing greater outlay than their smaller counterparts.[79] Little empirical evidence is available to test the accuracy of this relationship in respect of underdraining. However, estate size as a surrogate of potential landlord financial resources is an important and largely unexamined variable in the spread of draining in the nineteenth century.

The rate of return on landlord investment in the improvement has also been regarded as a significant factor influencing its adoption. The majority of estate studies has shown that the whole range of capital investment in agricultural improvements produced low levels of return in the form of increased rent to the landlord, rarely achieving more than 4 per cent in the third quarter of the nineteenth century.[80] Yet the interpretations of the effect of such returns, much lower than in other sectors of the economy, on the spread of underdraining have been varied. Collins and Jones have argued that the unproductive nature of draining as a landlord improvement was a major contribution to the small area of land drained and that such draining that was undertaken was a distinct investment miscalculation on the part of landowners, views that have been subsequently widely supported.[81] Despite the low return, others have detected an indirect financial benefit to landlords from capital expenditure on agricultural improvements, which would have encouraged their adoption. Interest on landlords' capital was a large element in rent, Holderness suggesting that interest probably accounted for three-fifths of average rents in the mid and late nineteenth century.[82] Underdraining may not have produced economic returns, but allowed growth in existing rental levels in periods of stable agricultural prices. Such outlays may have prevented excessive fall in rent in periods of depression and in a study of a sample of eight estates R. Perren identified a broad correlation between the amount of landlord investment in agricultural improvement and the extent to which rent levels were maintained after 1879.[83] Some have seen capital outlay in agricultural improvement as social investment by landlords, non-profit-making expenditure for the benefit of tenants. Thus, Thompson noted that improvement was undertaken in spite of its financial disadvantages rather than because of its profitability. For a landowner to improve was to discharge a duty towards his land, not to make an economic investment.[84] Such contrasting views indicate that far more needs to be known of the attitudes and expectations of landlords and their agents to the provision of capital for underdraining, key elements in the improvement's adoption.

Far less is known of the tenant contribution to the improvement. A number of studies has suggested the complete use of tenant capital in underdraining, especially in areas where the customs of tenant right prevailed, as in Lincolnshire, and after the introduction of the Agricultural Holdings Acts of 1875 and 1883, by which tenants were entitled to compensation for the unexhausted value of improvements they had undertaken. However, little evidence has been provided of the extent of such tenant-financed underdraining.[85]

If total tenant responsibility may have been rare, some form of tenant assistance was important in adopting draining, even with full landlord control of the improvement. Holderness has indicated a number of ways in which tenants contributed to permanent agricultural improvements.[86] Existing accounts with some reference to a tenant element in draining suggest that tenants either paid an annual rate of interest, averaging about 5 per cent, on the improvement's cost, which was added to the rent, or provided the labour costs.[87] The relative prevalence of these and other methods of tenant involvement is unknown, but the use of such resources implies a tenant choice and decision on having land drained. Thompson considered that the instigation of draining depended in great measure on the landlord's initiative, the work being supervised by the estate agent, with its farm-by-farm employment being determined by the requirements of individual farmers.[88] The applicability of this system to all estates has not been tested, but it points to the need for a closer examination of the role of the tenant in the adoption process of the improvement.

Underdraining was one of the few agricultural improvements that received both direct assistance and subsidy from the state during the nineteenth century. From 1840 onwards a body of legislation was enacted, culminating in the public money draining loans of 1846 and 1850 and in a series of land improvement companies, which established funds from which landlords could borrow for the purpose of underdraining, the sums to be repaid as a charge on the estate over a period of years.[89] The origins of such legislation have been variously seen as attempts to alleviate the difficulties that contemporaries considered that tenants for life under strict family settlement experienced in trying to improve their estates; as compensation to the landed interest after the repeal of the Corn Laws; and as a psychological measure to popularize agricultural improvement and increase agricultural output.[90]

Whatever the intention, the legislation provided a new source of capital for landowners for the improvement, defined in several studies, as discussed above, as the public component of draining activity. The loans were administered by the Inclosure Commissioners and their successors and the total amount borrowed by landowners for underdraining was published and is known: £8,995,000 in Great Britain between 1847 and 1899.[91] However, no detailed examination of the loans exists, leaving the precise role of such government assistance in the adoption of the improvement undetermined. A fuller understanding of the importance of such funds to different sized estates and the proportion that they formed of total estate expenditure would clarify the relationship between the public and private components in draining, a relationship so basic to many calculations of the amount of land drained in the second half of the nineteenth century.

The agricultural consequences of underdraining

The agricultural benefits and results to be derived from underdraining depended on the technical efficiency and permanence of the improvement. Underdraining was a valuable agricultural improvement only if it effectively removed excess soil water. More information is available in the literature on the technical development of underdraining in the nineteenth century than any other aspect of the improvement. This has already been summarized and there is little need to detail further these technical changes.

However, less consideration has been applied to the equally important aspect of the efficiency and permanence of various underdraining systems. While hollow and mole drains have been widely recognized as improvements largely the responsibility of the tenant and with a short life and limited effectiveness,[92] a divergence of opinion is evident on tile and pipe underdraining carried out from the 1840s onwards. Collins and Jones made the claim that much of such draining was ineffective, often ill done, falling short of expectation, and with a limited life.[93] On the other hand, others have noted that pipe and tile draining was an effective improvement, almost always paying its cost within a few years, and with a life expectancy of 25 to 30 years and more.[94] Such variation reveals the need to establish more carefully not only the extent and prevalence of particular draining systems and specific draining materials, factors fundamental to an understanding of technical effectiveness, but also their durability so that the permanence of the improvement in the nineteenth century can be more accurately judged.

As an agricultural improvement, underdraining sought to increase land productivity. The main assessment of that productivity has been in terms of yield increases in crops. However, in the literature it is too often assumed that the adoption of a new draining technique automatically implied increased productivity and there is little supporting evidence of the yield increments consequent on the improvement. In an attempt to calculate the financial benefit from investment in land improvement, P. A. David suggested that draining resulted in an increase in wheat yields of 15 to 20 per cent. His figures were based on a 'few and scattered contemporary reports of what seem to have been noteworthy increases in yields following drainage' and the representativeness of such figures is unknown.[95] Present-day studies of yield increases after draining are of the order of only 5 to 10 per cent.[96] There is a need to assemble a more extensive body of data on yield changes following draining and without adequate figures on crop yields it is difficult to estimate how important the adoption of underdraining would be to agricultural productivity.

The changes in agricultural practice that were produced by underdraining create another area of considerable divergence of opinion. Many have claimed that underdraining revalued agriculture on heavy lands and permitted the adoption of mixed-farming systems, providing the basis for direct competition with light-land agriculture: C. S. Orwin and E. H. Whetham noted, 'tile drains...would do for the clays what the turnip had done for the light land'.[97] Thus, the practice of bare fallowing disappeared to be replaced by an extension of root cultivation and the use of new rotations.[98] Others, not detecting such a complete transformation, have seen underdraining as a technique for the improvement and expansion of corn farming. T. L. Crosby emphasized Sir Robert Peel's linking of the public money draining loans to cereal production in his Corn Law repeal speech in 1846, while Moore saw underdraining as the means by which wheat output could be increased in the period of high farming.[99] The case has been made by Thompson that underdraining in conjunction with other improvements served to prolong the importance of cereal, especially wheat, farming in English agriculture because of the increased efficiency and productivity that it brought to that system.[100]

However, the very reverse of this pattern has also been identified as a result of the adoption of underdraining. Sturgess has argued that over the 1850s and 1860s 'there occurred a technical revolution on the clays of the north and west of the country which consisted of the conversion of cornland to an intensive grassland husbandry on newly-drained farms'.[101] In contrast to all these views, Collins and Jones have presented a negative picture of the agricultural changes resulting from underdraining. They considered that by 1880 underdraining had hardly begun to improve clayland agriculture. The attempts to adopt turnips on clays after draining were misjudged and the hope of introducing turnip husbandry unrealized: 'on the stiff plastic clays neither large root breaks nor the extensive overwintering of stock in the fields become possible...Bare fallows remained stubbornly the answer for foul land and prolific weed growth.' Overall changes in land use on claylands were slight, being no more than a series of minor adaptations to the relative price movements of cereals and livestock in a period which favoured the latter.[102]

Such contradictions indicate that the impact of underdraining on agricultural practice requires to be thoroughly re-assessed. Little evidence is provided in the literature to show not only how the existing patterns of land use affected the adoption of the improvement but also how the techniques effected changes in land use. Until such data are widely available, the degree of change that underdraining brought to agricultural systems remains in dispute.

Approach

The foregoing review has amply demonstrated how little is known about the significance of underdraining in nineteenth-century agriculture. Much of the confusion arises from the nature of the sources used. Too much reliance has been placed on secondary accounts traditionally employed in agricultural history, resulting in vague generalization, and on questionable surrogate measures of land drained. To clarify the issues, there is a need to provide precise detail of the amount of land drained, its regional distribution, the factors affecting its adoption and its effects on agricultural practice. Such primary material is to be found in the records of the various government and improvement companies' draining loans and of the draining activities of landed estates. These two data sources have largely formed the basis of the present study.

The loans for underdraining made by landowners under the land improvement acts have been regarded by both nineteenth-century agriculturalists and present-day scholars as a fundamental source to assess the improvement's importance.[103] They permit analysis of the problem at the national scale and provide detailed information throughout the country from 1847 to the end of the century on the amount, the distribution and chronology of underdraining and on the estates that borrowed. Examination of such data allows reliable assessments to be made of the pattern of draining activity, financed by the government and the improvement companies, for England as a whole.

Although providing precise data at a national scale, these draining loans cannot, however, cover all aspects raised by the use or adoption of the improvement. The representativeness of the underdraining carried out under the land-improvement legislation of all underdraining in the second half of the nineteenth century is unknown and the records yield little information on the organization, efficiency and agricultural consequences of the improvement. To overcome these difficulties, use has to be made of the other main primary source for nineteenth-century underdraining, estate records, which at a smaller scale are more detailed. Throughout the century, the estate emerged as the unit of management of permanent agricultural improvements.[104] Landowners held land in a variety of parishes and counties and their policies for agricultural improvement were transmitted through their agents to all parts of their estates. Existing studies have indicated the value of estate records for reconstructing the impact of underdraining on nineteenth-century agriculture.[105]

In the present study, records of groups of estates have been examined for three counties: Devon, Northamptonshire and Northumberland. In selecting these counties, there was a need to ensure that they represented different intensities and patterns of underdraining financed by government and improvement companies' loans and that they possessed diverse agricultural regimes and varied locations throughout the country. However, the counties were chosen only after extensive searches of estate material in county-record offices and private muniment rooms. Although far from complete, the available draining records of the landed estates in these three counties

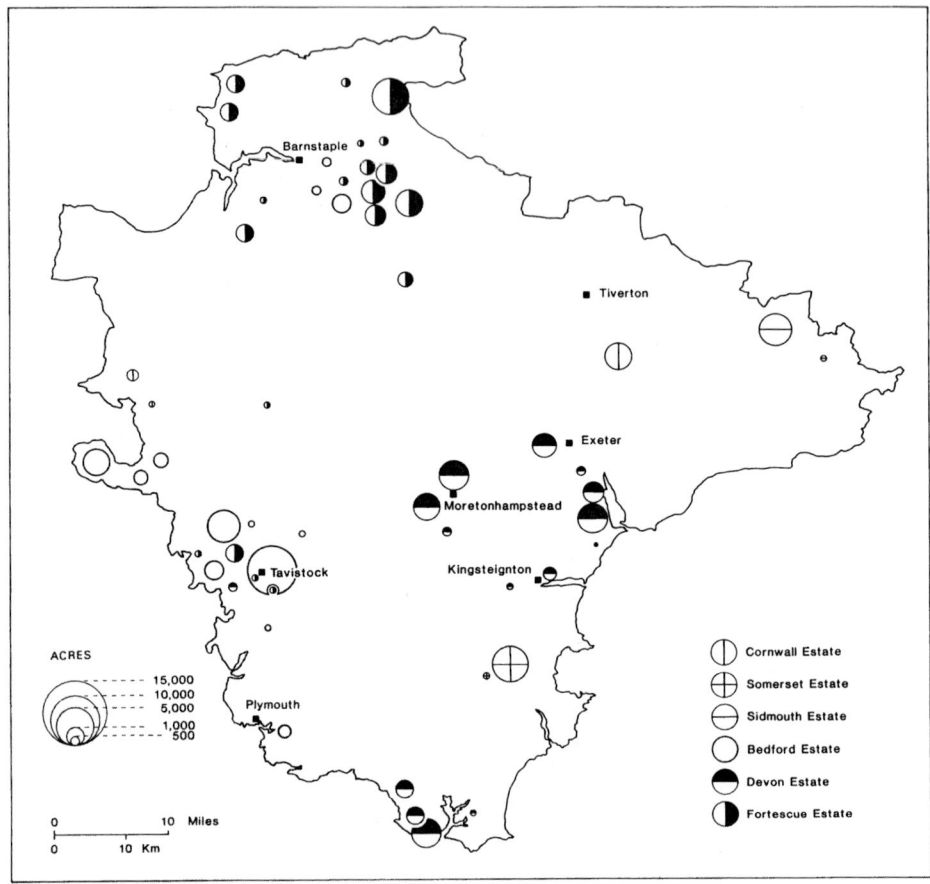

Figure 1.1 Location of sample estates in Devon, 1850–1875 (*Sources:* Bedford MSS:
Annual report for 1867, vol. 2; Cornwall MSS: Map of the honor of Bradninch by
W. Simpson, 1788; Valuation of the manor of Bradninch by R. Watt, 1855; Map of
the manor of Bradford, 1867; DRO, Courtenay MSS: 1508M/Estate papers/14/A/
III, Shelf III, Particular... of the estate of the earl of Devon, 1862; Fortescue MSS:
1262M/E29/58, List of estates in north Devon, 1864; 1262M/20/42, Rental
1849–50; 1262M/E22/10, 12, 16, 24, 51–7, Maps of north Devon estate, 1880;
1262M/E22/45, Map of land in south Devon, 1825; Michelmore, Lovey and Carter
MSS: 867B/Berry Pomeroy survey book *c.* 1850; Seymour MSS: 1392M/Estate
rentals/bundles 1 and 2, 1862–72; Sidmouth MSS: 152M/A general report of the
Upottery estate... by F. Thynne, 1850; PRO, 1R29 and 30/9/280,429, Tithe maps
and apportionments for Membury and Upottery)

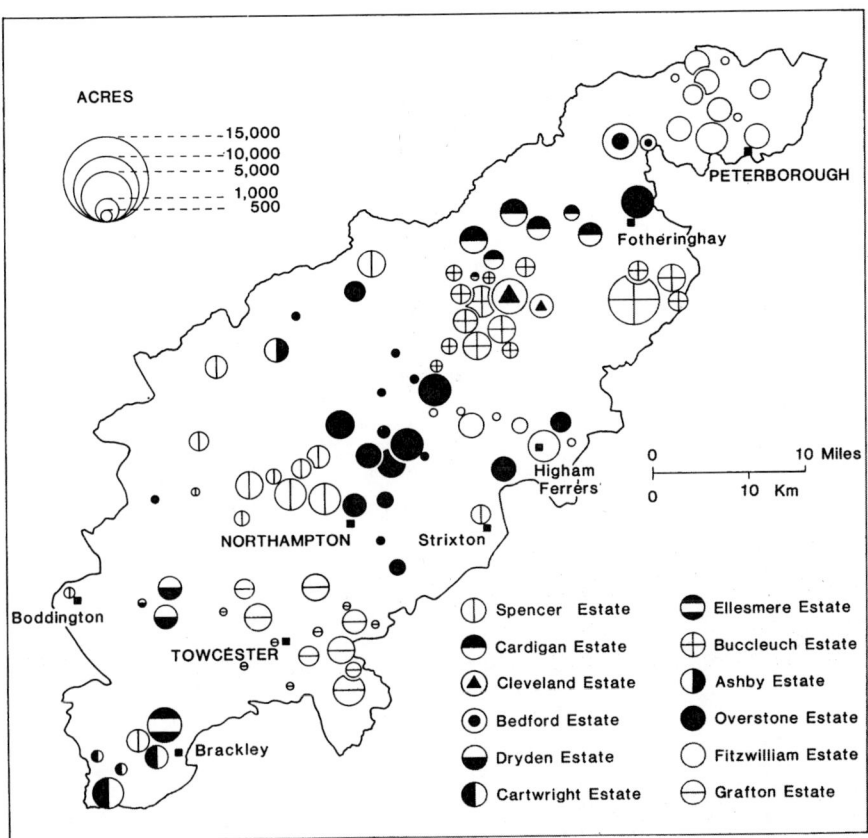

Figure 1.2 Location of sample estates in Northamptonshire, 1850–1875 (*Sources:* Bedford MSS: Annual report for 1857; Buccleuch MSS: Particular of estates within the Boughton collection, 1834; Numerical reference to the Boughton estate, 1896; Raby MSS: Plans of the Brigstock and Sudborough estates of the duke of Cleveland, 1855; Spencer MSS: Reference to the estate of the earl Spencer, 1859; NRO, Brudenell MSS: ASR/163, Terrier of the Cardigan estates in Northampton-shire ... 1871; Buccleuch MSS: Misc. Ledgers, 137, A particular of the Barnwell estate, 1860; Cartwright MSS: C(A) 3849–55, 3871a, Rental accounts, 1854–71; C(A) 4914, Report on the Cartwright estate, 1893; Dryden MSS: D (CA) 446, Valuation of estate, *c*. 1847; Ellesmere MSS: Box X.461, Rentals, 1836–75; Fisher-Sanders MSS: FS1/34, Terrier of the Ashby estate, 1860; Fitzwilliam MSS: Misc. vols., 656, Valuation of estate around Great Harrowden, 1857; Misc. vols., 738, Surveys of estates around Higham Ferrers, *c*. 1860; Misc. vols., 753, Reference to the Milton estate, constructed 1841, corrected 1855; Misc. vols., 450, Valuation of land in Castor, 1835; Grafton MSS: G. 1558, Farms, tenants and parishes, 1822; Overstone MSS: Ov. vol. 4, Terrier of Lord Overstone's estates, 1877; Ov. map 342, Map of the Northamptonshire estates of Lewis Loyd, 1850)

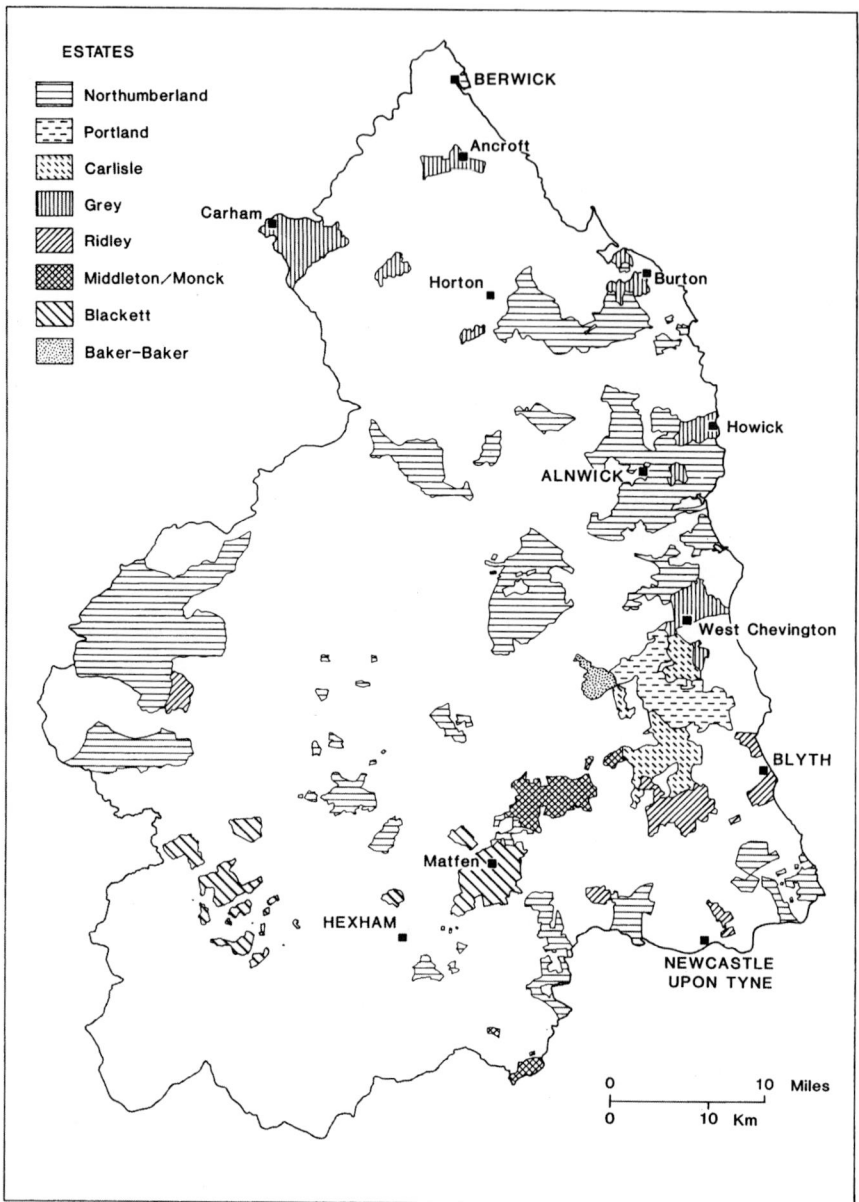

Figure 1.3 Location of sample estates in Northumberland, 1850–1875 (*Sources:* Grey MSS: Maps of Ancroft, Chevington and Howick estates, 1844–98; Northumberland MSS: T. Bell and sons, Survey and terrier of the duke of Northumberland's estate, 1850: map indices to bailiwicks; DPD, Baker-Baker MSS: 119/19, Plan of estate in Stanton, 1847; Grey MSS: Estate cropping book, 1845–1878; Howard MSS:

Table 1.1 *Sample estates by size and acreage, c. 1875*

Estate size in acres	No. of estates	Total acreage	Percentage of total estate area
Devon			
999 and under	–	–	–
1,000–2,999	–	–	–
3,000–9,999	3	12,224	4.7
10,000 and over	3	64,658	21.3
Total	6	76,882	5.1
Northamptonshire			
999 and under	–	–	–
1,000–2,999	3	6,503	1.2
3,000–9,999	4	20,179	20.0
10,000 and over	5	87,527	49.2
Total	12	114,209	19.3
Northumberland			
999 and under	–	–	–
1,000–2,999	1	2,067	1.2
3,000–9,999	1	9,061	3.6
10,000 and over	6	234,503	39.4
Total	8	245,631	20.6

–: No estates in these size categories included in the samples
Sources: As for Figs. 1.1–1.3; J. Bateman, *The Great Landowners*, 1883, 501–11;
F. M. L. Thompson, *English Landed Society*, 1963, 32, 113–17

Fig. 1.3 (cont.)
N99/2, Survey of Northumberland estates, 1886; NdRO, Belsay MSS: Part III
(Supplemental), S.18, Estate plans, 1847–1861; Blackett MSS: ZBL/4/10, Indenture
with Lands Improvement Company, 1861; ZBL/269/12, 20, 26, 34, 43, Maps of the
Matfen and Fallowfield estates, 1846–78; Ridley MSS: ZR1/44/4, Miscellaneous
farm accounts, 1847–82; ZR1/49/11 and 12, Sir M. W. Ridley's fieldbook, 1889;
Sample MSS: ZSA/18/2/1 and 2, A schedule of the Bothal estate belonging to the
duke of Portland, 1861; ZSA/51/29, Map of the Bothal estate (n.d.); Newcastle
Central Library, L.622.33, J. T. W. Bell's plans of the Newcastle Coal District, 1847,
and of the Blyth and Warkworth Coal District, 1851; PRO, 1R29 and 30/25/66, 82,
86, 124, 154, 244, 264, 288, 309, 319, 365, 376, 426, 459, 460, Tithe maps and
apportionments for Bradford, Budle, Burton, Coldmartin, Downham, Henshaw,
Horton, Learmouth, Melkridge, Milfield, Presson, Ridley, Thorngrafton, Wark,
Wark and Sunnilaws; M. Hughes, 'Lead, land and coal as sources of landlord income
in Northumberland between 1700 and 1850', 1963, vol. 1, 338, 353, 363; vol. 2, 60,
100–1)

appeared the most promising for present purposes. From these, a fuller picture of the total amount and chronology of underdraining can be constructed, allowing an assessment of the representativeness of the trends determined from the draining loans under the land improvement acts. The detail that is available at the estate level provides the opportunity to analyse the processes by which the improvement was adopted on particular estates and to identify individual fields that had been drained, the techniques employed and land use before and after draining. Judgments on the organization and efficiency of the improvement and the resulting changes in agricultural practice can be made only from such detailed data. Examination of estate material is an essential complement to the national coverage contained in the government and improvement companies' draining loans.

The extent and location of the estates whose records have been consulted have been depicted for each county in the period 1850–75, the only time when a full cover of estate maps and surveys is available (Figs. 1.1–1.3). There is no complete map coverage for the estates in Devon and Northamptonshire and landownership units have been indicated on a parish basis, holdings being plotted on the same scale as the county as a whole. As map coverage is more comprehensive for the estates in Northumberland, their exact location in the third quarter of the nineteenth century has been marked. The estates form but a proportion of each county. In 1873, they represented respectively 5.1, 19.3 and 20.6 per cent of the total area excluding waste occupied by estates in Devon, Northamptonshire and Northumberland (Table 1.1). Moreover, using the functional estate-size categories devised by Bateman and modified by Thompson, far more information is extant for the larger than for the smaller estates.[106] Such limitations are the product of the known differential survival of records from estates in these size categories and the availability of essentially private material.[107]

In the absence of complete estate cover, it will never be possible to produce an absolute record of draining activity in the nineteenth century. However, the combined analysis of data at a national level from the draining loans under the land improvement acts and of the detail arising from the estate material of these counties is likely to create real progress in the understanding of the role of underdraining as a nineteenth-century agricultural improvement and its contribution to heavy-land agriculture.

2

The need for underdraining in the nineteenth century

In the assessment of the impact of underdraining on nineteenth-century agriculture, a measure of the area that would benefit from the improvement and of its distribution is a priority. Few attempts were made by nineteenth-century agriculturalists to calculate the amount of land in need of draining in England. The dichotomy between heavy- and light-land farming was widely reported in the contemporary literature at both the national and local scale: as William Marshall explained in 1818, 'the most natural division, and at the same time the best agricultural distinction, is into strong and light lands'.[1] And many commentators reiterated the view of Léonce de Lavergne in 1855 that 'the draining away of superabundant water, especially upon stiff soils, has always been the chief difficulty in English agriculture'.[2] Although aware of the problem, most nineteenth-century agriculturalists relied on a qualitative judgment of the extent of the improvement's need, exemplified by that of Josiah Parkes in 1845 that 'a most enormous and untold quantity' of land required to be drained.[3]

In general, those agricultural writers who added some quantification to such statements gave no indication of either the basis of calculation of the areas or their distribution: nevertheless, the results achieved currency in the contemporary literature. Thus Philip Pusey claimed in 1841 and 1842 that one-third of England, 10 million acres, required underdraining.[4] This figure was reported in the *Farmer's Magazine* in 1843, was included by Henry Hutchinson in 1844 in a study of underdraining, and corresponded to the acreage that James Caird considered in need of draining in 1873, although he had applied it to Scotland and Wales as well as England.[5] Others suggested much larger areas: a young J. Bailey Denton in 1842 proposed that 10 million of the 12 million acres of arable in Great Britain should be drained.[6] This scale of activity was adopted by Joshua Trimmer in 1847, who added that 15 million acres of pasture should also be treated, making a total of 25 million acres, about 75 per cent of the cultivated area of Great Britain.[7] Although these high values may have emphasized the contemporary view of the need for the improvement, their range is a reflection of the lack of any thorough survey in their formulation.

The only detailed nineteenth-century analysis of the area requiring draining was made by Denton, which appeared in 1855 and with modifications in 1883.[8] For individual and for groups of counties, he calculated the area of wet land which 'has been drained or remains to be drained with advantage'.[9] In 1855 he concluded that 15.337 million acres of England, 48.1 per cent of total area, needed draining. The distribution of wet land was not uniform throughout the country and at county level he identified much of the southwest and East Anglia as below average, although in these counties wet land occupied between 30 and 40 per cent of total area (Table 2.1). With Essex, the west midland counties of Cheshire, Staffordshire and Warwickshire possessed the largest concentrations of wet land, forming in each over 60 per cent of total area.

By 1883, Denton had increased the area of wet land to 16.452 million acres, 50.5 per cent of England (Table 2.1). He had expanded the wet-land acreage of the northern counties, Devon and Somerset, and the east midland counties of Cambridgeshire, Huntingdonshire, Leicestershire, Lincolnshire and Nottinghamshire. The resulting pattern revealed a distinct variation in the need for draining throughout the country. The counties of Cheshire, Staffordshire, Warwickshire and Essex still retained their high concentrations of wet land, but similar levels were now recorded for Durham and Nottinghamshire (Fig. 2.1). There had been some reduction in the number of counties with the lowest proportions of wet land, being limited to Norfolk and Suffolk in the east, Cornwall in the southwest and a group of five southern counties. Elsewhere the need for draining was relatively uniform, with wet land forming between 40 and 59 per cent of all counties.

Denton's total of wet land would suggest a large proportion of the cultivated area of England needed underdraining: wet land in 1883 formed 69.1 per cent of the average annual cultivated area of the country (23.810 million acres) in the decade 1870–9.[10] However, an element of exaggeration entered Denton's calculations. His figures were derived from the solid geology of different parts of the country and the wet-land areas represented the total acreage of geological formations with drainage difficulties. No distinction was made between cultivated and uncultivated land within these divisions. The wet-land acreage consequently incorporated significant areas that were not cultivated either in 1855 or in 1883 and that as upland would not have warranted cultivation. Denton's results provide an over-assessment of the amount of cultivated land in need of underdraining. Such enhancement of the area that would benefit from the improvement may not have been unintentional. Although Denton wrote extensively on draining matters, he was not an impartial observer. He played a leading role in establishing in 1849 the General Land Drainage and Improvement Company, a major aim

Table 2.1 *Wet-land areas in England in 1855 and 1883, after J. B. Denton*

County or county groups	1855 Total acreage	1855 Wet-land acreage	1855 Percentage of total	1883 Total acreage	1883 Wet-land acreage	1883 Percentage of total
Cheshire	673,280	500,000	74.3	700,000	490,000	70.0
Cornwall	851,200	255,000	30.0	870,000	255,000	29.3
Derbyshire	657,920	300,000	45.6	661,700	300,000	45.3
Durham	702,080	400,000	57.0	650,000	500,000	76.9
Essex	981,120	650,000	66.3	1,015,000	650,000	64.0
Leicestershire	500,000	250,000	50.0	514,000	280,000	54.5
Northamptonshire	650,240	375,000	57.7	632,000	375,000	59.3
Nottinghamshire	535,680	300,000	56.0	530,000	350,000	66.0
Rutland	95,360	40,000	41.9	100,000	50,000	50.0
Staffordshire	757,760	550,000	72.6	650,000	550,000	73.3
Warwickshire	574,080	350,000	60.9	570,000	360,000	63.2
Bedfordshire Buckinghamshire	768,640	330,000	42.9	760,000	330,000	43.4
Berkshire Dorset Hampshire Oxfordshire Wiltshire	3,223,840	1,200,000	37.2	3,253,800	1,210,000	37.2
Cambridgeshire Huntingdonshire Lincolnshire	2,467,600	1,318,000	53.4	2,591,560	1,468,000	56.6
Cumberland Lancashire Northumberland Westmorland Yorkshire	7,405,120	3,612,000	48.8	7,774,000	4,112,000	52.9
Devon Somerset	2,857,200	1,050,000	36.7	2,991,870	1,347,000	45.0
Gloucestershire Herefordshire Shropshire Worcestershire	2,666,500	1,497,000	56.1	2,715,200	1,450,000	53.4
Hertfordshire Middlesex	583,680	300,000	51.4	580,000	300,000	51.7
Kent Surrey Sussex	2,420,480	1,300,000	53.7	2,407,800	1,300,000	54.0
Norfolk Suffolk	2,254,960	700,000	31.0	2,235,360	700,000	31.3
Unspecified	290,320	60,000	20.7	294,480	75,000	25.5
Total	31,917,060	15,337,000	48.1	32,596,770	16,452,000	50.5

Sources: J. B. Denton, *The Under-Drainage of Land*, 1855, 3–5, and *Agricultural Drainage*, 1883, 33–6.

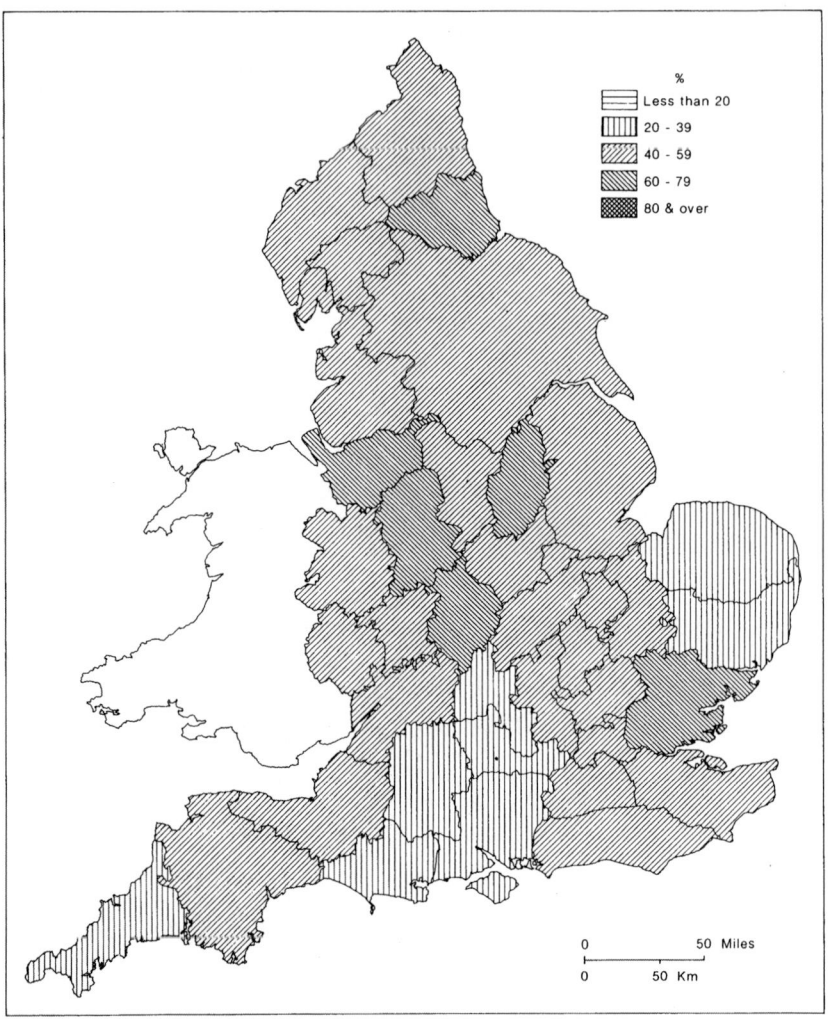

%
Less than 20
20 - 39
40 - 59
60 - 79
80 & over

0   50 Miles
0   50 Km

Figure 2.1 Distribution of wet land by county as estimated by J. B. Denton in 1883
(*Source:* J. B. Denton, *Agricultural Drainage*, 1883, 33–6)

of which was both to finance and undertake draining projects, becoming its
principal engineer, a position which he held into the 1880s.[11] Clearly,
Denton's own interests would have been well served by inflating the extent
of the need for draining. Nevertheless, his figures represented for nineteenth-
century agriculturalists the best contemporary framework in which to judge
the development and progress of the improvement.

In recent times, Denton's analysis has been superseded as increasingly precise estimates of the amount and distribution of land in need of draining have been developed, based on calculations of the area of clayland defined by geological divisions, surveys of the drainage status of fields and the soil maps of England and Wales published by the Soil Survey. As these assess the need for draining in terms of the physical conditions of soils, they provide an absolute measure of the quantity of land that could have benefited from underdraining in the nineteenth century.

The first of these more recent estimates related drainage requirement to the area occupied by claylands as determined by geological divisions, working on the premise that claylands had long been recognized as being greatly in need of underdraining. In 1946, H. H. Nicholson calculated the acreage of land lying on clay formations on published solid geology maps. The resulting estimate of 4.4 million acres represented 17 per cent of the then cultivated area. As Nicholson used only solid geology maps, he excluded from his reckoning Boulder Clay areas marked on drift geology maps, and consequently undervalued greatly the extent of claylands in the country.[12]

However, H. C. Darby's 1964 map of all the main clayland areas below 1,000 ft, and therefore largely cultivated land, renders possible a more accurate assessment of the clayland acreage.[13] From the original map at the scale of 1:1,000,000 both the total area and county distribution of clayland have been computed[14] (Fig. 2.2a and b). In all, clayland covered 9.572 million acres, 29.8 per cent of the total area of England (Table 2.2).

The distribution of clayland on a county basis reveals a concentration in the midlands, with Bedfordshire, Buckinghamshire, Huntingdonshire, Leicestershire, Northamptonshire, Rutland, Warwickshire and Worcestershire all possessing 45 per cent or more of their area as such, while high proportions are also located in Cheshire, Essex and Suffolk (Fig. 2.2b). The lowest densities are recorded in Cornwall, Devon and Herefordshire, each with less than 10 per cent of its area as clayland. Yet, the distribution of claylands does not offer a complete picture of underdraining need in the country. The clays included in the present dicussion vary in structure from heavy soils to lighter loams where the need for draining is slight.[15] More significantly, many soils other than those based on clays require draining and present-day drainage designers record the improvement's need on soils as various as clay soils, clay loams, sand loams and silty loams, with varying intensity and according to land use.[16] The value of underdraining on soils besides clays was also appreciated by nineteenth-century agriculturalists. Although clayland was of little extent in Cornwall, Devon and Herefordshire, the authors of the Royal Agricultural Society of England's prize essays on the farming of those counties, W. F. Karkeek, Henry Tanner and Thomas Rowlandson, could all note the necessity for extensive underdraining on a variety of soil types.[17] Clearly, the physical need for underdraining extended to an area greater than that encompassed by claylands.

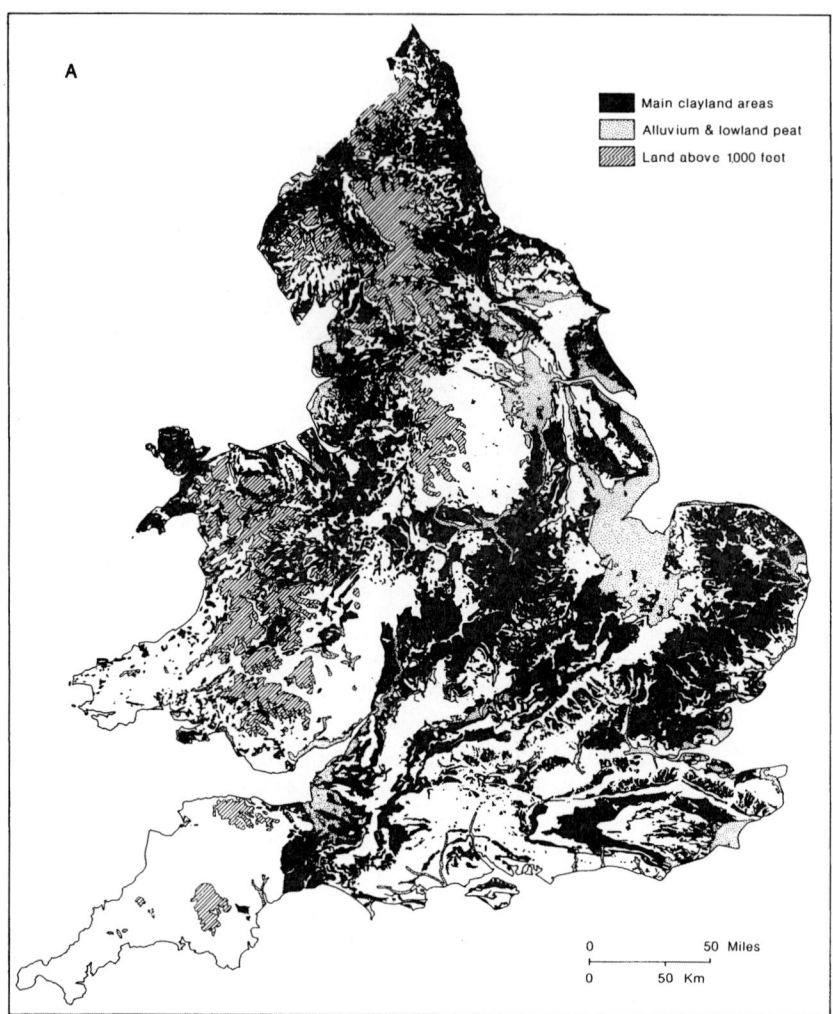

Main clayland areas
Alluvium & lowland peat
Land above 1,000 feet

0   50 Miles

0   50 Km

Figure 2.2 A. Distribution of claylands

    This conclusion is substantiated by the results of a survey into drainage
needs instituted by the Ministry of Agriculture in 1968–9.[18] In this, the
drainage status of a random 5 per cent sample of fields throughout the
country was assessed jointly by Ministry of Agriculture drainage advisers
and farmers. Of the cultivated area in England at the time, 23.55 million
acres, it was estimated from this sample that 6.67 million acres had been
drained and that a further 6.35 million acres were still in need of the
improvement. Although the distribution of this land is not detailed, the

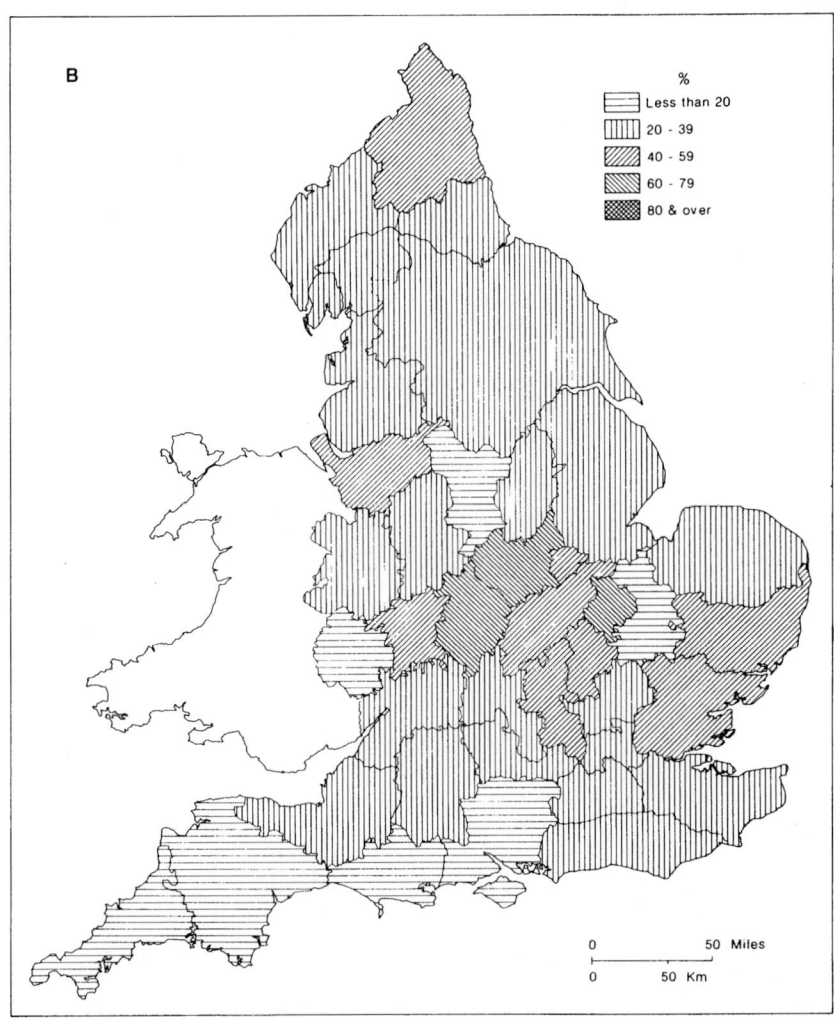

Figure 2.2 B. Density of claylands by county (*Source:* As for Table 2.2)

results of the sample suggest that in all underdraining was physically necessary on 13.02 million acres in the country, 55 per cent of the cultivated area.[19]

The publication by the Soil Survey of the *Soil Maps of England and Wales* in 1975 and 1983 allows greater clarification of the amount of land in need of underdraining. These maps depict both the area and distribution of soils with drainage difficulties. At the national and county level, they provide the most reliable and complete data to calculate the need for underdraining in terms of the physical conditions of soils.

Table 2.2 *Distribution of main clayland areas, after H. C. Darby*

County	1873 acreage	Clayland acreage	Percentage of county
Bedfordshire	295,516	175,340	59.3
Berkshire	455,035	122,738	26.9
Buckinghamshire	468,574	239,100	51.0
Cambridgeshire	547,427	87,670	16.0
Cheshire	715,835	344,304	48.1
Cornwall	857,608	0	0
Cumberland	973,510	325,176	33.4
Derbyshire	642,794	84,482	13.1
Devon	1,657,749	133,896	8.1
Dorset	628,225	108,392	17.3
Durham	699,626	231,130	33.0
Essex	994,608	573,840	57.7
Gloucestershire	810,995	180,122	22.2
Hampshire	1,027,673	168,964	16.4
Herefordshire	540,539	41,444	7.7
Hertfordshire	390,828	109,986	28.1
Huntingdonshire	230,486	146,648	63.6
Kent	1,002,972	205,626	20.5
Lancashire	1,205,037	400,094	33.2
Leicestershire	511,428	380,966	74.5
Lincolnshire	1,725,641	385,748	22.4
Middlesex	178,466	55,790	31.3
Norfolk	1,352,291	494,140	36.5
Northamptonshire	633,286	310,830	49.1
Northumberland	1,236,655	495,280	40.1
Nottinghamshire	529,281	148,242	28.0
Oxfordshire	467,306	105,204	22.5
Rutland	92,696	41,444	44.7
Shropshire	852,493	261,416	30.7
Somerset	1,043,879	280,544	26.9
Staffordshire	729,248	270,980	37.2
Suffolk	943,166	479,794	50.8
Surrey	479,921	148,242	30.9
Sussex	925,076	216,784	23.4
Warwickshire	565,448	393,718	69.6
Westmorland	508,115	103,610	20.4
Wiltshire	869,233	208,814	24.0
Worcestershire	463,730	216,784	46.7
Yorkshire	3,858,624	894,234	23.2
Total	32,111,020	9,571,516	29.8

Sources: H. C. Darby, 'The draining of the English clay-lands', 1964, 191; W. G. Hoskins and L. D. Stamp, *The Common Lands*, 1963, 92–3

The 1975 map at the scale of 1:1,000,000, based on detailed soil maps covering 20 per cent of the total area, reconnaissance soil surveys, geological and relief maps and mean-maximum, soil-moisture deficit data, is divided into 71 units, each representing a distinct soil group.[20] The characteristics of the dominant and associated soils of each of these soil-group units are recorded, including a description of drainage quality. A fivefold classification of these 71 soil groups may be devised, based on their drainage qualities: alluvium and lowland peat, with high groundwater levels mainly controlled by ditches and pumps; upland soils and peat, mostly over 1,000 ft; well-drained soils on chalks, limestones, loams and sands with no drainage problems; well or moderately well-drained loamy, sandy and silty soils associated with clayey and loamy soils with impeded drainage which experience drainage difficulties; and clayey and loamy soils with impeded drainage which require underdraining[21] (Fig. 2.3a). By computation, the areas of alluvium and lowland peat and of upland soils and peat amounted to 2.315 and 3.193 million acres, respectively 7.2 and 9.9 per cent of England. Well-drained soils covered 8.457 million acres, 26.4 per cent of the total, while soils with impeded drainage occupied 12.540 million acres, 38.6 per cent of the total. The extent of soil groups of well-drained soils associated with clayey and loamy soils with impeded drainage was 5.538 million acres, 17.9 per cent of England. The exact proportion that the associated soils with impeded drainage formed of the last category cannot be determined. However, spatially they were of secondary importance and for purposes of quantification the assumption has been made that such soils represented 10 per cent of the total area of this group. Taking this assumption into account, the total area of clayey and loamy soils with impeded drainage amounted to 13.094 million acres, 40.8 per cent of the country.

This area of land with drainage difficulties is corroborated by the *Soil Map of England and Wales* of 1983 published at the larger scale of 1:250,000.[22] For this map a new description of the drainage characteristics of soil groups was adopted, impeded drainage being replaced by the term slowly permeable, seasonally waterlogged. However, the correspondence of the total area of clayey and loamy soils with impeded drainage with that of slowly permeable, seasonally waterlogged clayey and loamy soils is very high, the latter occupying 39.8 per cent of England and Wales.[23]

The 1975 and 1983 soil maps reveal similar patterns in the distribution of soils with problems of drainage throughout the country. Using data computed from the 1975 map, the density of clayey and loamy soils with impeded drainage was highest on a country basis in the east midlands, forming 65 per cent and over of the total area (Fig. 2.3b; Table 2.3). Such soils were also significant in area in both the west-midland and northern counties with the exception of Westmorland. In general, they formed a smaller proportion of the total area of eastern, southeastern and southern

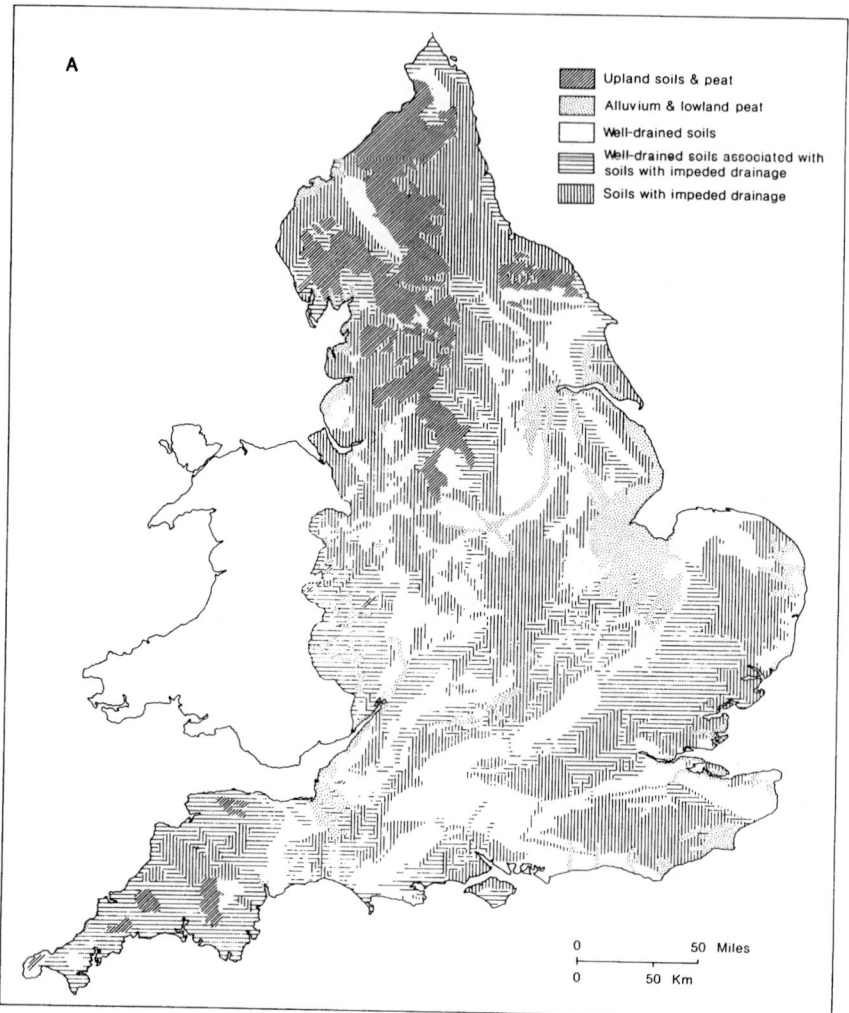

Figure 2.3 A. Distribution of soil types

counties, although relatively high levels were recorded for Suffolk, Middlesex, Surrey and Sussex. The southwestern counties possessed the lowest quantities of clayey and loamy soils with impeded drainage.

The extent of soils with drainage difficulties varied in the three counties where draining activity has been examined on individual estates. These soils occurred most widely in Northamptonshire and Northumberland, occupying respectively 64.8 and 53.1 per cent of each county, while in Devon they formed a much smaller area, 28.9 per cent. Use has been made of the larger-scale 1983 *Soil Map of England and Wales* to plot the distribution of soil

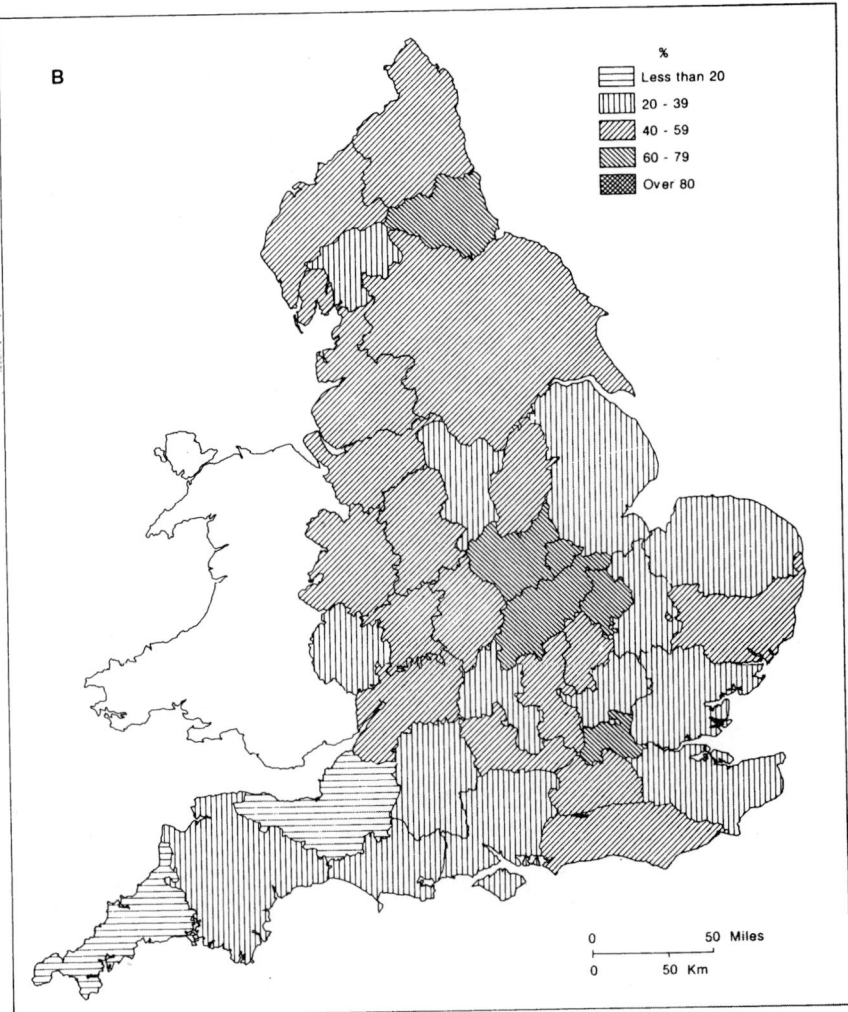

Figure 2.3 B. Density of soils with impeded drainage by county (*Source:* As for Table 2.3)

types in the three counties, the results indicating clear differences between them in the spatial pattern of slowly permeable, seasonally waterlogged clayey, loamy and silty soils in need of underdraining (Figs. 2.4–2.6). Large districts of both Northamptonshire and Northumberland were dominated by such soils, but in Devon their distribution was broken, being mixed with other soil types. Comparison of the location of the sample estates with the distribution of soils with differing drainage characteristics reveals that the estates encompassed all soil types in each county, permitting close

Table 2.3 *Distribution of clayey and loamy soils with impeded drainage*

County	Area of soil groups of clayey and loamy soils with impeded drainage (acres)	10 per cent of area of soil groups of well-drained soils associated with clayey and loamy soils with impeded drainage (acres)	Total area of clayey and loamy soils with impeded drainage (acres)	Total as percentage of 1873 county area
Bedfordshire	157,486	2,047	159,533	53.9
Berkshire	195,665	10,011	205,676	45.2
Buckinghamshire	178,059	13,589	191,648	40.9
Cambridgeshire	116,105	2,007	118,112	21.6
Cheshire	422,343	4,295	426,638	59.6
Cornwall	42,880	65,178	108,058	12.6
Cumberland	428,344	9,735	438,079	45.0
Derbyshire	128,558	15,427	143,985	22.4
Devon	421,068	57,855	478,923	28.9
Dorset	212,340	1,885	214,225	34.1
Durham	421,874	6,996	428,870	61.2
Essex	328,220	53,709	381,929	38.4
Gloucestershire	372,247	1,865	374,112	46.1
Hampshire	369,963	17,470	387,433	37.7
Herefordshire	82,162	33,243	115,405	21.3
Hertfordshire	58,624	21,886	80,510	20.6

Huntingdonshire	162,723	0	162,723	70.6
Kent	320,952	13,039	333,991	33.3
Lancashire	578,417	8,435	586,852	48.7
Leicestershire	393,800	2,557	396,357	77.5
Lincolnshire	465,923	18,982	484,905	28.1
Middlesex	114,218	6,425	120,643	67.6
Norfolk	440,847	10,227	451,074	33.4
Northamptonshire	401,503	9,119	410,622	64.8
Northumberland	656,664	0	656,664	53.1
Nottinghamshire	214,359	0	214,359	40.5
Oxfordshire	126,173	11,215	137,388	29.4
Rutland	63,033	0	63,033	68.0
Shropshire	323,947	28,985	352,932	41.4
Somerset	167,021	37,580	204,601	19.6
Staffordshire	422,965	2,188	425,153	58.3
Suffolk	458,379	3,773	462,152	49.0
Surrey	249,559	7,679	257,238	53.6
Sussex	525,443	2,580	528,023	57.1
Warwickshire	265,761	5,089	270,850	47.8
Westmorland	148,878	5,843	154,721	30.4
Wiltshire	219,916	1,825	221,741	25.5
Worcestershire	262,935	7,095	270,030	58.2
Yorkshire	1,620,622	54,021	1,674,643	43.4
Total	12,539,976	553,855	13,093,831	40.8

Source: B. W. Avery et al., *Soil Map of England and Wales, 1:1,000,000*, 1975

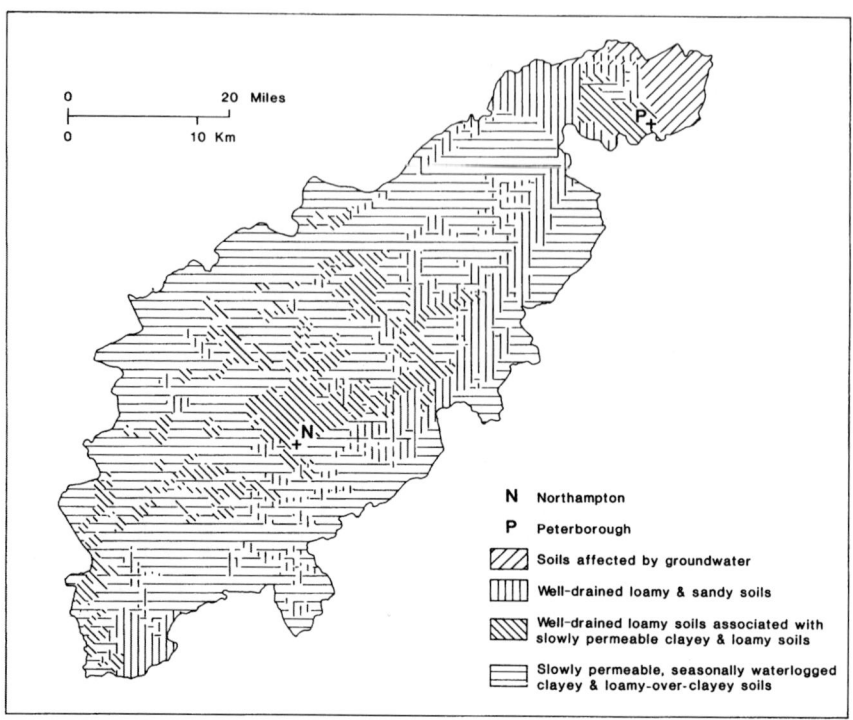

Figure 2.4 Soil types in Northamptonshire (*Source:* Soil Survey of England and Wales, *Soil Map of England and Wales, 1:250,000,* 1983)

examination of the relationship between the physical need for draining and the occurrence of the improvement.

This analysis of drainage qualities based on the two soil maps of England and Wales has indicated that soils with problems of drainage, which are likely to need underdraining, covered just over 13 million acres in England. The close correspondence of this figure to that derived from the other most recent source, the Ministry of Agriculture's 1968–9 sample survey of the drainage status of fields, would point to the basic reliability of the estimate. The results obtained from these maps have been adopted in the present study as the best measures of the absolute amount and distribution of land in the country that could benefit from underdraining. Although smaller than Denton's estimates, the area of soils with drainage problems represents about 55 per cent of the average annual cultivated acreage of the country in the decade 1870–9 and demonstrates the fundamental importance of under-draining to half of English agriculture in the nineteenth century.

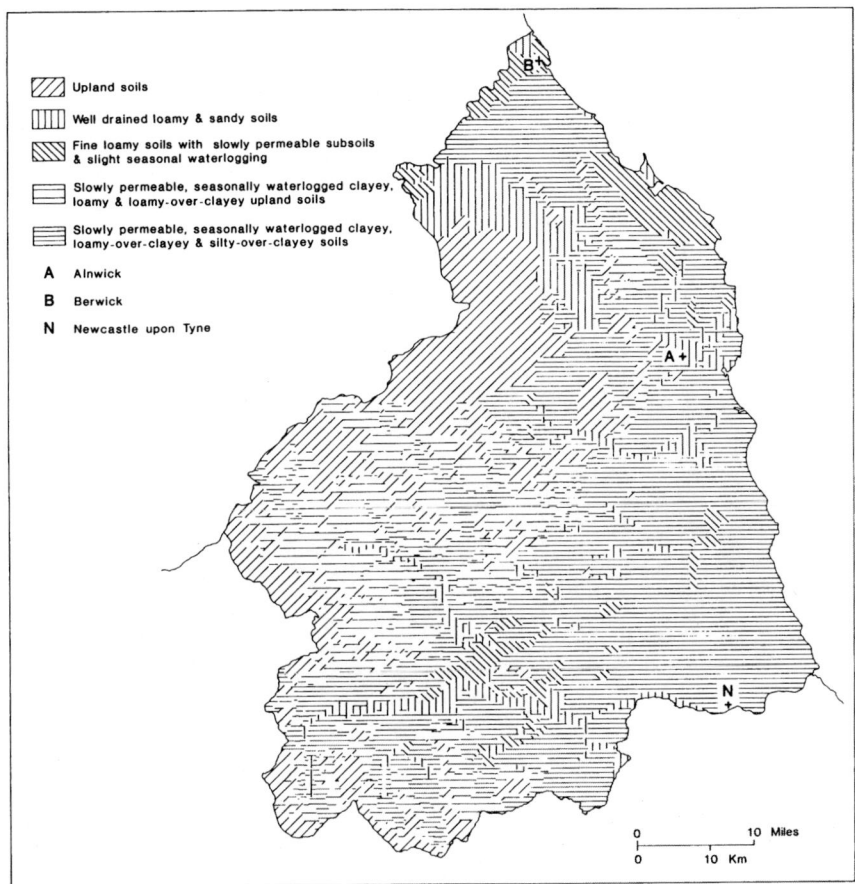

Figure 2.5 Soil types in Northumberland (*Source:* As for Fig. 2.4)

The extent of underdraining at the beginning of the nineteenth century

No precise evidence is available at the national level of the extent to which this area had been underdrained by the beginning of the nineteenth century. In the absence of such material, the wealth of agricultural literature that appeared during the Napoleonic Wars may be used to provide a picture of draining activity in the country around 1800. The series of *General Views* of the agriculture of each county sponsored by the Board of Agriculture are the most valuable of these literary sources. These county accounts were written to a uniform plan which included a chapter on improvements in which draining was to be discussed. Two editions of these *General Views* exist, an earlier, provisional quarto edition dating from 1793–6 in which the plan was

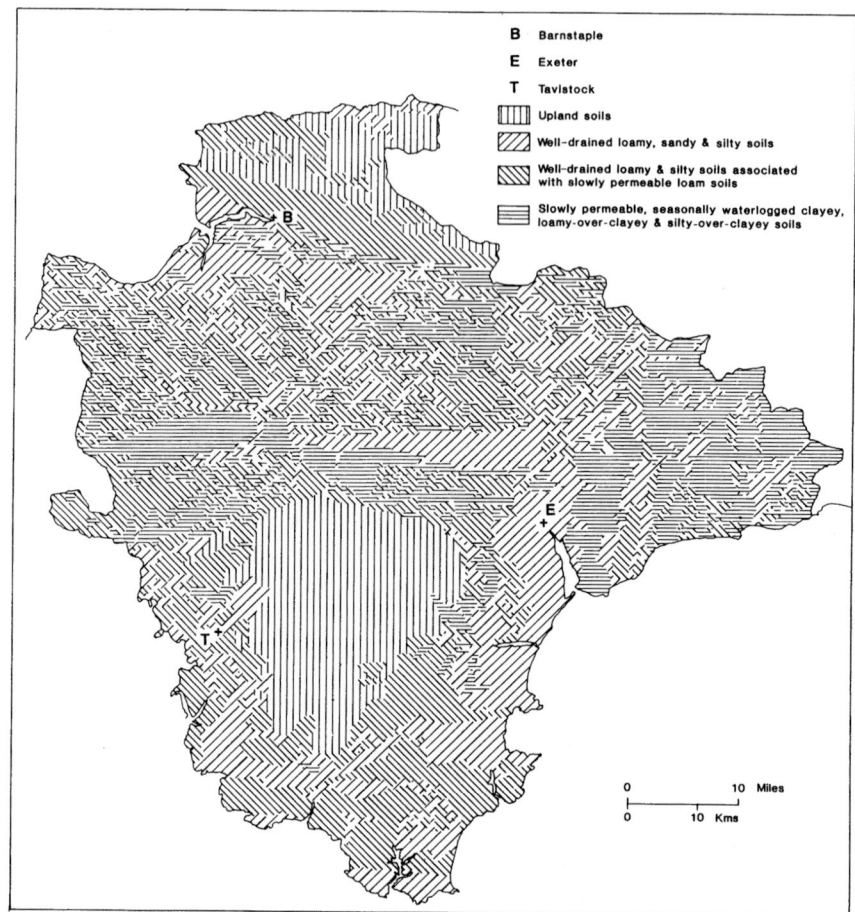

Figure 2.6 Soil types in Devon (*Source:* As for Fig. 2.4)

less rigidly maintained and from which observations on draining were occasionally lacking, and a later, more thorough octavo edition, published between 1794 and 1817, the work usually of different authors and with a separate review of draining as an agricultural improvement. Together, these accounts allow an assessment to be made of the amount of land drained throughout the country about 1800.[24]

This material has been supplemented by detailed information on underdraining reported in the *Annals of Agriculture*, which ran from 1784 to 1815; the early numbers of the *Farmer's Magazine*, which began in 1800; the *Communications to the Board of Agriculture*, dating for the period 1801–6; the report by the Board of Agriculture on *The Agricultural State of the*

Kingdom in...1816; and a number of draining treaties, particularly John Johnstone's account of Joseph Elkington's system of spring draining. William Marshall's studies of the rural economy of various districts of England published between 1787 and 1798 have also been used, but in these draining is treated less consistently and in less detail than in the *General Views*.[25]

Distinct limitations should be recognized in the use of this literary evidence. First, the material is not of a uniform quality and there is considerable variation in the level of information available for each county. In the production of the *General Views* of the Board of Agriculture, a large number of authors were involved, 30 alone contributing to the octavo editions.[26] Despite the set plan for these volumes, this range of authors ensured that the topic of draining was discussed in great variety, depending on the reporter's competence and attention to detail. John Farey provided much information on the practice of underdraining in Derbyshire, listing not only places drained but also the names of professional drainers and the location of drain-tile manufacturers, while in neighbouring Cheshire Henry Holland made no comment on the specific incidence of the improvement.[27] Such differences in detail and reliability of information cannot be remedied. Secondly, the greater part of the agricultural literature on draining is practical and technical in approach. The authors were primarily concerned in describing the types and methods of draining land, in identifying best practice and in exhorting its adoption to increase overall productivity. Far less attention was paid to the spatial aspects of the improvement, such as the distribution and the amount of land drained, these being treated where mentioned largely in general terms. As a result, the picture of draining activity derived from this body of material is not as complete as could be wished.

An indication of the distribution of underdraining in the country around 1800 may be obtained by plotting references from the above sources to places that had been drained, distinguishing separately the system of tapping springs developed by Joseph Elkington (Fig. 2.7). The resulting map, however, should be regarded as no more than a very tentative and incomplete picture of underdraining activity. This assessment is a reflection of the varying detail recorded in these diverse late eighteenth- and early nineteenth-century accounts. In addition, it recognizes the difficulty in attempting to bring precision to a qualitative source: there is not necessarily a direct relationship between the number of places mentioned and the extent and intensity of underdraining. Thus, H. E. Strickland in his report of the agriculture of the East Riding of Yorkshire wrote of underdraining being widely practised in the area but identified only one example of the improvement.[28] Nevertheless, while acknowledging these problems, the map would point to a broad distinction in draining activity between a group of

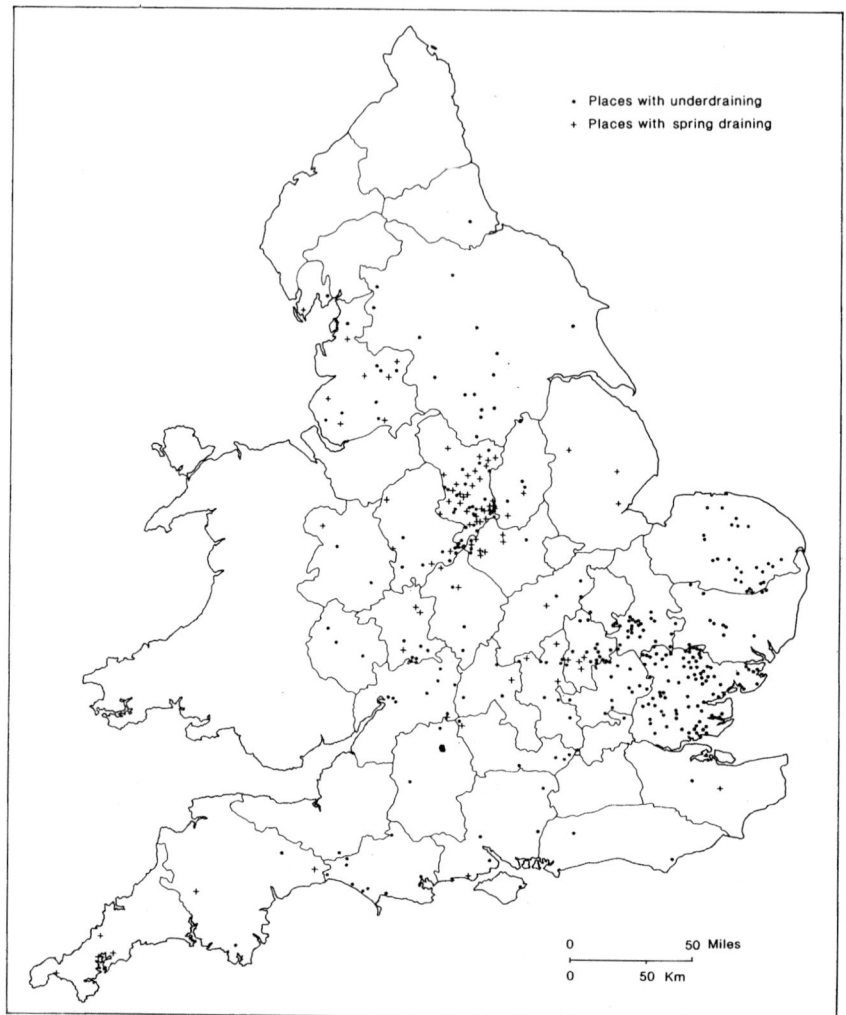

Figure 2.7 References to underdraining around 1800 (*Sources:* The county *General Views* of the Board of Agriculture; *Annals of Agriculture*, 1784–1815, vols. 1–45; *Farmer's Magazine*, 1800–16, vols. 1–15; W. Marshall's *Rural Economies* of parts of England, 1787–98; J. Johnstone, *An Account of the Mode of Draining Land*, 2nd edn, 1801; Board of Agriculture, *The Agricultural State of the Kingdom in... 1816*, 1816)

eastern counties and other parts of the country. The concentration of references in eastern England would suggest that the improvement was widely adopted in that region, while in the rest of the country the use of underdraining would seem in general less intense, although the level of application varied between individual counties.

Amplification of this pattern may be achieved by examining the comments of contemporary agriculturalists on draining practice. In the north, John Bailey and George Culley recorded that the improvement had only recently been introduced into Cumberland, Durham and Northumberland and that in consequence little land had been drained.[29] A similar lack of underdraining was noted in Westmorland and in parts of Yorkshire.[30] More activity was reported in Lancashire by both John Holt and R. W. Dickson but it was largely associated with the reclamation of mossland in which underdraining formed but a part.[31]

A distinction was drawn for all the west-midland counties from Cheshire south to Gloucestershire between the existence of large areas of wet land and the small amount of land drained. William Pitt observed in Staffordshire that, although the practice of underdraining had recently spread, much remained to be done, while William Marshall wrote more simply in Gloucestershire 'much underdraining is wanted'.[32] Although an element of variation in the intensity of draining was reported in the east midlands, the general view was that the improvement was little adopted. Arthur Young ascribed the absence of underdraining in Lincolnshire to its recent introduction.[33] While 'the necessity of draining wet lands has of late years been much better understood and attended to', as Robert Lowe found in Nottinghamshire, 'the very great neglect' of the improvement, that Richard Parkinson observed particularly in Huntingdonshire, was widely noted.[34] Only John Farey's meticulously detailed report for Derbyshire deviated from the general view, asserting 'draining of land has been so much practised'.[35]

A relatively uniform level of draining activity was recorded for the southern counties. The authors of the *General Views* of Devon and Somerset agreed on the general want of the improvement in those counties.[36] William Marshall and Thomas Davis indicated that the use of underdraining was more widespread in Wiltshire but were aware that large areas of the county remained untouched.[37] The lack of extensive underdraining was a theme common to the reports on all the southeastern counties, the Rev. Arthur Young observing that in Sussex, where the improvement was greatly needed, the methods were not thoroughly understood and the practice was confined to a few individuals.[38]

Throughout the East Anglian counties, a higher level of draining activity was reported, the practice being noted to be widely understood and implemented. In the *General Views* of Cambridgeshire, Charles Vancouver and William Gooch listed 26 parishes where underdraining had been carried out.[39] Arthur Young considered that in Norfolk the improvement was widely undertaken and in Suffolk the practice was 'general on all the wet lands of the county [and was] too well-known to need a particular description'.[40] Similar levels of draining were noted in Essex and Hertfordshire, there being no part of the latter county where the improvement was 'not well understood and practised'.[41] Nevertheless, some exaggeration should be recognized in these assessments and, despite Arthur Young's enthusiasm for detail, only eleven specific examples of draining were recorded in Suffolk and thirteen in Hertfordshire.[42]

The general picture provided by contemporary agricultural literature was that little underdraining had been undertaken in the country around 1800 while much land remained in need of the improvement. Some divergence from this pattern was identified in a group of eastern counties where the improvement was observed as well established and amount of land drained would seem greater. Little opportunity exists to quantify these views, but where instances are given the acreages involved tend to be small. For example, of the fifteen certified cases of the success of Joseph Elkington's system of spring draining cited by Sir John Sinclair in 1795 in his report as president of the Board of Agriculture to the House of Commons supporting the award of a government grant to Elkington, acreages are noted in thirteen.[43] The total area recorded as being drained in these thirteen cases was 779 acres, an average of 60 acres at each location. The literary evidence would suggest that the absolute amount of land underdrained, even in eastern England, must have represented only a small proportion of the area in need.

This judgment is reinforced by an examination of the draining systems in use throughout the country at the time. In the literature, there was general acceptance that wetness in soils arose from two sources: the product of springs and the accumulation of rain water on impervious surfaces.[44] Commentators were aware that the former source of soil wetness tended to affect specific point locations and required treatment by methods of spring tapping pioneered by Elkington but that the influence of the latter was more extensive, creating problems for agriculture in large parts of the country, particularly areas of clayland and heavy loams, and being remedied by the use of a variety of underdraining systems.[45]

Descriptions of techniques in the northern counties concentrated on spring draining. No methods of underdraining to remove surface water were recorded for Cumberland and Westmorland[46] and they were given little mention in either Durham or Lancashire. Mole draining was reported at single locations in both counties, while forms of turf underdraining were noted in Lancashire.[47] More attention was paid in the midlands to underdraining systems which aimed to deal with surface water. The use of the mole plough was observed in Gloucestershire, Herefordshire, Bedfordshire and Huntingdonshire, while throughout the midlands stone, turf and wood underdrains were frequently described.[48] Nevertheless, references to methods of spring draining predominated revealing the popularity of this draining system, and a large number of examples of Elkington's schemes were located in this area, ranging from Stretton upon Dunsmore, Warwickshire, where he first applied the technique in 1764, and Flitwick, Bedfordshire, on the duke of Bedford's estate, to Madeley, Staffordshire, on the estates of the Crewe family, where he died in 1806 at the age of 67.[49]

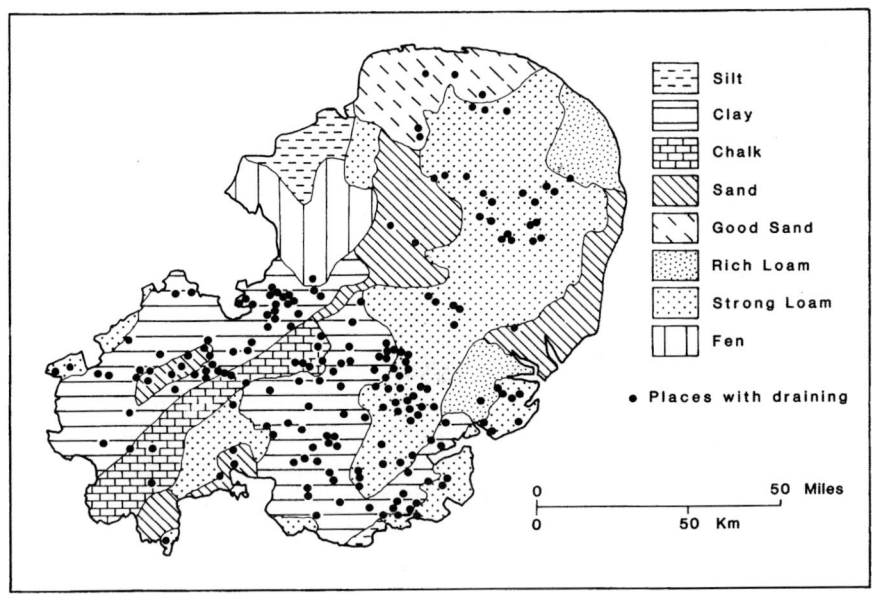

Silt
Clay
Chalk
Sand
Good Sand
Rich Loam
Strong Loam
Fen

• Places with draining

0   50 Miles

0   50 Km

Figure 2.8 References to underdraining in eastern England around 1800 in relation to soil regions identified in the *General Views* (*Source:* As for Fig. 2.7)

A similar pattern was reported in the southern counties. Stone, turf and wood underdrains had been noted at various locations from Cornwall through Hampshire to Kent, while mole ploughs were at work in Berkshire, Hampshire, Surrey, Sussex and Wiltshire.[50] However, far more detail was given of methods and examples of spring draining. The relative use of the two types of draining system was appreciated by Charles Vancouver in Devon, where he wished 'as much pains [had] been bestowed in endeavouring to carry the water from off the ground as in most cases have been exerted to prevent it coming on'.[51]

A more integrated approach to removing excess soil water was recorded in the eastern counties. The system of underdraining in use was intended, as Lord Petre noted in Essex and Hertfordshire, 'not only to carry off the water from springy, spewy soils but as a melioration of stiff loamy clays'.[52] Greater heed was given to underdraining as a means to remove surface water than in other parts of the country. This emphasis is confirmed by a comparison of the distribution of places recorded as having been drained about 1800 with the soil regions marked on the maps accompanying the octavo editions of the *General Views* of seven counties in eastern England (Fig. 2.8): underdraining activity was predominantly located on clay and loam soils. Throughout the area underdrains were described as being most commonly filled with bushes, straw, thorns, wood and occasionally stones, while mole draining was reported in all the counties.[53]

The concentration on methods of spring draining that was noted in all counties with the exception of those in East Anglia provides support for the view that the large areas of heavy soils suffering from surface water had been little touched by underdraining at the beginning of the nineteenth century. Indeed, the importance of some of the underdraining methods described in use on these soils may have been overestimated. The existence of mole draining was reported from many counties, the ploughs most often mentioned being those of Adam Scott and R. Lumbert. Yet, besides the large number of horses required to pull both, the Royal Society of Arts recognized in 1795–6 that the former possessed technical deficiencies which hampered its adoption, while William Gooch noted in 1813 in his *General View* of Cambridgeshire that throughout the country only forty of the latter plough were at work.[54]

A contributory factor in the neglect of underdraining was the distinction that can be detected in the literature for the responsibility for implementing the different draining systems. The dramatic changes described as resulting from the adoption of Elkington's mode of spring draining, transforming areas of bog into cultivable land, attracted both the attention and capital of landowners ranging from Charles Towneley at Towneley, Lancashire, and the marquess of Donegall at Fisherwick, Staffordshire, to the earl of Radnor at Coleshill, Berkshire.[55] The systems of underdraining to remove surface water on already cultivated land produced less startling effects. Moreover, the design of these underdrains and the materials used as fill resulted in systems with a limited life. The depths of underdrains were reported in general to be shallow, averaging in East Anglia between 20 and 30 in and for mole drains from 18 to 20 in, and rendering them susceptible to disturbance particularly in arable areas, while in the absence of the widespread use of tiles, the most frequently cited fill materials – from bushes to stones – were liable to easy blockage of the drainage channel.[56] Although Arthur Young found instances of underdrains lasting twenty-five years and more in Suffolk, a much shorter expectancy of life was more usually recorded.[57]

Such underdraining with its lack of permanence was identified in the *General Views* as forming part of the tenant's commitment to agriculture.[58] Many commentators were of the opinion that tenant responsibility restricted the spread and effectiveness of the improvement.[59] William Stevenson could note in Surrey that tenant responsibility 'too often renders the draining bestowed incomplete either in the width and depth or in the number of the drains or in the kind of materials used to fill them'.[60]

In effect, it would seem that few tenants were prepared to accept the responsibility for adopting the improvement, for, according to John Lawrence in 1801, 'the common error in the management of clays [was] the omission of drainage'.[61] Indeed, certain clays were considered too retentive to be satisfactorily underdrained. Thomas Rudge observed this to be the case in the clay vales of Gloucestershire, and similar soils were identified in Essex and Sussex.[62] In such circumstances, tenants resorted to alternative and simpler methods of removing surface water from heavy lands, and throughout the country the drainage of these areas was accomplished by methods of surface draining, either in the form of ridge and furrow or by the creation of surface cuts and grips.[63] Despite their limited efficiency and the loss of cultivated land, extensive use of these systems was described in Durham, Lancashire, Northumberland and the North Riding of Yorkshire.[64] Thomas Rudge expressed the view that no other method could deal with the claylands in Gloucestershire, while in Bedfordshire and Leicestershire Thomas Batchelor and William Pitt saw ridge and furrow as the means of effectively draining areas of strong loams and clayey soils.[65] Abraham and William Driver recorded that in Hampshire much of the arable was simply drained by throwing the land into ridges.[66] The use of ridge and furrow was recommended to remove surface water in Berkshire, Kent and Middlesex, John Boys believing the practice on heavy lands in Kent to be as effective as underdraining.[67] Even in East Anglia, such techniques were employed, being found by Arthur Young on clay soils in Essex, Hertfordshire and Norfolk.[68] The distribution maps compiled under the guidance of W. R. Mead of present-day survival of ridge and furrow in six midland counties and in Cambridgeshire and Kent provide an indication of how extensive these forms must have been on heavy lands.[69] This material and the numerous descriptions of surface draining in the agricultural literature must be regarded as further, if indirect, evidence of the lack of underdraining around 1800.

A detailed literature records the methods and systems of draining practised in England around 1800. It reveals that spring-draining techniques had been generally successful, although highly localized and applied to individual springs. Other material suggests that the use of underdraining systems on heavy lands to remove surface water had not been widely adopted and that considerable recourse had been made to forms of surface draining. The dominant impression arising from these literary sources remains that at the beginning of the nineteenth century effective underdraining of soils with drainage difficulties had hardly begun.

3

The intensity and location of underdraining, 1845–1899

For the period 1845–99, statistical data are available to examine the spread of underdraining in England as a whole. The loan capital that was established from 1840 onwards by the body of land-improvement legislation provided landowners with an alternative source to private funds for financing underdraining. From 1847 to 1899, under this legislation, representing the public component of draining activity, £8,995,000 was lent for the purpose of underdraining in England, Scotland and Wales.[1] The records of these loans have in the main survived and may be used to construct an index of the amount and distribution of land drained in England over the period. Although the political and administrative framework of this land-improvement legislation has been described by F. M. L. Thompson and D. Spring,[2] a review of its development is essential for an understanding of the nature and value of this data source on underdraining in the second half of the nineteenth century.

The development of government- and improvement-company-financed draining

The various land improvement acts arose from efforts to overcome the limitations that contemporaries perceived were placed on owners of settled estates in undertaking agricultural improvement.[3] Although the extent of settled estates and the effect of settlement on agricultural improvement remain unclear,[4] a series of acts was passed from 1840 onwards which aimed at relieving the restrictions supposedly experienced on settled estates by allowing tenants for life to borrow money for agricultural improvement and to establish a rentcharge on the lands improved to redeem the capital and interest over a number of years. Although eventually applied to a range of improvements, underdraining was the first and most important of the improvements that these acts permitted. Because in theory these measures were for the use of the tenant for life, part of the legislation had to protect the reversionary interest of the settled estate by ensuring the security and efficiency of the improvements undertaken. As part of the machinery for this protection, administrative bodies were employed to supervise the loans and the improvements, and it is from their records that data on underdraining may be obtained.

The principle that tenants for life might borrow for agricultural improvement on the security of their estate, repaying capital and interest in a rentcharge over a period of years, was established in the 1840 act to enable owners of settled estates to defray the expense of draining by way of a mortgage.[5] However, the value of the act in encouraging draining was limited. To be allowed to use the act, a landowner had to petition the court of either Chancery or Exchequer, detailing the amount and cost of draining to be done. A surveyor was appointed to produce a report on the proposed improvement, with which the master of the court decided whether or not to permit such a loan to cover the cost of draining to be made.[6] The procedure was clumsy and petitioners complained of the delay involved.[7] In addition, the process was costly and expenses, amounting in some cases to nearly 10 per cent for every £1,000 borrowed, could not be charged on the estate.[8] The rentcharges resulting from loans made under the act were also high. Tenants for life could not borrow at a rate of interest greater than 5 per cent and the period to amortize the loan was restricted to between 12 and 18 years. With money borrowed at $4\frac{1}{2}$ per cent, the repayment of a loan over 18 years necessitated an annual rentcharge of £8.4 per cent on the outlay, an amount greater than the £7 per cent increase in the value of land that the surveyor had to report was likely to arise from the adoption of the improvement.[8] Applications for loans under this act were few, only eleven being recorded by 1845, and evidence exists of only four loans being taken up.[10] The act was amended in 1845 to reduce the cost of expenses, but the amendment was rendered ineffective by subsequent legislation.[11]

However, the principle embodied in these two acts was developed more effectively in the Public Money Draining Act, 1846, by which £2 million was made available from the Treasury to landowners to borrow for underdraining. In 1850 a further £2 million was sanctioned.[12] The loans applied to England, Scotland and Wales, a separate sum being allocated for Ireland. Under these acts, money borrowed by landowners was repaid over 22 years and, as the Treasury provided the funds at $3\frac{1}{2}$ per cent interest, resulted in a rentcharge of £6.5 per cent on the outlay.[13] The administration of the loans was given to the Inclosure Commissioners and, with the exception of two acts discussed below, subsequent nineteenth-century land-improvement legislation in England retained that body and their successors, the Land Commissioners and the Board of Agriculture, as administrators of such improvement loans.

Landowners wishing to borrow to drain their estates under these acts applied to the Inclosure Commissioners for an advance. The Inclosure Commissioners arranged for the land to be examined by an assistant commissioner or inspector, as they became known, who reported on the proposed draining. If a favourable report was received, the commissioners issued a provisional certificate to the amount of the proposed loan so that an applicant could obtain credit to carry out the improvement. When completed, an inspector checked the draining to ensure that it had been properly undertaken. If satisfied, the commissioners ordered an absolute certificate to be issued by the Treasury to the amount spent by the landlord.[14] This procedure was common to all draining loans made under subsequent land-improvement legislation involving the Inclosure Commissioners. It attempted to secure that the draining was satisfactorily executed, lasting at least as long as the duration of the rentcharge and increasing the value of the land improved to at least the level of the rentcharge, and that as a result the rentcharge to redeem the loan should not become a burden on the estate.

Based on the same machinery and principles, the Private Money Draining Act, 1849, was introduced to allow landowners to borrow from private sources, as distinct from the government, for the purpose of underdraining, repaying loans by means of a 22-year rentcharge on the lands improved. Attempting to achieve a low level of rentcharge, the borrowing of funds was limited to rates of interest at 5 per cent and less. As money could not be obtained in the market at the rate offered in the Public Money Draining Acts, the rentcharges under this act were inevitably higher, and assuming an interest rate of $4\frac{1}{2}$ per cent they amounted to £7.25 per cent on the outlay.[15] In addition, there was a reluctance to make such private funds available, for the rentcharge provided annual repayment of both interest and loan, and involved individuals who had lent capital in a continuous, annual task of re-investing.[16] This act was repealed in 1864, its terms being incorporated into the Improvement of Land Act, 1864. Retaining all the machinery of the earlier, the latter act expanded the range of improvements for which private funds could be borrowed. The rentcharge period for repayment was extended to twenty-five years, although the interest rate limit of 5 per cent was maintained for money borrowed. With money available at $4\frac{1}{2}$ per cent interest, this extension resulted in a lowering of the annual rentcharge to £6.7 per cent on the outlay.[17]

Two forms of loan capital for underdraining have been identified in the land-improvement legislation – a central government fund from which landowners could draw money directly, and the means by which landowners could borrow from private sources. A third form was also introduced through the establishment by act of land-improvement companies which bridged the gap between the first two systems. These improvement companies obtained funds from insurance companies at current rates of interest and made them available to landowners.

Attempts to apply such collective capital to the draining of estates dated back to 1843 with the proposal to establish the Yorkshire Land Draining Association under the guidance of the agriculturalist J. H. Charnock.[18] Although in itself unsuccessful, the model was used for the creation of the five improvement companies which by act were empowered to lend money for draining and other improvements in England and to issue rentcharges on the land improved to redeem such loans. In 1847, the act incorporating the Landowners' Drainage and Inclosure Company, centred mainly in Cheshire, Lancashire and north Wales, was passed to be followed in 1848 by the West of England and South Wales Land Draining Company's act.[19] Of the five improvement companies, these two deviated from the usual methods of charging the cost of the improvement on the land. Under both acts, the loans, improvements and creation of rentcharges did not need the approval of the Inclosure Commissioners, that body being involved only when a remainderman on a settled estate opposed the improvement, or when the limited owner was a minor or a lunatic. Where these conditions were inapplicable, the tenant for life had merely to advertise in a local newspaper his intention to drain and then obtain a certificate from two justices of the peace that such a notice had been served for the applicant to charge the land with the cost of the improvement. The period of repayment under these two company acts could be either a lien in perpetuity or limited to a distinct number of years.[20]

Subsequent improvement company acts resorted to the use of the Inclosure Commissioners to sanction loans as a more satisfactory means of safeguarding the reversionary interest and, moving from the principle of a perpetual charge for an agricultural improvement, prescribed specific periods for the repayment of monies. Under the General Land Drainage and Improvement Company's act of 1849, the length of the rentcharge for draining could range from 22 to 50 years, the most frequent period adopted by landowners being 31 years. With money at $4\frac{1}{2}$ per cent, this period of redemption produced a rentcharge of £6.01 per cent on outlays of over £2,000. The company levied a commission of $5\frac{1}{2}$ per cent on the amount of the loan, which became part of the sum charged on the estate.[21] The Lands Improvement Company and the Land Loan and Enfranchisement Company were established by acts passed respectively in 1853 and 1860. Both were restricted to repayment periods for draining loans of twenty-five years, which resulted in a rentcharge of £6.7 per cent on outlays of over £1,000, with money borrowed at $4\frac{1}{2}$ per cent. The level of their commission, also added to the gross sum charged on the estate, was at 5 per cent of the value of the loan, slightly less than that of the General Land Drainage Company.[22]

Having detailed the legislation that made available loan capital for draining purposes, we may now review the surviving records of such loans. No central list is extant of the two sets of reports submitted by inspectors to the Inclosure Commissioners on draining loans. By 1873 there were forty-six inspectors at work and they represented the leading land agents of their time at both the regional and national level. In Devon, for example, Charles Gordon, W. B. Johnson, J. C. Knollys, A. S. Parker and Thomas Webber, all agents from Devon and Somerset, had been employed by the Inclosure Commissioners as draining inspectors in the period 1848–66 together with the nationally known Hewitt Davis and Josiah Parkes.[23] As the first report of an inspector commented on the proposed draining, describing the land to be drained, the planned system of draining, the estimated cost and the probable increase in the value of land after draining, while the second ensured that the draining had been carried out in the proposed manner, the material would be of considerable value in assessing draining as an agricultural improvement. At present few of these reports have come to light, the main exception being those of Andrew Thompson, agent to Ralph Sneyd's Staffordshire estate. Thompson was appointed an inspector to the Inclosure Commissioners in 1847 and remained one until his death in 1869. Copies of Thompson's reports to the Inclosure Commissioners exist for the period 1857–68 and, although but a sample, present a clear insight into loan-financed draining.[24]

Far more data are available of the financial transactions of the loans. The most useful and complete of these records are the registers of certificates of loans. Although the detail contained in these registers varies between both acts and companies, in common they furnish the name of the landowner who borrowed, the sum borrowed, the location of the lands to be charged with the repayment of the loan, the amount, duration and assignment of the rentcharge, and the date of the loan. A full record is extant of the loans made under the two Public Money Draining Acts and the Improvement of Land Act, 1864. Of the improvement companies, a similarly complete cover can be obtained of all loans of the three largest: the General Land Drainage, the Lands Improvement and the Land Loan Companies.[25] This body of material provides the basis for the present analysis at a national level of the intensity and location of draining in the second half of the nineteenth century.

No registers of certificates are extant for the three remaining sources of loan capital. The Landowners' Drainage Company never made use of its powers to charge lands with the cost of improvements. It deveoped into a company that only implemented agricultural improvements, and by 1851 it had ceased to exist.[26] However, the West of England Company did provide loans for draining which became a rentcharge on the land. As the Inclosure Commissioners were incidental to the company's general functioning, that body maintained no record of its loan certificates. But for the same reason the sums lent by the company were never included in the total amounts reported by the Inclosure Commissioners as having been borrowed for draining under the land-improvement legislation of the middle of the century.[27] In any case, this gap is not as important as it may seem, for the total lent by the company was small in comparison to the other sources of loan capital. In 1873, the company secretary, F. Brodie, reported that, since establishment by act in 1848, it had advanced £300,000 for agricultural improvement, of which £200,000 had been for draining.[28] This amount represents but 2.7 per cent of the £7,500,319 lent for draining in England, Scotland and Wales under the other land improvement legislation by 1873.[29] No record is available of loans resulting from the Private Money Draining Act, 1849. The act was unsuccessful in its aim of encouraging the investment of private funds in draining, partly from the high level of rentcharge for the borrower and partly from the difficulty of finding private capitalists to lend money on such terms for a 22-year period.[30] By the time of the act's repeal in 1864, only £272,717 had been lent under its powers in England, Scotland and Wales.[31] Although this sum was incorporated into the reports of the Inclosure Commissioners of the monies lent under the land improvement acts, it formed 3 per cent of the total lent in England, Scotland and Wales by the end of the century. The loan capital for draining provided by the West of England Company and under the Private Money Draining Act involved only small sums and the absence of such material should not significantly affect the conclusions on the amount and distribution of draining in England derived from the surviving data.

There are, however, difficulties in the use of the draining-loan data that exist. The registers of certificates of loans had a common function, the detailing of a financial transaction. They recorded that the Inclosure Commissioners had sanctioned an improvement, that a certain sum had been lent to a landowner on a specific date for a particular improvement, that the land described in the certificate was to be charged with the cost of the improvement, and that an annual rentcharge was to be levied from these lands to redeem the loan. The acreage drained is not noted and the registers of loans do not allow the reconstruction of the area drained under the land-improvement legislation. The amount and intensity of draining resulting from loan capital has to be expressed in terms of expenditure.

The Inclosure Commissioners approved only new draining schemes with loan capital as distinct from repairing existing draining systems.[32] Nevertheless, some of the loans embody an administrative fee which cannot always be isolated. The loans recorded under the Public Money Draining Acts incorporate a charge levied by the commissioners for the work of their inspectors. The rate varied inversely with the sum borrowed and on average amounted to 2 per cent of the loan.[33] Similarly, the loans recorded in the registers of the Lands Improvement Company included the 2 per cent fee of the Inclosure Commissioners and the 5 per cent commission levied by the company for arranging the loan.[34] The registers of loans under the Improvement of Land Act and of the General Land Drainage and Land Loan Companies excluded such expenses and distinguished the amount borrowed solely for the improvement. In examining the pattern of draining expenditure throughout the country, the inclusion of an expense element in some loans should be borne in mind.

The registers of loans provide detailed evidence on the location of draining activity. A rentcharge to redeem a loan could be raised only on land that had benefited from the improvement outlay, a relationship not altered until 1899.[35] The land described as subject to a rentcharge in the loan certificates represented the location of draining undertaken with such loan capital. The description took the form of a schedule which normally gave the name and area of the farms on the estate involved and always the parish and county in which they lay. However, the breakdown of the loan between the constituent farms and parishes was not recorded. As a result the distribution of draining in the country financed by loan capital may be demonstrated at a parish level by plotting parishes subject to draining rentcharge, but the intensity of such draining expenditure cannot be similarly expressed on a parish basis.

Draining loans were provided on an estate basis: a landowner applied for a loan to improve his estate. In the absence of detailed maps of estate location in the nineteenth century, the county forms the smallest areal unit by which to examine the spatial variation in the intensity of draining outlay under the land-improvement legislation and has been adopted in the present study. In support of its use, the great majority of estates lay within the confines of a single county. However, some estates straddled county boundaries and some landowners held estates in two or more counties. A number of draining loans were made for estates in several counties. As the schedules of lands charged did not record the distribution of a loan between parishes, the amount allocated to each county cannot be precisely determined and a division of the monies has been devised to maintain the county basis of analysis. In these cases, the loan has been distributed proportionally between the counties involved according to the amount of land in each county recorded in the schedule of lands subject to the rentcharge.

A final difficulty in the data arises from the use to which loans were put. The Public and Private Money Draining Acts provided loan capital solely for draining. Under the Improvement of Land Act and the improvement companies, landowners were able to borrow for improvements other than draining and often made combined loans for draining and other agricultural improvements. In these cases, with the exception of those for the Land Loan Company, the loan registers recorded a single sum, not distinguishing the amounts lent for the individual improvements. In order that the draining element of such combined loans should not be lost for analysis, the assumption has been adopted that half the amount of such loans was for draining, the remainder being used for the other named improvements. Support for this division may be obtained by examining in detail the applications for such joint-improvement loans. When a contract was made by a landowner with the General Land Drainage Company and under the Improvement of Land Act for a loan for draining and other improvements, a record survives of the amounts proposed to be spent on each of the improvements. Such material is also available for contracts with the Lands Improvement Company but only for the period 1861–9.[36] These proposals formed the basis for the improvements sanctioned by the Inclosure Commissioners and for the loans granted by the two companies and under the Improvement of Land Act. The proportion that draining represented of these joint-improvement proposals may therefore be reasonably applied to the loans actually implemented.

Details exist of 159 contracts made by landowners with the General Land Drainage Company over the period 1852–99 for loans to finance draining and other improvements. Of the £833,616 proposed to be borrowed, 58 per cent was to be spent on draining. The proportion was higher in the 110 contracts agreed between landowners and the Lands Improvement Company between 1861 and 1869 for such loans. Expenditure on draining formed 76 per cent of the £329,874 intended to be borrowed for combined-improvement schemes. Only in the 41 contracts for loans for draining and other improvements made under the Improvement of Land Act in the period 1864–99 did the proportion occupied by draining fall below half, forming 45 per cent of the £193,715 projected to be borrowed. From the available material, landowners entered contracts proposing to borrow £1,357,205 for combined-improvement schemes involving draining. Of this sum, £821,418, some 61 per cent, was to be devoted to draining. This evidence would suggest that the assumption adopted in the present study, that in loans for draining and other agricultural improvements half the sum was employed for draining, should not be regarded as too arbitrary and probably under-estimates the amount spent on draining in such loans.

Despite these problems presented by the data, the legislation which set up this body of loan capital for draining incorporated few restrictions or biases. Although the legislation developed in response to the perceived need of owners of settled estates, the use of the loans was open to all landowners, from parsons holding glebeland and institutional landowners to owners of estates in fee simple. The acts applied to the whole country with no spatial limitations being placed on their availability. Spring has suggested that an attempt was made to introduce an equitable distribution amongst counties of that part of the second Public Money Draining Act loan used for draining in England[37] but in practice such a scheme was not implemented. The sums applied for by landowners recorded in the *London Gazette* ranged widely between counties from £200 for Rutland to £516,616 for Yorkshire[38] (Fig. 3.1a). In sanctioning loans to landowners under the Public Money Draining Acts, the Inclosure Commissioners did not rectify this variation and the county totals of sums borrowed reveal considerable divergence, reflecting

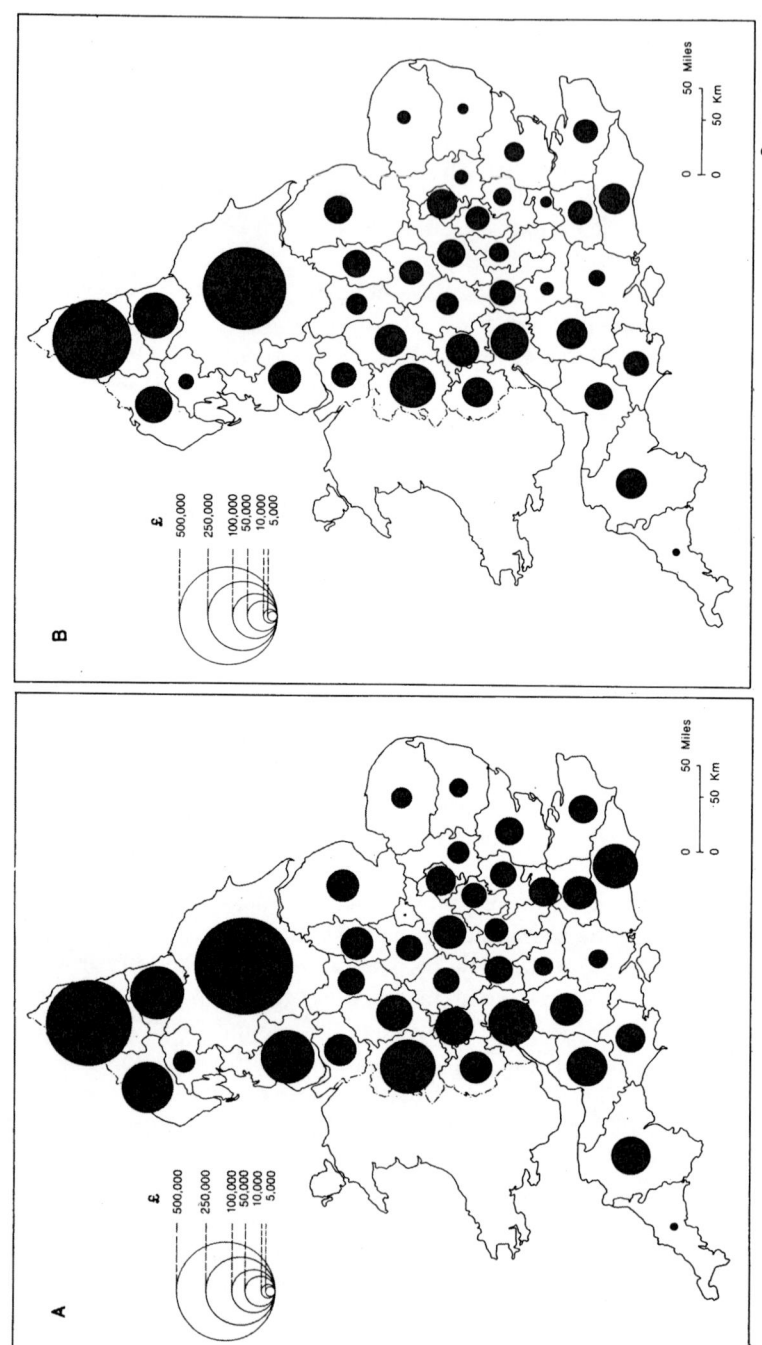

Figure 3.1 A. Applications for loans under the Public Money Draining Acts, 1846 and 1850; B. Loans granted under the Public Money Draining Acts

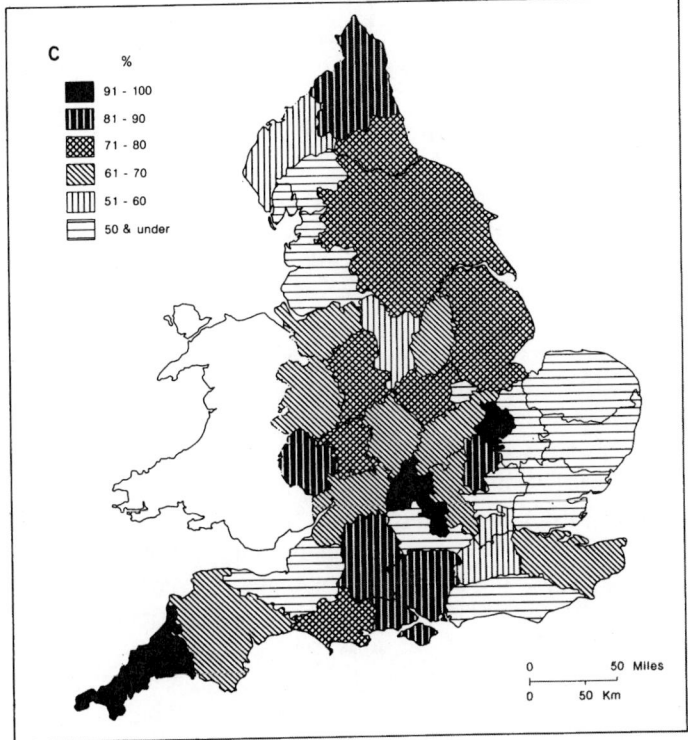

Figure 3.1 C. Loans granted as a percentage of applications (*Sources:* PRO, 1R3/6–38; *London Gazette*, 1846–78, *passim*)

the differential use made by estates of these funds throughout the country (Fig. 3.1b).

Constraints on the amounts that could be borrowed existed only on loans under the Public Money Draining Acts. These arose from the fact that the £4 million represented a finite sum which, provided at the below-market interest rate of $3\frac{1}{2}$ per cent, was in demand from landowners. Initially, under the 1846 act no limit was placed on the amount borrowable; by 1847 a maximum of £10,000 per landowner was instituted, which in 1850 was lowered to £5,000.[39] This restriction led some landowners to abandon their proposed loans: of the 21 English landowners who applied for loans in excess of £9,000 under the 1846 act, five were unprepared to accept a reduction.[40] These controls on the size of loans under the two Public Money Draining Acts meant that the sums applied for were never met in the allocation of funds, loans granted under the acts forming 67 per cent of the £2,975,579 requested by landowners. This proportion varied widely between counties, being for example 36 per cent in Lancashire, 45 per cent in

Westmorland, 53 per cent in Cumberland and 81 per cent in Northumberland (Fig. 3.1c), a fact which serves further to emphasize the unreliability of R. W. Sturgess' use of the applications for loans under the Public Money Draining Acts published in the *London Gazette* as a measure of the location and intensity of underdraining in the second half of the nineteenth century.[41]

Landowners were not limited in the size of loan by the other land-improvement legislation. However, the funds provided by these sources were not available at the low-interest rates of the Public Money Draining Acts loans. The capital used by the improvement companies and under the Improvement of Land Act came in the main from insurance companies[42] and its cost reflected trends in the money market. Although the size of loan had a slight inverse effect on the interest rate, there was little variation between the companies in the rate at which they lent to landowners, as can be seen from the close correspondence of the interest rates most frequently cited for new loans of under £1,000 made annually by the Land Loan Company in the period 1863–99 with those of the Lands Improvement Company on new loans of between £1,000 and £2,000 from 1854 to 1899 and with those of the General Land Drainage Company on new loans of over £2,000 between 1852 and 1895 (Fig. 3.2). As the number of loans under the Improvement of Land Act was much smaller, a continuous series of interest rates on money borrowed cannot be constructed, but for those years when data are available the rates mirrored those of the improvement companies. From the early 1850s to the middle of the 1870s, the interest rate on such loans was stable at $4\frac{1}{2}$ per cent. This level fell to 4.37 per cent in the late 1870s and to 4.25 per cent in the early 1880s. From 1892, the rate sagged markedly, the companies offering funds to landowners at an interest rate of $3\frac{1}{2}$ per cent in 1899. The rates at which money was lent to landowners for draining by the improvement companies were not fixed, so lessening their attraction as sources of capital, but followed the long-term movement in the availability of funds, a pattern demonstrated by comparison with the average annual percentage yield on 3 per cent consols in the second half of the nineteenth century[43] (Fig. 3.2).

The rate at which funds could be obtained from the improvement companies throughout the period differed little from that of other sources of borrowed capital, such as loans or mortgages from insurance companies. Indeed, some insurance companies refused to lend money to the improvement companies because the proffered rate of interest was too low.[44] As the rentcharge arising from land-improvement loans ensured the amortization of both principal and interest, the annual charge was always higher than the simple-interest payment under a mortgage- or insurance-company loan, which did not include an annual redemption element. However, no landowner wishing to repay both principal and interest could privately

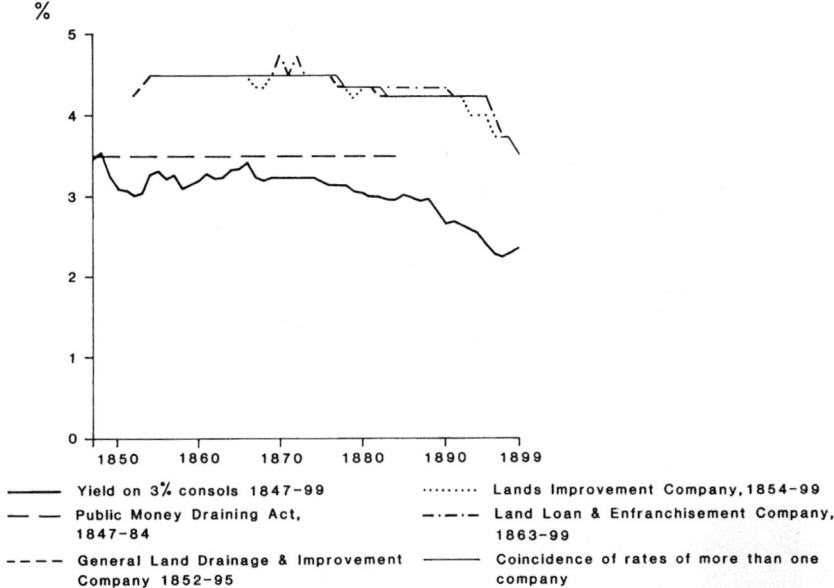

Figure 3.2 Interest rate on loans granted under the land-improvement legislation (*Sources:* PRO, 1R3/6–38; MAF 66/1–2, 4–6, 13–22, 25–39, 43–7; S. Homer, *A History of Interest Rates*, 1963, 195–7)

secure a loan at lower interest rates than those offered by the improvement companies. And the advantage of such loans to the landowner was that they provided instant, usable capital which was fully repaid by means of a rentcharge in a stated number of years.[45] The broad correspondence of the level of annual rentcharges established under all the legislation, save for the Private Money Draining Act, at around £6.5 per cent on loan outlay annually, irrespective of redemption period, with the anticipated rate of tenant contribution towards the cost of draining reported by agriculturalists at 5 or 6 per cent on outlay[46] may have provided further confirmation to landowners of the value of such funds.

The land-improvement legislation offered uniform treatment, administratively, financially and legally, to all landowners in England applying for loan capital. At the same time, no inherent bias can be detected in the adoption of such loans by landowners. Although, once a decision to borrow had been made, a preference may have existed for one source of funds over another and although the amount borrowed varied according to individual landowners' needs and intentions, estates of all sizes in all parts of the country employed this loan capital for draining purposes. These loans represent, therefore, a coherent and virtually complete statistical source on a scale unavailable elsewhere for the analysis of the distribution and intensity of draining in England in the second half of the nineteenth century.

The intensity and location of loan-financed underdraining

The first draining loan in England was made under the Public Money Draining Act in 1847. The provision of funds under these acts peaked in 1852 but continued until 1884 (Fig. 3.3). With the decline in the amount of capital available from this source, landowners turned to the improvement companies, and draining loans were first granted by the General Land Drainage Company in 1852, being maintained until the end of the century. Sums provided by the Lands Improvement Company were consistently larger than by the other two companies (Fig. 3.3). Loan capital for draining under the Improvement of Land Act was first authorized in 1866, but the amounts were on a smaller scale than from the other sources and from the late 1880s became irregular in occurrence.

The pattern of spatial adoption of draining financed by this body of capital from 1847 to 1899 may be determined by compiling a series of maps of the parishes subject to draining rentcharge. Save for the period 1847–9, parishes with new draining rentcharges have been recorded on a 10-yearly basis. Although few in number, a distinction has been made for parishes which possess a rentcharge only for draining combined with other agricultural improvements. As each map is cumulative, incorporating parishes with draining rentcharges from previous periods, the series allows the complete distribution of draining implemented under the land-improvement legislation to be identified.

As would be expected with the loan system in the process of establishment, few parishes became subject to draining rentcharges in the period 1847–9 (Fig. 3.4a). However, the four northern counties and north Yorkshire contained more landowners who were original adopters of loan capital than any other part of the country and most early draining activity was concentrated in this area. A number of landowners in Devon and Shropshire had also been quick to take advantage of loans, but in the rest of England there had been little recourse to these funds.

The expansion in the number of parishes subject to rentcharge between 1850 and 1859 presents direct evidence of a considerable diffusion in the use of loan capital by landowners for draining purposes in the country (Fig. 3.4b). With the exception of Rutland, by 1859 draining loans had been employed in all counties. Nevertheless, distinct patterns can be detected in their spatial adoption. The early concentration in the northern counties was maintained, there being a major intensification of draining activity in Northumberland, Durham and north Yorkshire, while the system of

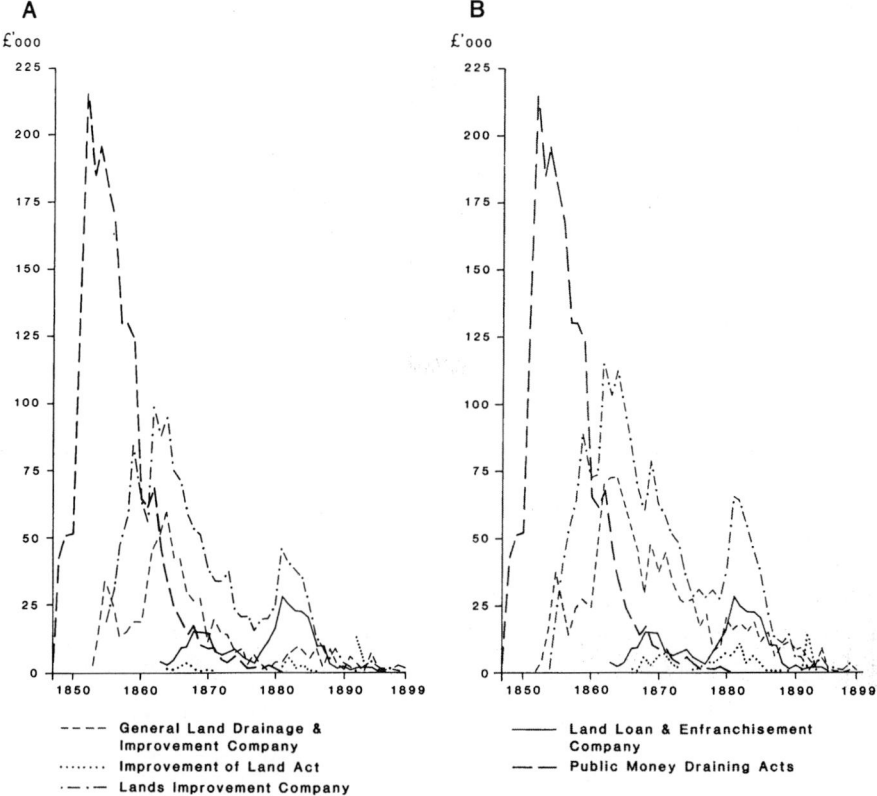

A

£'000

B

£'000

Figure 3.3 A. Supply of loan capital for draining alone; B. Supply of total loan capital for draining, incorporating the assumption that half the amount of combined loans was for draining (*Sources:* PRO: 1R3/6–38; MAF 66/1–6, 8–9, 13–22, 25–38, 43–7)

draining spread within Lancashire and other parts of Yorkshire. Only in Westmorland was the spatial extension of draining less marked. A similar growth in loan-financed draining occurred in Shropshire, the practice being adopted equally intensively in the neighbouring west-midland counties of Cheshire, Herefordshire and Worcestershire. Expansion from the remaining early adopting centre, Devon, was slight, not rivalling the level in the other two areas. However, by 1859 there were some parts of the country where little use had been made of loan capital for draining, especially in Cornwall, the southern counties of Berkshire and Hampshire, the East Anglian counties of Cambridgeshire, Norfolk and Suffolk, and Rutland and Derbyshire. In the remaining counties, although the density was less than in the north and west midlands, the adoption of draining loans was widespread.

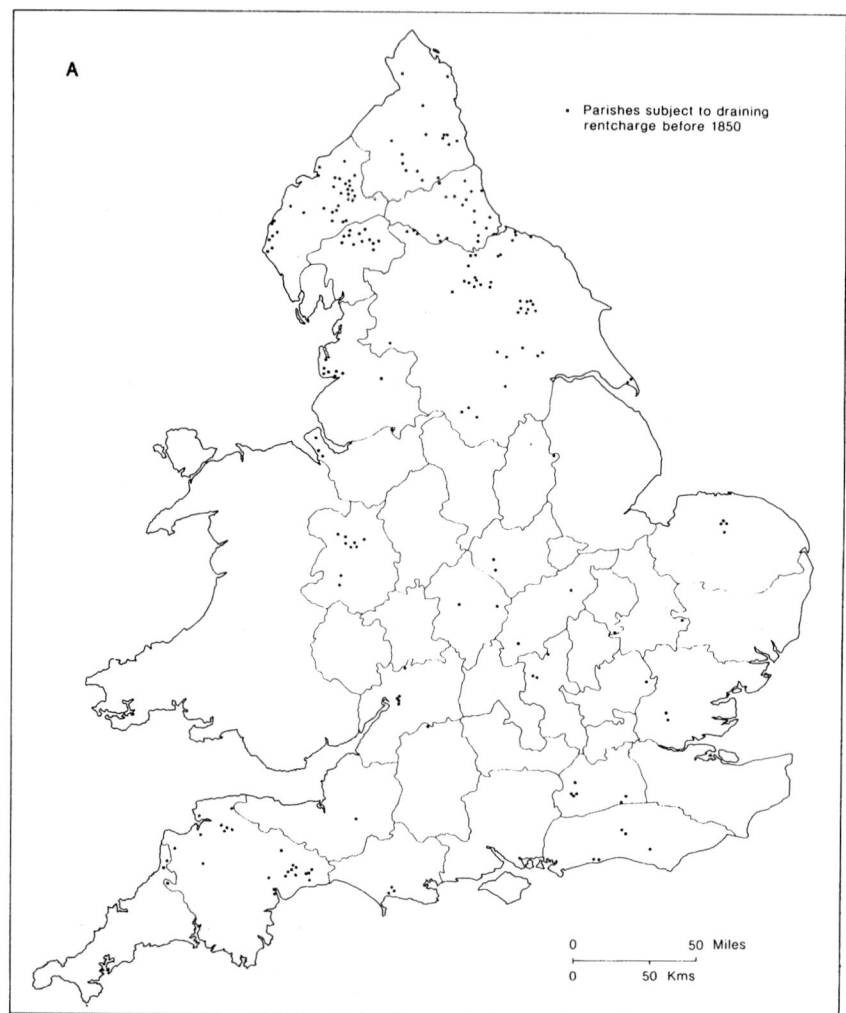

Figure 3.4 A. Parishes subject to draining rentcharge, 1847–1849

The decade 1860–9 exhibited a less vigorous spatial development of draining, the total number of parishes that became subject to draining rentcharges being much less than in the previous period (Fig. 3.5a). The pattern of draining activity was little changed in the north, there being a slight increase in density. In most midland counties a marked spatial extension in the adoption of draining loans can be recognized, notably in Derbyshire, Herefordshire, Northamptonshire, Nottinghamshire and Warwickshire. A similar expansion also took place in several southern counties,

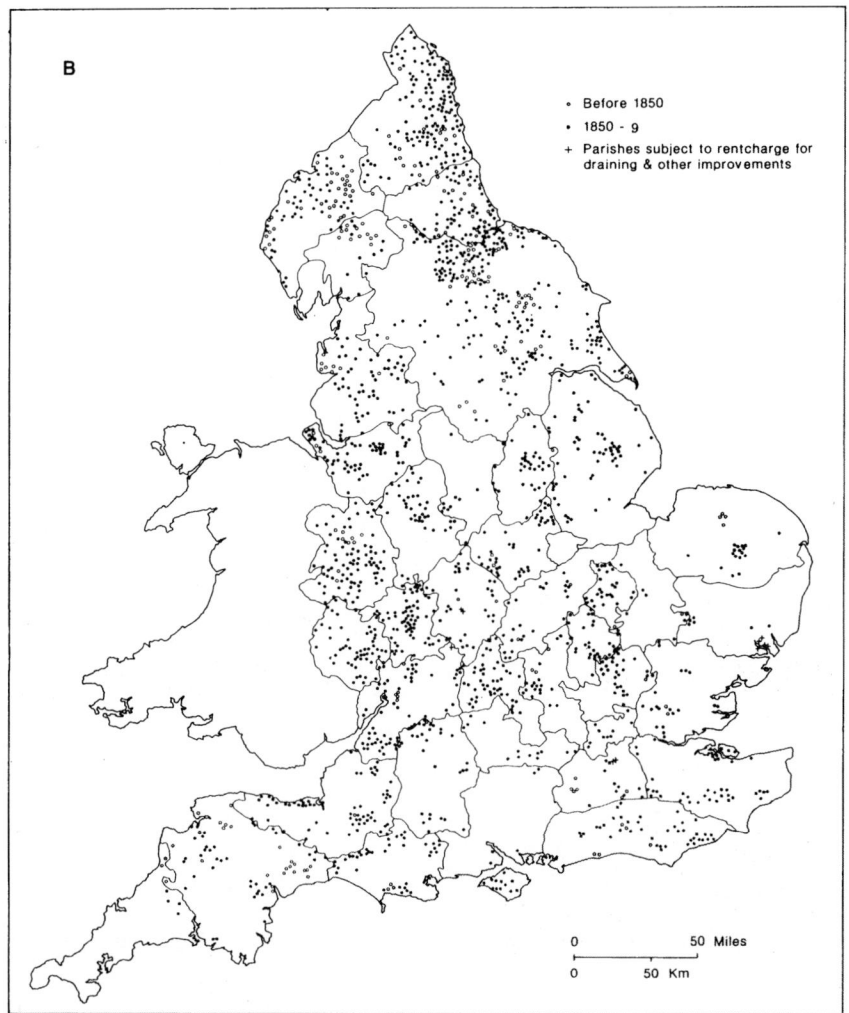

B

∘ Before 1850
· 1850 - 9
+ Parishes subject to rentcharge for
draining & other improvements

0   50 Miles
0   50 Km

Figure 3.4 B. Parishes subject to draining rentcharge, 1850–1859 (*Source:* As for
Fig. 3.3)

being most evident in Dorset, Hampshire and Wiltshire. There still existed by
1869 parts of the country with a lack of use of loan capital by landowners for
draining, a situation found in Cornwall, Rutland and the East Anglian
counties of Norfolk and Suffolk.

Little alteration was registered to the pattern of spatial adoption of
draining between 1870 and 1879 (Fig. 3.5b). Few new parishes became
subject to draining rentcharges in the decade and most of those that were
added infilled and made more dense the existing distribution, as for example

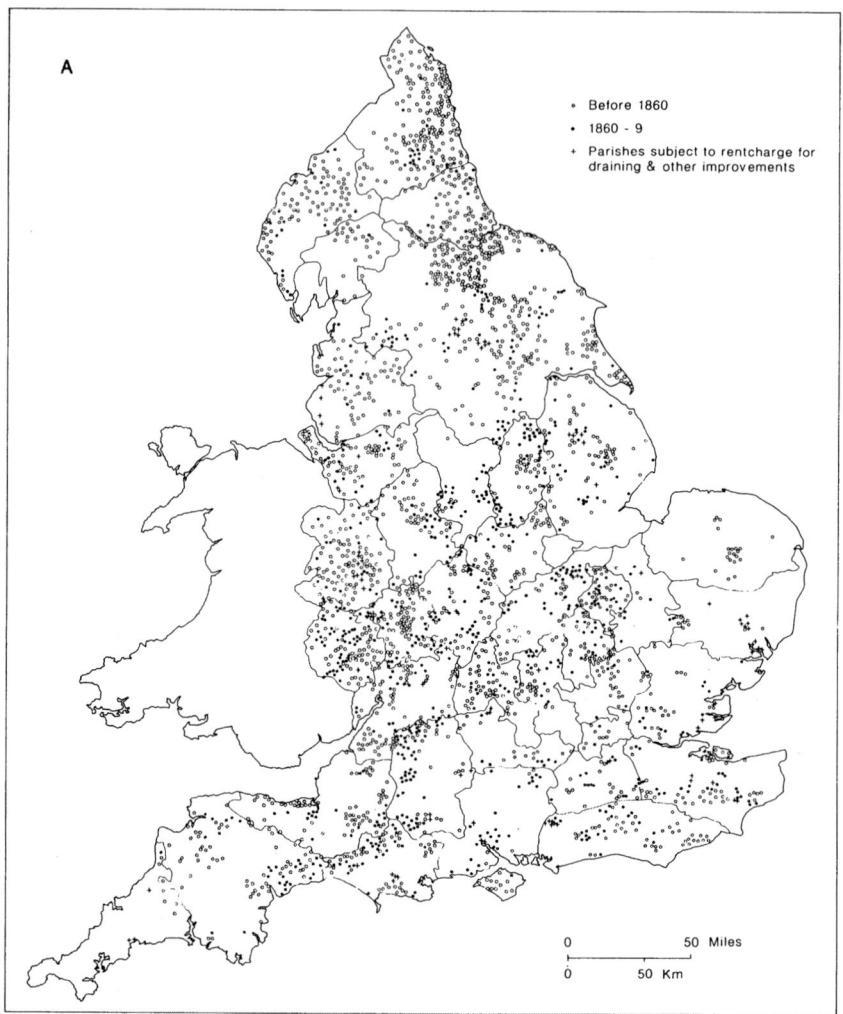

Figure 3.5 A. Parishes subject to draining rentcharge, 1860–1869

in north Devon, Dorset, Kent and Northamptonshire. Some extension in the use of draining loans can be identified in Essex and in this period they were first introduced in Rutland. An increase in the number of new parishes with draining rentcharges is recorded in the decade 1880–9 (Fig. 3.6a). These were predominantly located in midland counties and took the form of intensifying the existing pattern of draining loans. This trend was particularly evident in Leicestershire and Lincolnshire but was also found in Northamptonshire, Nottinghamshire, Oxfordshire, Staffordshire and Warwickshire. The spread

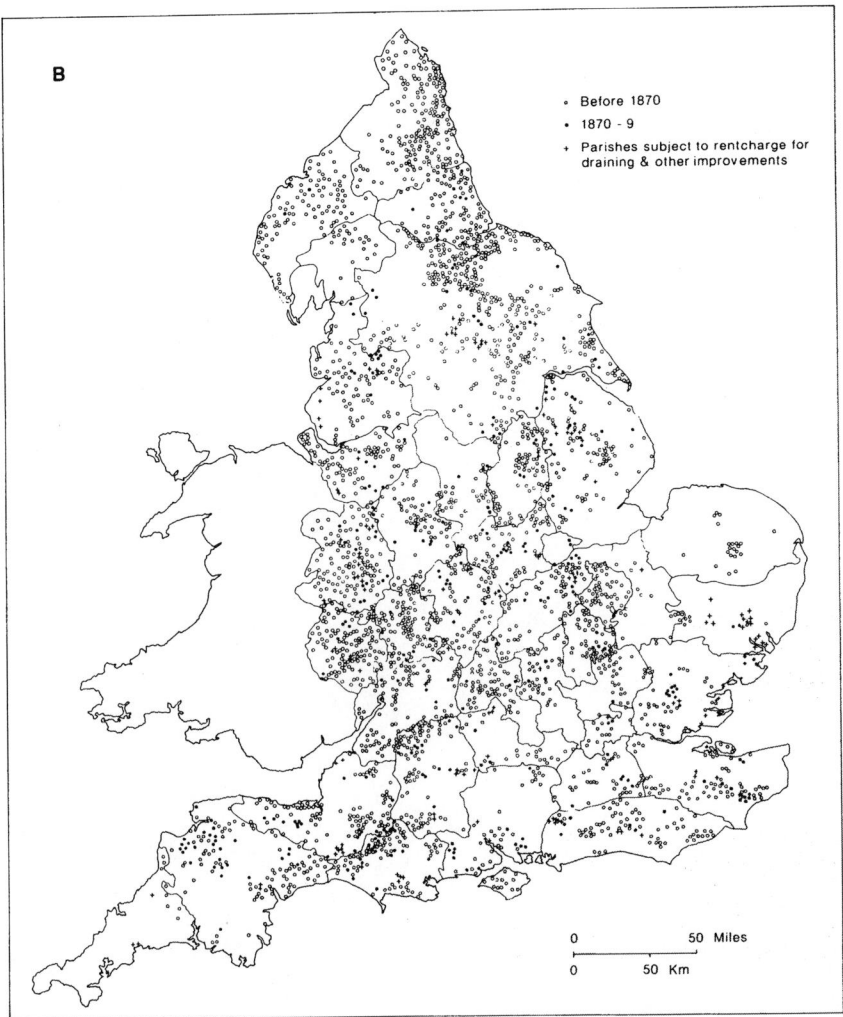

B

Before 1870

1870 - 9

Parishes subject to rentcharge for
draining & other improvements

0   50 Miles

0   50 Km

Figure 3.5 B. Parishes subject to draining rentcharge, 1870–1879 (*Source:* As for Fig. 3.3)

of loans continued in Essex and Rutland, while a revival in activity can be detected in Cambridgeshire and Norfolk. Few new parishes were added to the stock of those possessing draining rentcharges in counties in the south and the north in this period.

In the final decade of the nineteenth century, as only a small number of parishes became subject to draining rentcharge, there was little spatial extension or intensification of the pattern of draining loans recorded in 1889. Nevertheless, the map of the location of parishes with draining rentcharges

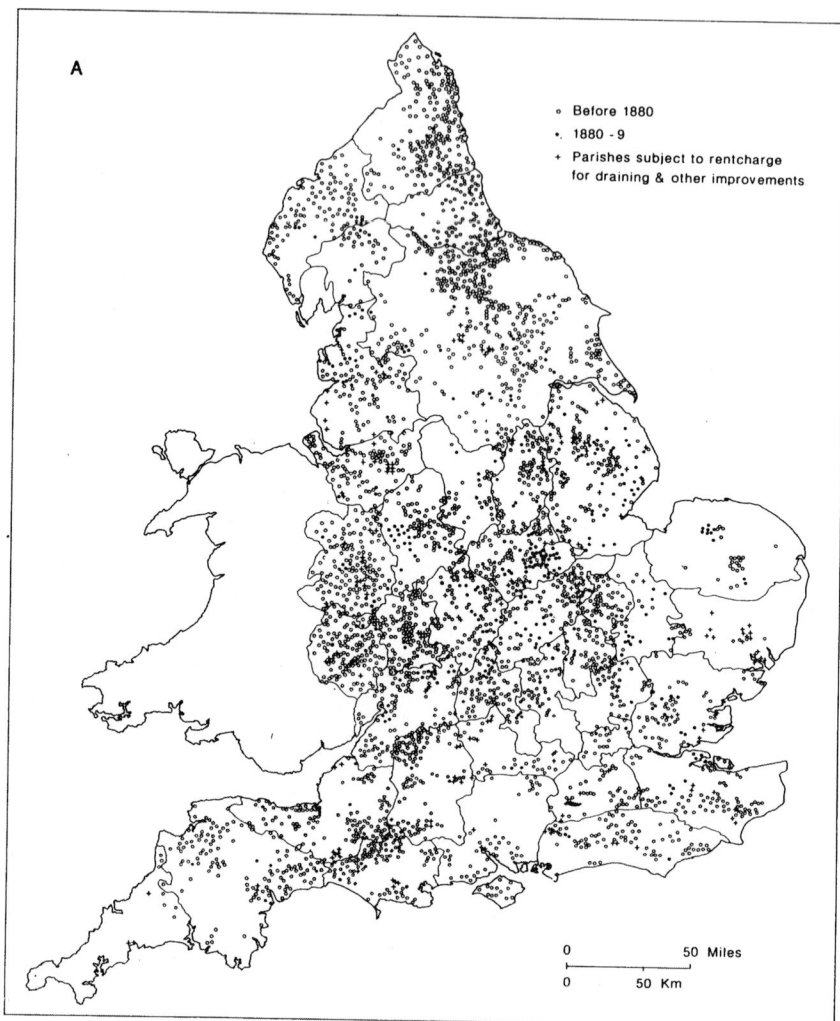

A

○ Before 1880

. 1880 - 9

+ Parishes subject to rentcharge
   for draining & other improvements

```
0                    50  Miles
├────────────────────┤
0         50  Km
```

Figure 3.6 A. Parishes subject to draining rentcharge, 1880–1889

by 1899 provides an opportunity to examine the overall distribution of draining financed by loan capital in the period 1847–99 (Fig. 3.6b). It is clear that draining loans were not used uniformly throughout the country. Two concentrations of draining activity emerged by 1899 – the northeastern counties of Northumberland, Durham and north Yorkshire and the west-midland counties of Herefordshire, Shropshire and Worcestershire. Draining loans were adopted widely and densely in both areas. The spatial pattern was established earlier in the former area, largely complete by 1860,

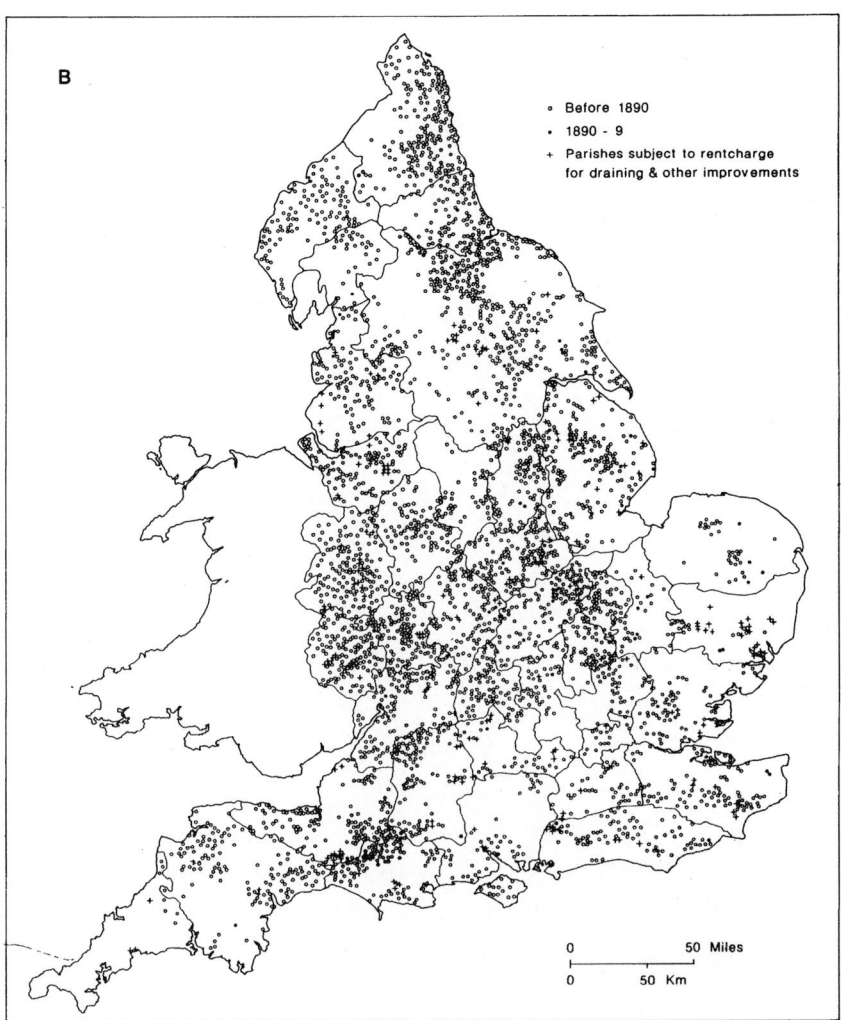

Figure 3.6 B. Parishes subject to draining rentcharge, 1890–1899 (*Source:* As for Fig. 3.3)

a state not achieved in the latter until 1870. Away from these areas, the density of the distribution of parishes with draining rentcharges declined. The use of draining loans was spatially more even and widespread in the remaining midland counties than in the northern counties abutting the northeastern centre. However, in both adjacent areas the development of the pattern took longer periods of time, in the case of the midland counties up to 1889. Irregularity marked the distribution of draining loans in the southwest. Parts of Devon, Dorset, Somerset and Wiltshire possessed distinct

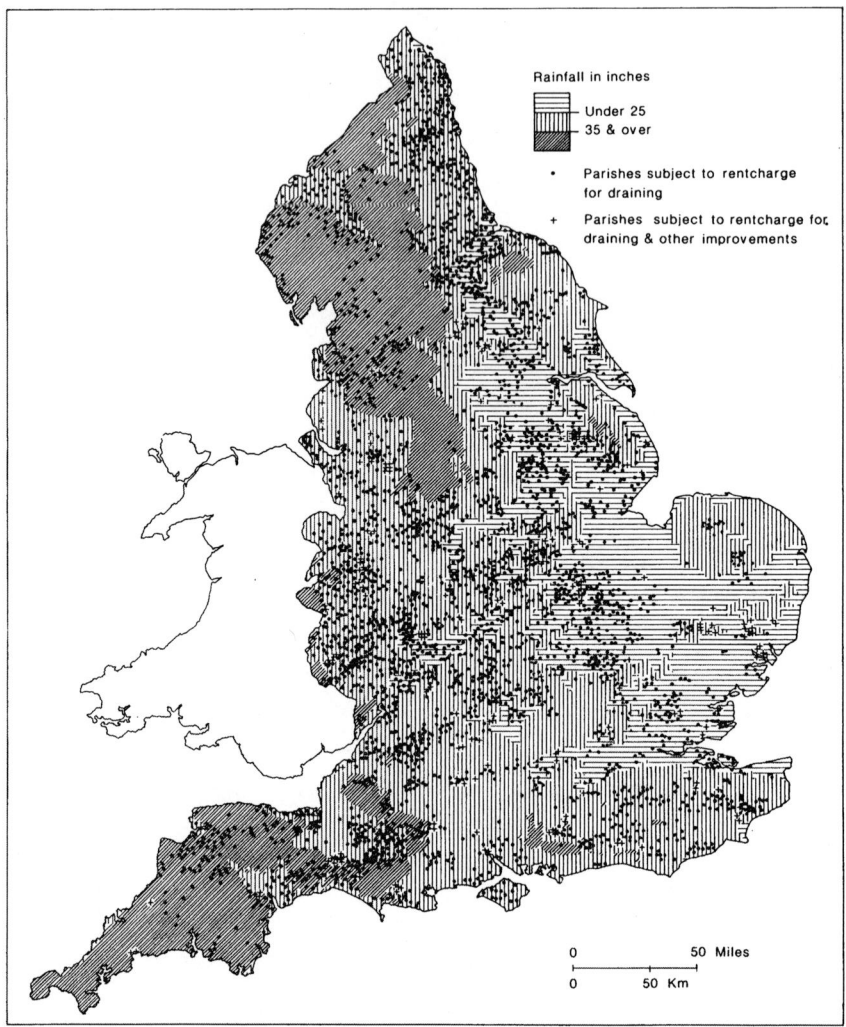

Figure 3.7 Parishes subject to draining rentcharge, 1847–1899, and average annual rainfall, 1881–1915 (*Sources:* Ministry of Town and Country Planning, *Rainfall: Annual Average, 1881–1915, 1:625,000,* 1949; as for Fig. 3.3)

concentrations of draining activity, while in other districts parishes with draining rentcharges were absent. Over the whole period, the adoption of draining loans by landowners in Cornwall had been neglected. Evidence of a similar, if less extreme, neglect can be found throughout southeastern and more particularly East Anglian counties, and in these areas the spatial diffusion of draining loans had made little advance between 1847 and 1899.

Some explanation of the location of draining financed by loan capital may be provided by relating the pattern to the two most significant physical factors theoretically responsible for the need for the improvement, rainfall and soil type. However, draining activity would not seem a direct product of rainfall quantity. Using rainfall data for the period 1881–1915,[47] no precise correspondence can be identified between the distribution of annual average rainfall and the location of parishes with draining rentcharges (Fig. 3.7). The pattern of draining was not at its most dense in those districts of highest rainfall in the northwest and southwest of the country, as M. Robinson has suggested.[48] Indeed, areas of high rainfall largely correspond to uplands with poor quality agricultural land, where extensive investment in draining would have been difficult to justify. At the same time, areas with low levels of rainfall were not uniformly marked by a lack of the improvement. Thus, there were few parishes with draining rentcharges in East Anglia, while draining loans had been widely employed in Bedfordshire and Huntingdonshire, counties with comparable annual rainfalls. Draining as represented by parishes with draining rentcharges was mainly located in areas with annual rainfall ranging from 25 to 35 in, but even in such areas variation was considerable, as a comparison of Norfolk with coastal Northumberland and Durham demonstrates.

A more satisfactory relationship to the location of loan-financed draining is to be found in the drainage requirements of different soil types throughout the country. Landowners were aware of the varying spatial need for such capital and only a small proportion of parishes with draining rentcharges were located in areas of well-drained soils (Fig. 3.8). Draining loans were little employed to upgrade upland areas, only the occasional parish over 1,000 ft having a draining rentcharge. The lack of draining in certain parts of counties from Derbyshire northwards may be largely attributed to the existence of extensive upland. Similarly, loans were not adopted on the moorland areas of the southwest. Little loan-financed draining also occurred in the extensive areas of alluvium and lowland peat, such as the Fens and the Somerset Levels. In such cases, land improvement was sought predominantly by measures of arterial drainage.[49] The lack of fall, the lowering of the surface and the possibility of inundation in these districts would have created severe technical problems for the introduction of underdraining, while the rentcharge associated with the improvement would have made heavier the rates already levied for existing arterial drainage schemes. The application of loan capital was concentrated in areas that possessed clayey and loamy soils with impeded drainage, either as the dominant soil type or in association with well- or moderately drained soils: landowners essentially made use of such funds to aid heavy-land agriculture. However, although the distribution of loan-financed draining reflected the occurrence of clayey and loamy soils with impeded drainage, draining loans were not adopted

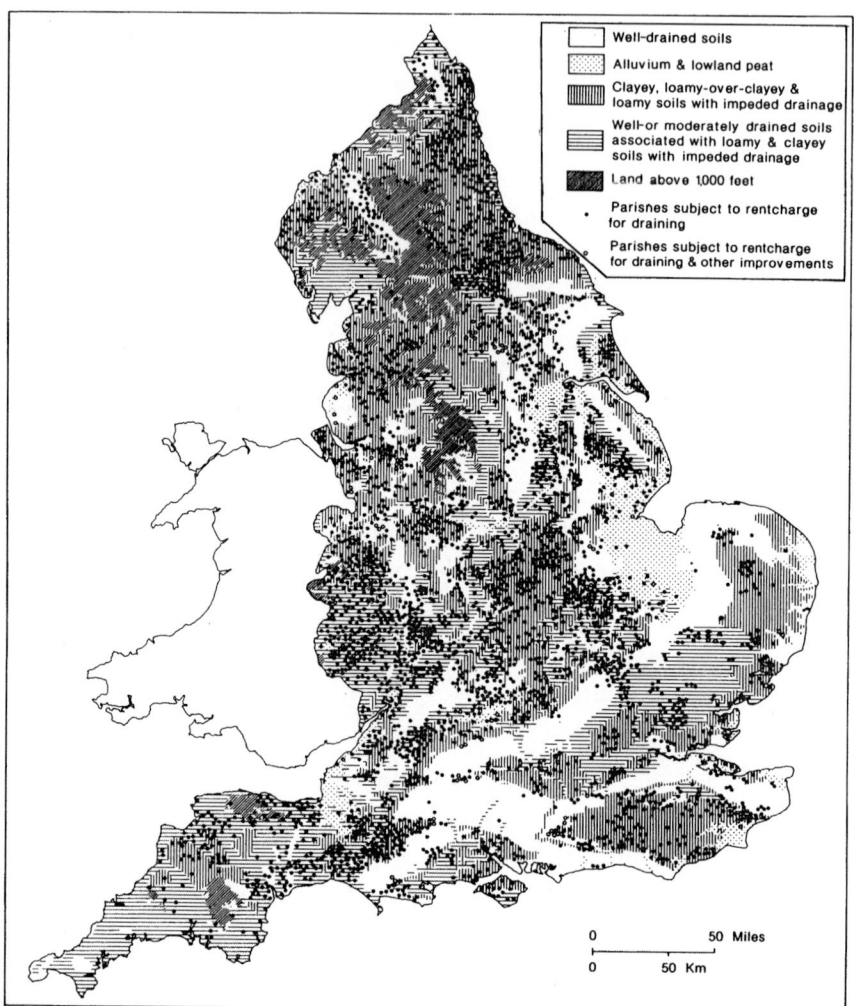

Figure 3.8 Parishes subject to draining rentcharge, 1847–1899, and soil types (*Sources:* As for Figs. 2.3 and 3.3)

uniformly by landowners on such soils and distinct spatial variation in the density of draining activity can be distinguished. For example, parishes with draining rentcharges were widespread and dense on clayey and loamy soils in the northeastern counties but on soils similarly classified in the southwest and East Anglia there were few. A comparable discrepancy in draining may be identified on clayey and loamy soils associated with well- or moderately drained soils found in west-midland counties and in Essex and Suffolk.

This spatial variation in the use of loan capital may be quantified by determining on a county basis the sums borrowed by landowners. These amounts were obtained by aggregating for each county the monies borrowed by individual landowners recorded in the registers of certificates of loans under the various land-improvement acts, incorporating the constraints already discussed. This exercise necessitated the examination of over 15,000 certificates of loans. Where duplicate and complementary series of registers were available, as in the case of loans under the Public Money Draining Acts, the General Land Drainage, the Lands Improvement and Land Loan Companies, these were cross-checked to ensure the accuracy of the figures recorded in the certificates. Although loans were made by landowners in Scotland and Wales from these sources, these sums have been excluded from the present study. The detailed draining-loan data derived from the registers of certificates used in this study relate solely to England, where the importance of the improvement has been subject to most debate.

Between 1847 and 1899, loans for draining alone amounted to £4,503,000 (Fig. 3.3a; Table 3.1). In addition to this capital solely for draining, loans totalling £1,993,300 were issued by the General Land Drainage and Lands Improvement Companies and under the Improvement of Land Act for draining jointly with other agricultural improvements (Table 3.2). Invoking the assumption that expenditure on draining constituted half these combined loans, a sum of £5,499,600 was borrowed for draining in England in this period from the loan sources for which data are extant (Fig. 3.3b; Table 3.3). Of this total, most came from the Public Money Draining Acts, 36.5 per cent, closely followed by the Lands Improvement Company with 35.7 per cent. The General Land Drainage Company provided 20.2 per cent of the draining loan capital, loans from the Land Loan Company and under the Improvement of Land Act forming much smaller proportions, 5.5 and 2.1 per cent respectively. There was close correspondence throughout all counties in the amounts lent for draining alone and for draining and other improvements, the Spearman rank correlation coefficient producing a value of 0.81 between the amount of loans solely for draining and the amount of combined loans for draining and other improvements in counties over the period 1847–99.[50] Indeed, the inclusion of the draining element from such combined loans produces little change in the relative county distribution of the funds solely for draining purposes, merely serving to reinforce and intensify that pattern (Figs. 3.9a and b).

Draining-loan capital in England averaged £141,017 per county, although the range was wide, from £4,000 borrowed by landowners in Cornwall to £771,609 in Northumberland (Fig. 3.9b; Table 3.3). The major concentration of funds was located in the northeastern counties of Northumberland, Durham and Yorkshire, where landowners absorbed 32.5 per cent of the total. Covering a slightly larger area, a group of counties from Lancashire south to Wiltshire, mainly in the west midlands and including Cheshire,

Table 3.1 *Loan expenditure solely on draining, 1847–1899*

	Source of loan capital (£)					
County	Public Money Draining Acts	General Land Drainage Co.	Lands Improve- ment Co.	Land Loan Co.	Improve- ment of Land Act	Total
Bedfordshire	27,261	16,965	9,991	5,593	295	60,105
Berkshire	8,940	1,045	6,325	279	0	16,589
Buckinghamshire	19,133	8,415	30,205	1,271	0	59,024
Cambridgeshire	10,311	1,922	16,570	647	0	29,450
Cheshire	32,982	110,683	23,646	1,093	0	168,404
Cornwall	2,773	0	0	0	0	2,773
Cumberland	73,460	816	33,351	18,689	3,372	129,688
Derbyshire	21,786	7,896	21,676	1,217	0	52,575
Devon	48,096	20,406	6,152	5,648	0	80,302
Dorset	31,865	17,858	30,485	7,227	0	87,435
Durham	111,134	37,082	78,759	10,774	201	237,950
Essex	18,793	8,922	7,384	2,923	2,996	41,018
Gloucestershire	76,454	6,981	43,158	2,633	0	129,226
Hampshire	15,217	26,387	12,475	22,391	0	76,470
Herefordshire	46,404	5,748	70,294	10,084	520	133,050
Hertfordshire	16,604	8,094	2,292	4,102	0	31,092
Huntingdonshire	43,857	23,419	3,663	15,507	0	86,446
Kent	29,608	17,354	21,895	136	2,681	71,674
Lancashire	55,082	34,438	26,588	11,860	0	127,968
Leicestershire	28,243	17,863	33,335	9,499	14,588	103,528
Lincolnshire	39,093	14,542	39,791	8,160	2,699	104,285
Middlesex	5,842	9,162	3,836	598	0	19,438
Norfolk	10,257	2,002	0	1,092	0	13,351
Northamptonshire	37,446	29,713	25,444	3,723	0	96,326
Northumberland	323,202	5,214	343,795	22,822	3,753	698,786
Nottinghamshire	37,995	753	22,083	6,022	351	67,204
Oxfordshire	37,828	3,570	31,320	3,321	0	76,039
Rutland	0	549	11,507	0	0	12,056
Shropshire	110,424	30,596	81,858	11,094	0	233,972
Somerset	40,559	25,435	13,813	4,735	0	84,542
Staffordshire	51,566	19,830	54,469	4,812	0	130,677
Suffolk	6,388	1,144	0	499	0	8,031
Surrey	31,504	3,542	11,694	414	0	47,154
Sussex	47,975	16,130	13,505	10,397	0	88,007
Warwickshire	25,944	27,743	46,774	25,095	756	126,312
Westmorland	11,468	499	2,814	21,053	2,650	38,484
Wiltshire	47,877	10,102	19,901	32,633	1,735	112,248
Worcestershire	56,085	25,985	63,714	2,566	1,322	149,682
Yorkshire	369,347	69,272	222,077	10,735	210	671,641
Total	2,008,803	668,077	1,486,639	301,344	38,139	4,503,002

Sources: PRO, 1R3/6–38; MAF 66/1–6, 8, 9, 11, 13–22, 25–39, 43–7

Table 3.2 *Expenditure on combined loans for draining and other improvements,
1847–1899*

| County | Source of loan capital (£) | | | |
	General Land Drainage Co.	Lands Improvement Co.	Improvement of Land Act	Total
Bedfordshire	23,769	2,760	0	26,439
Berkshire	18,284	6,218	0	24,502
Buckinghamshire	2,198	24,503	0	26,701
Cambridgeshire	11,602	4,021	0	15,623
Cheshire	79,145	6,296	40,201	125,642
Cornwall	2,455	0	0	2,455
Cumberland	830	12,837	0	13,677
Derbyshire	0	8,147	1,739	9,886
Devon	4,460	34,954	0	39,414
Dorset	52,206	31,956	5,264	89,426
Durham	1,334	42,188	768	44,290
Essex	15,402	10,377	0	25,779
Gloucestershire	5,150	24,779	11,737	41,666
Hampshire	12,798	18,945	0	31,743
Herefordshire	38,828	48,452	622	87,902
Hertfordshire	10,292	1,897	0	12,189
Huntingdonshire	24,944	238	1,918	27,100
Kent	24,185	8,330	6,796	39,311
Lancashire	51,402	50,170	25,200	126,772
Leicestershire	35,821	10,248	2,348	48,417
Lincolnshire	7,295	77,812	1,854	86,961
Middlesex	14,169	4,957	0	19,126
Norfolk	1,108	0	0	1,108
Northamptonshire	10,563	33,908	11,006	55,477
Northumberland	9,012	130,365	6,270	145,647
Nottinghamshire	3,304	30,126	250	33,680
Oxfordshire	7,088	23,212	0	30,300
Rutland	0	0	0	0
Shropshire	67,180	74,096	6,298	147,574
Somerset	55,810	10,784	0	66,594
Staffordshire	12,540	35,824	3,150	51,514
Suffolk	33,427	0	0	33,427
Surrey	18,310	11,672	0	29,982
Sussex	23,848	13,662	0	37,510
Warwickshire	21,968	14,628	0	36,596
Westmorland	0	3,990	0	3,990
Wiltshire	104,618	14,940	27,772	147,330
Worcestershire	13,594	21,098	0	34,692
Yorkshire	62,830	104,394	5,648	172,872
Total	881,779	952,694	158,841	1,993,314

Sources: PRO, MAF 66/1–3, 11, 25–39, 43–7

Table 3.3 *Total loan expenditure on draining, 1847–1899*

Source of loan capital (£)

County	Public Money Draining Acts	General Land Drainage Co.	Lands Improvement Co.	Land Loan Co.	Improvement of Land Act	Total	Percentage of overall total
Bedfordshire	27,261	28,849	11,326	5,593	295	73,324	1.33
Berkshire	8,940	10,187	9,434	279	0	28,840	0.52
Buckinghamshire	19,133	9,514	42,456	1,271	0	72,384	1.32
Cambridgeshire	10,311	7,723	18,580	647	0	37,261	0.68
Cheshire	32,982	150,255	26,794	1,093	20,100	231,224	4.20
Cornwall	2,773	1,227	0	0	0	4,000	0.07
Cumberland	73,460	1,236	39,769	18,689	3,372	136,526	2.48
Derbyshire	21,786	7,896	25,749	1,217	869	57,517	1.05
Devon	48,096	22,636	23,629	5,648	0	100,009	1.82
Dorset	31,865	43,961	46,463	7,227	2,632	132,148	1.40
Durham	111,134	37,749	99,853	10,774	585	260,095	4.73
Essex	18,793	16,623	12,572	2,923	2,996	53,907	0.98
Gloucestershire	76,454	9,556	55,547	2,633	5,868	150,058	2.73
Hampshire	15,217	32,786	21,947	22,391	0	92,341	1.68
Herefordshire	46,404	25,162	94,520	10,084	831	177,001	3.22
Hertfordshire	16,604	13,240	3,240	4,102	0	37,186	0.69
Huntingdonshire	43,857	35,891	3,782	15,507	959	99,996	1.82

County						
Kent	29,608	29,446	26,060	136	6,079	91,329	1.66
Lancashire	55,082	60,139	51,673	11,860	12,600	191,354	3.48
Leicestershire	28,243	35,773	38,459	9,499	15,762	127,736	2.32
Lincolnshire	39,093	18,189	78,697	8,160	3,626	147,765	2.69
Middlesex	5,842	16,246	6,314	598	0	29,000	0.53
Norfolk	10,257	2,556	0	1,092	0	13,905	0.25
Northamptonshire	37,446	34,994	42,398	3,723	5,503	124,064	2.26
Northumberland	323,202	9,720	408,977	22,822	6,888	771,609	14.03
Nottinghamshire	37,995	2,405	37,146	6,022	476	84,044	1.53
Oxfordshire	37,828	7,114	42,926	3,321	0	91,189	1.66
Rutland	0	549	11,507	0	0	12,056	0.22
Shropshire	110,424	64,186	118,906	11,094	3,149	307,759	5.59
Somerset	40,559	53,340	19,205	4,735	0	117,839	2.14
Staffordshire	51,566	26,100	72,381	4,812	1,575	156,434	2.84
Suffolk	6,388	17,857	0	499	0	24,744	0.45
Surrey	31,504	12,697	17,530	414	0	62,145	1.13
Sussex	47,975	28,054	20,336	10,397	0	106,762	1.94
Warwickshire	25,944	38,727	54,088	25,095	756	144,610	2.63
Westmorland	11,468	499	4,809	21,053	2,650	40,479	0.74
Wiltshire	47,877	62,411	27,371	32,633	15,621	185,913	3.38
Worcestershire	56,085	32,782	74,263	2,566	1,332	167,028	3.03
Yorkshire	369,347	100,687	274,274	10,735	3,034	758,077	13.78
Total	2,008,803	1,108,962	1,962,981	301,344	117,558	5,499,648	

Sources: PRO, 1R3/6–38; MAF 66/1–6, 8, 9, 11, 13–22, 25–39, 43–7

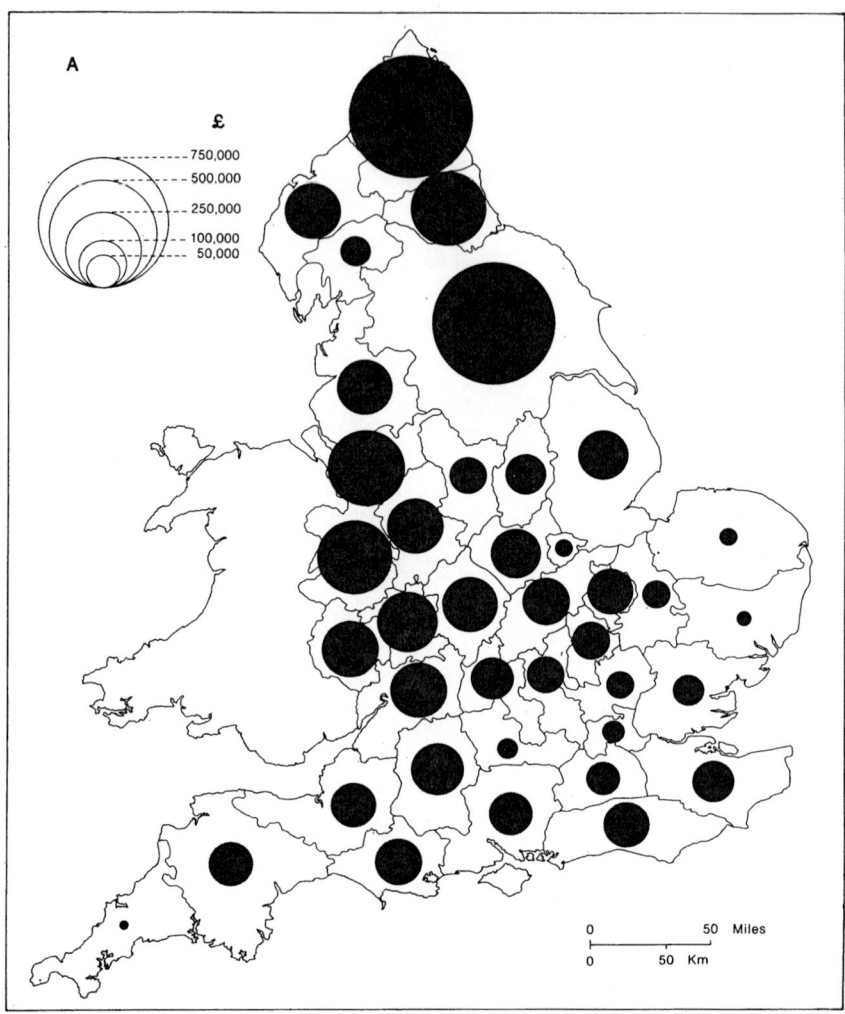

Figure 3.9 A. Loan expenditure solely on draining, 1847–1899

Gloucestershire, Herefordshire, Shropshire, Staffordshire, Warwickshire and Worcestershire, represented a further centre with 31.1 per cent of all loan capital. Small amounts were employed in eastern England and the six adjoining counties of Cambridgeshire, Essex, Hertfordshire, Middlesex, Norfolk and Suffolk accounted for 3.6 per cent of all loan-financed draining. In the remaining counties, save for Lincolnshire, the quantities of loan capital were also below the national average but, with the exception of Berkshire, Cornwall, Rutland and Westmorland whose joint outlay attained no more than 1.5 per cent of the total, at a higher level than that recorded in the eastern counties.

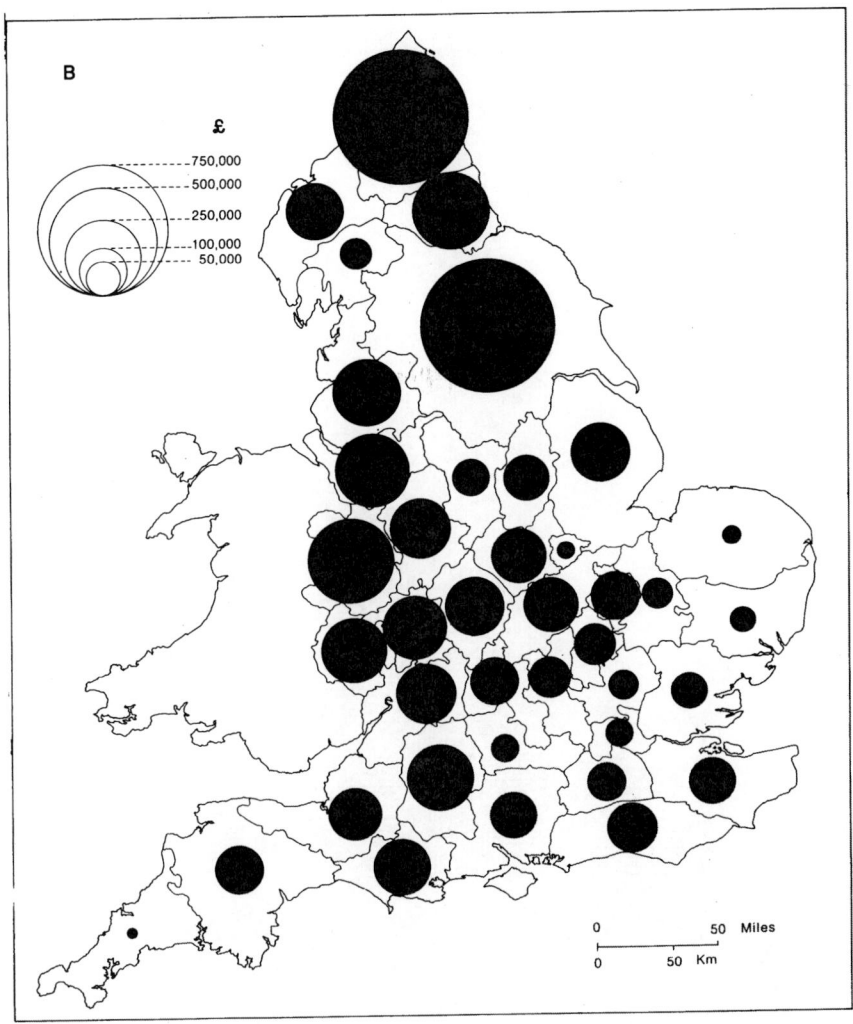

Figure 3.9 B. Total loan expenditure on draining, 1847–1899 (*Source:* As for Fig. 3.3)

The county distribution of total expenditure provides an imprecise measure of the relative impact of loan-financed draining on agriculture throughout the country, and for agricultural purposes this may be expressed more effectively by examining outlay per cultivated acre. Using average annual cultivated acreages for the decade 1870–9,[51] total draining-loan capital from 1847 to 1899 represented an average outlay of £0.23 per cultivated acre in England (Fig. 3.10; Table 3.4). However, considerable disparity existed in the intensity of draining outlay on agricultural land, per cultivated acre expenditure in Norfolk and Suffolk, for example, being but 2

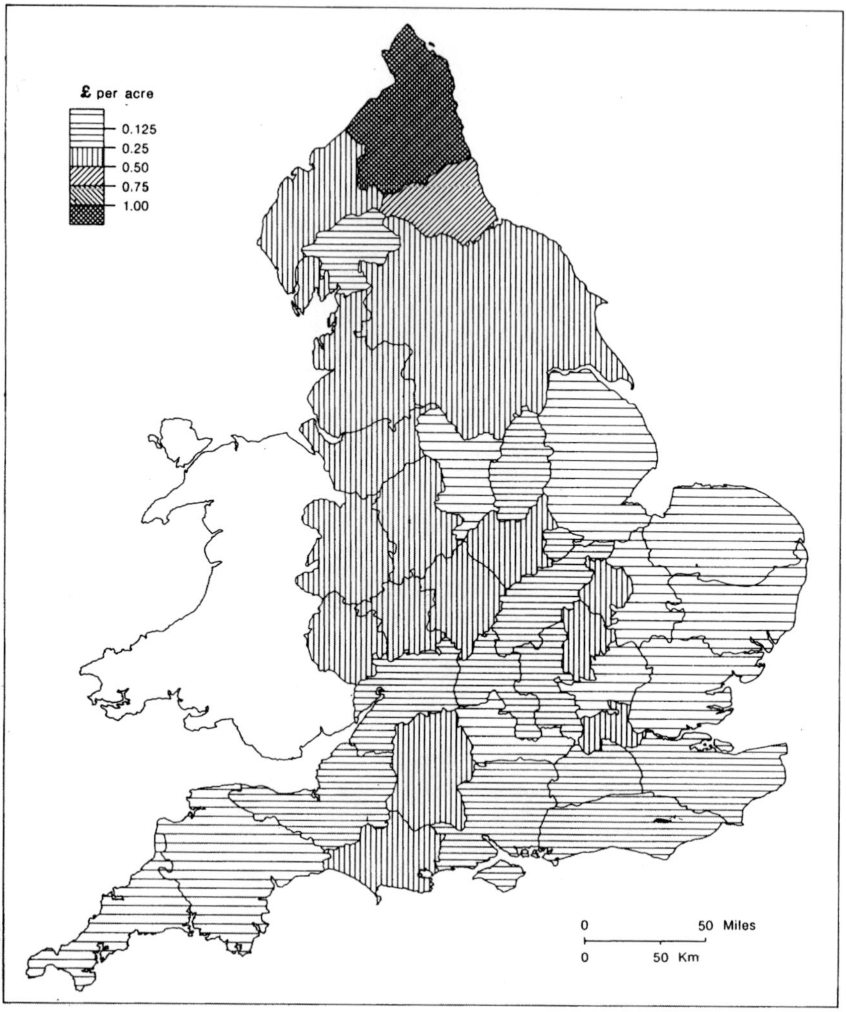

Figure 3.10 Total draining-loan expenditure, 1847–1899, per average annual cultivated acre, 1870–1879 (*Source:* As for Table 3.4)

per cent of that in Durham and Northumberland. A broad distinction may be drawn in the pattern of loan investment in the draining of agricultural land between counties in the north and the west midlands, with outliers in Bedfordshire and Huntingdonshire and in Dorset and Wiltshire, and counties in the east, southeast and southwest and the remaining parts of the midlands. In the former group, investment levels per cultivated acre were in the main above average. The highest outlays on agricultural land were

Table 3.4 *Total draining loan expenditure, 1847–1899, per cultivated acre and per acre of clayey and loamy soil with impeded drainage*

County	Average annual cultivated acreage 1870–79	Expenditure per cultivated acre (£)	Acreage of clayey/ loamy soils with impeded drainage	Expenditure per acre of soil with impeded drainage (£)
Bedfordshire	257,340	0.28	159,533	0.46
Berkshire	370,443	0.08	205,676	0.14
Buckinghamshire	400,748	0.18	191,648	0.38
Cambridgeshire	480,093	0.08	118,112	0.32
Cheshire	520,005	0.44	426,638	0.54
Cornwall	521,362	0.01	108,058	0.04
Cumberland	547,348	0.25	438,079	0.31
Derbyshire	500,360	0.11	143,985	0.40
Devon	1,082,010	0.09	478,923	0.21
Dorset	472,396	0.28	214,225	0.62
Durham	409,125	0.64	428,870	0.61
Essex	818,595	0.07	381,929	0.14
Gloucestershire	645,897	0.23	374,112	0.40
Hampshire	697,466	0.13	387,433	0.24
Herefordshire	433,267	0.41	115,405	1.53
Hertfordshire	336,102	0.11	80,510	0.46
Huntingdonshire	208,034	0.48	162,723	0.61
Kent	730,085	0.13	333,991	0.27
Lancashire	754,083	0.25	586,852	0.33
Leicestershire	467,926	0.27	396,357	0.32
Lincolnshire	1,470,034	0.10	484,905	0.30
Middlesex	115,761	0.25	120,643	0.24
Norfolk	1,066,092	0.01	451,124	0.03
Northamptonshire	554,347	0.22	410,622	0.30
Northumberland	680,384	1.13	656,664	1.18
Nottinghamshire	444,063	0.19	214,359	0.39
Oxfordshire	412,956	0.22	137,388	0.66
Rutland	84,936	0.14	63,033	0.19
Shropshire	692,282	0.44	352,932	0.87
Somerset	825,769	0.14	204,601	0.58
Staffordshire	589,049	0.27	425,153	0.37
Suffolk	762,910	0.03	462,152	0.05
Surrey	296,298	0.21	257,238	0.24
Sussex	653,690	0.16	528,033	0.20
Warwickshire	482,905	0.30	270,850	0.53
Westmorland	237,025	0.17	154,721	0.26
Wiltshire	739,611	0.25	221,741	0.84
Worcestershire	390,150	0.43	270,030	0.62
Yorkshire	2,658,713	0.29	1,674,643	0.45
Total	23,809,660	0.23	13,093,831	0.42

Sources: As for Tables 2.3 and 3.1; BPP, 1870, LXVIII; 1871, LXIX; 1872, LXIII; 1873, LXIX; 1874, LXIX; 1875, LXXIX; 1876, LXXVIII; 1877, LXXXV; 1878, LXXVII; 1878–79, LXXV, 'Agricultural returns, 1870–9'

recorded in Durham and Northumberland, the latter with £1.13 per cultivated acre. Although values in Cheshire, Herefordshire, Shropshire and Worcestershire were lower, they were nevertheless about twice the national average. The latter group of counties was generally marked by below average draining expenditure per cultivated acre. Investment in draining from loan capital was least important on agricultural land in Devon and Cornwall and in six eastern counties from Lincolnshire southwards to Essex, where levels were on average less than £0.10 per cultivated acre.

The use of cultivated acreage provides a standard measure to assess the relative importance of loan-financed draining on agricultural land throughout the country and facilitates comparison with other capital investments in agriculture. However, its employment leads to an under-estimation of draining intensity because not all cultivated land required underdraining. Draining activity financed by loan capital has been shown to be predominantly located in areas of clayey and loamy soils with impeded drainage and the relationship of loan amounts to such soils by county provides a more specific indication of draining intensity. As the area occupied by these soils formed only 55 per cent of the cultivated acreage in the 1870s, the intensity of investment must be expected to be greater, and on average throughout the country £0.42 was spent per acre of clayey and loamy soil with impeded drainage by means of draining loans (Table 3.4). The results produce a more coherent pattern of loan-financed draining than that based on cultivated acreage. Although a statistical relationship may be established between the total loan expenditure on draining and the area of clayey and loamy soils with impeded drainage in each county, the Spearman rank correlation coefficient being 0.54, significant at the .001 level,[52] the relatively low value clearly reveals that landowners provided a varied response to the draining needs of such soils, and counties fall broadly into four distinct groups of draining outlay (Fig. 3.11). Investment in such soils was most intense in four parts of the country: Durham and Northumberland; the west-midland counties of Cheshire, Herefordshire, Oxfordshire, Shropshire, Warwickshire and Worcestershire; Huntingdonshire and to a lesser extent Bedfordshire in the east midlands; and Dorset, Somerset and Wiltshire. Per acre values were highest in Herefordshire and Northumberland at £1.53 and £1.18 respectively and in general these counties corresponded to those areas with well-above-average loan-financed outlay per cultivated acre. Although levels were lower, usually less than the national average, a broad uniformity in the intensity of loan-financed draining was found on clayey and loamy soils with drainage difficulties throughout the rest of the northern and midland counties, with the exception of Rutland. Counties in eastern, southeastern and southwestern England generally displayed the lowest draining expenditure on such soils. And, within that general area, the values in Cornwall, Norfolk and Suffolk were particularly notable, no greater than

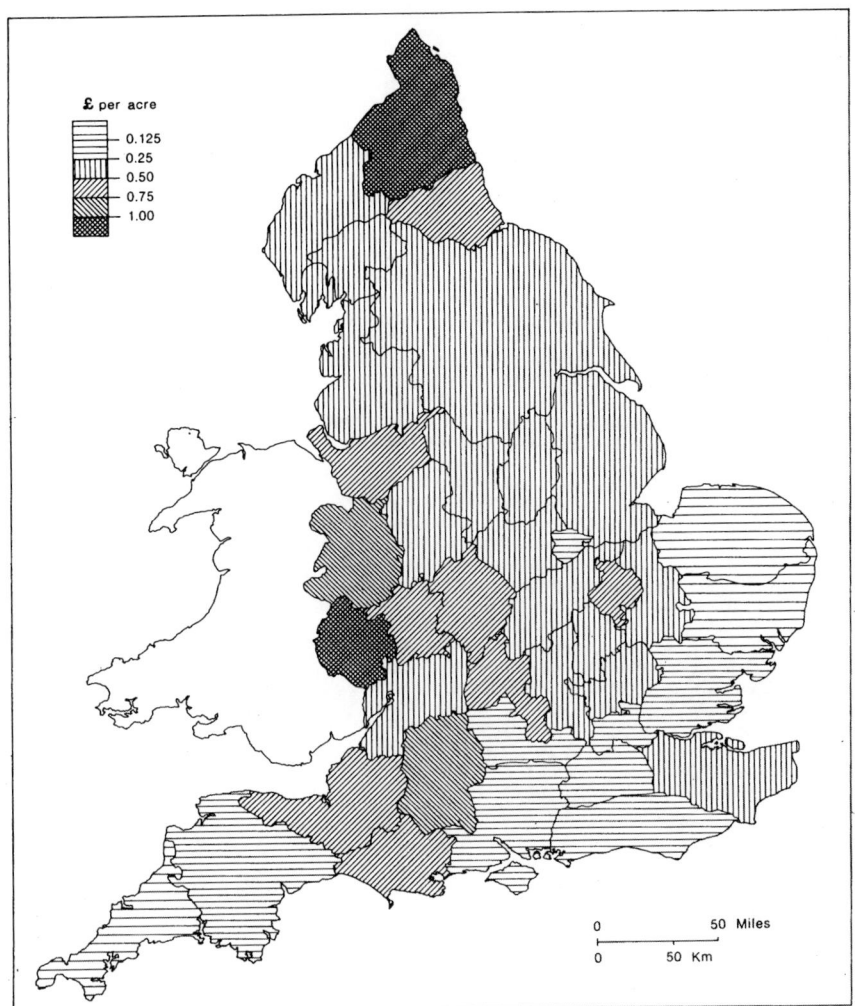

£ per acre

	0.125
	0.25
	0.50
	0.75
	1.00

0   50 Miles

0   50 Km

Figure 3.11 Total draining-loan expenditure, 1847–1899, per acre of clayey and loamy soil with impeded drainage (*Source:* As for Table 3.4)

£0.05 per acre of clayey and loamy soil, indicating that draining financed by loan capital had made little impact on the heavy lands of those counties.

These patterns of draining intensity may have been modified by funds from the Private Money Draining Act, the only source of loan capital administered by the Inclosure Commissioners for which data are not extant. However, although registers of certificates of loans are not available, notices of applications for loans under the act had to be published in the *London Gazette* and these proposals may be used to indicate the amount and location

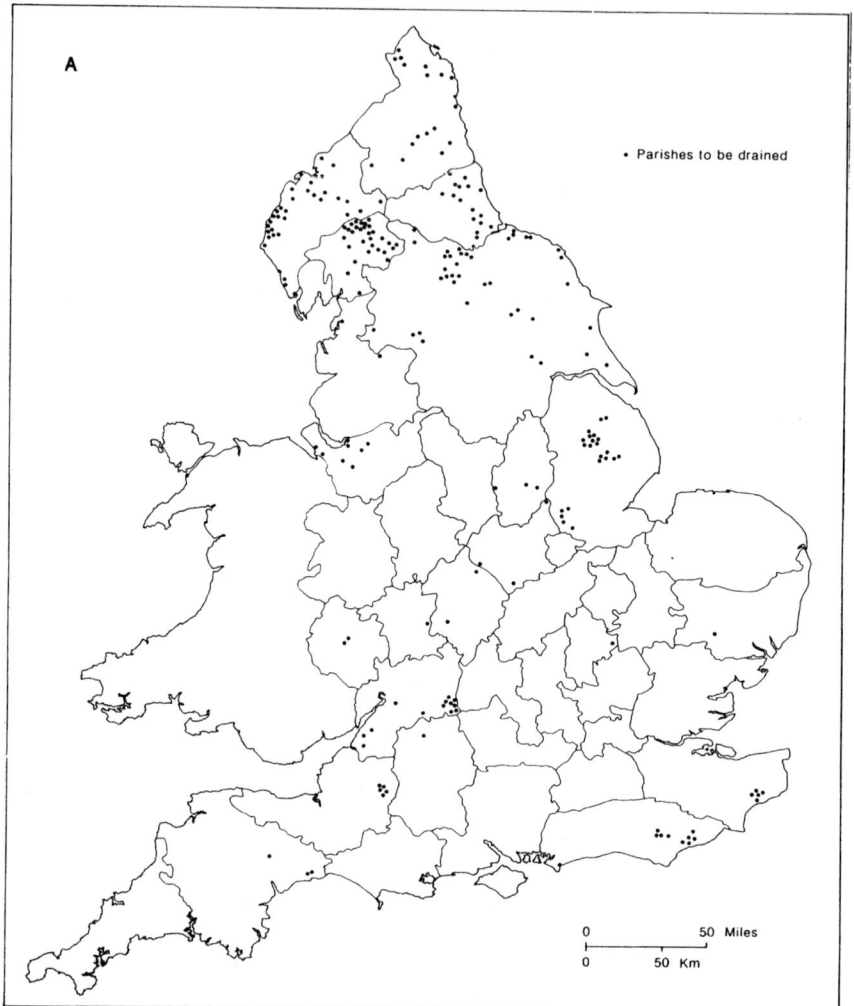

A

• Parishes to be drained

0   50 Miles

0   50 Km

Figure 3.12 A. Parishes to be drained under the Private Money Draining Act, 1849

of draining capital generated from this source.[53] In England, applications for loans totalling £339,768 were made, exceeding the amount actually lent under the act in the whole of England, Scotland and Wales by some 25 per cent. Parishes proposed to be drained were not located in areas neglected by other sources of loan capital but replicated and intensified the existing distribution (Fig. 3.12a). Intended expenditure was to be greatest in the five northern counties and in Lincolnshire, amounts ranging from £24,294 in Northumberland to £82,610 in Yorkshire (Fig. 3.12b). Yet, with the exception of Westmorland, the inclusion of these proposed sums would not

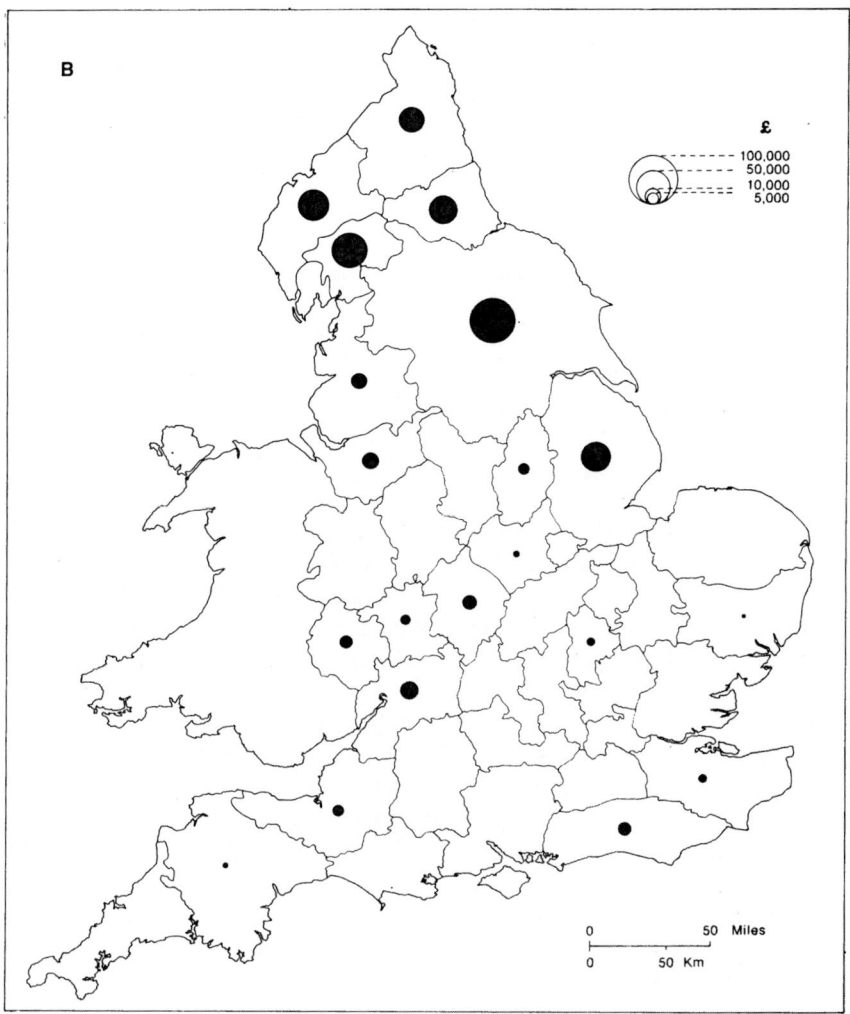

Figure 3.12 B. Amounts applied to be borrowed under the Private Money Draining Act, 1849 (*Source: London Gazette,* 1849–1864, *passim*)

have altered for any county the class interval of draining outlay per cultivated acre and per acre of clayey and loamy soil shown in Figs. 3.10 and 11. The sum intended to be used in Westmorland would have raised that county to the same category as neighbouring Cumberland and Lancashire for draining expenditure per cultivated acre, but would have produced no change to the category that it occupied for draining outlay on clayey and loamy soils. The absence of data on loans granted under the Private Money

Draining Act would not warrant any significant revision of the patterns of loan-financed draining already identified.

Conversion of these levels of investment into precise acreage equivalents is difficult because the cost of draining was not uniform in the second half of the nineteenth century, varying both in time and space. Contemporary estimates ranged from £4 to £8 per acre over the period,[54] but perhaps the most reliable large-scale guide to average acreage draining costs financed by loan capital may be found in the surviving reports that Andrew Thompson made as an inspector to the Inclosure Commissioners. Between 1857 and 1868, Thompson reported on 133 estates of all sizes, covering some 367,464 acres, mainly located in the midlands,[55] the owners of which had applied for draining loans from one of the sources of capital established by the land-improvement legislation of the middle of the century and administered by the Inclosure Commissioners (Fig. 3.13). On these estates, a sum of £473,381 was proposed to be borrowed to drain 76,411 acres, resulting in an average outlay to the nearest pound of £6 per acre.[56] The adoption of this figure as the standard acreage rate of loan-financed draining allows an areal assessment of land drained in the country from 1847 and 1899 by means of these funds.

At this rate the loan capital would have accomplished the draining of 916,000 acres in England. This represented but a small proportion of the cultivated area, 4 per cent of the national average annual acreage for the decade 1870–9, and of the land in need of draining, 7 per cent of the total acreage of clayey and loamy soils with impeded drainage. However, the use of a draining equivalent does emphasize the degree of spatial variation in the intensity of loan-financed draining throughout the country. In terms of cultivated area, recorded outlay would have drained, for example, 19 per cent of Northumberland, 7 per cent of Cheshire and Shropshire and 0.4 per cent of Norfolk and Suffolk, while as a proportion of soils with impeded drainage it would have resulted in 16 per cent of Durham and Northumberland, 12 per cent of Cheshire, Herefordshire, Shropshire, Warwickshire and Worcestershire and less than 1 per cent of Cornwall, Norfolk and Suffolk being improved.

The loans arising from the land-improvement legislation provide a coherent source for the analysis of draining in England in the second half of the nineteenth century. Although the amount of land drained with such capital was small, the data reveal distinct patterns in the location and intensity of loan-financed draining. Yet these funds were but one source of capital available to landowners for draining purposes and the representativeness of loan-financed draining to all draining undertaken in the second half of the nineteenth century has not been determined. While disagreeing on the proportion that loan-financed draining formed of total draining, estimates running from a half, a third and a quarter to higher fractions, contemporary agriculturalists, such as James Caird and J. Bailey Denton, were of the

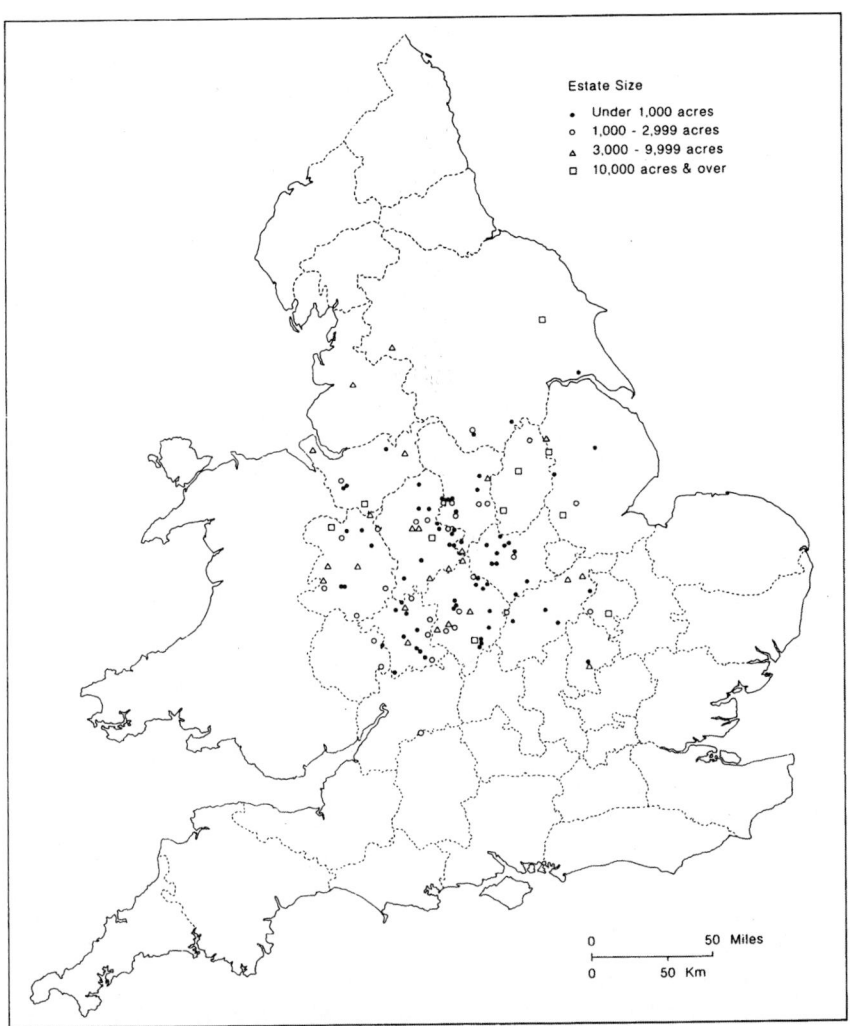

Figure 3.13 Estates inspected by Andrew Thompson, 1857–1868 (*Source:* KU, Sneyd MSS: Thompson's reports, vols. 4–8)

opinion that such draining typified all draining activity in the country.[57] However, before accepting the pattern of loan-financed draining as representative of general, national trends, the relationship between the two must be more closely assessed. To this end, in three counties with considerable range in the intensity of loan-financed draining in terms of both cultivated land and soils in need of drainage, Devon, Northamptonshire and Northumberland, the amount and distribution of draining have been examined on a number of estates for which relevant material is extant. The data yielded by these county groups of estates may be regarded as a sample to test the applicability and reliability of the patterns of loan-financed draining.

The amount of underdraining on estates in Devon, Northamptonshire and Northumberland

While providing precise detail, limitations should be appreciated in the records of draining activity on these estates. Whereas the draining loans are nearly complete, continuous series of accounts or other data indicating draining over the period 1845–99 are not available for all estates in the sample, being restricted in the main to the largest and best organized. Of the twenty-six estates examined in the three counties, nine possess draining records for the whole period and a further six for forty or more years. For the remaining eleven estates, data relate to smaller time periods. As with the loan-capital material, draining activity on these estates has to be expressed primarily in financial terms from sums spent on the improvement listed under various headings in estate accounts. Few estates maintained a separate ledger identifying fields and acreages drained or plotted the information in map form. However, the outlays recorded in estate papers represent essentially the landlords' investment in draining, which may not in all cases coincide with the total expenditure on the improvement. Unlike draining loans where the Inclosure Commissioners sanctioned only new draining schemes, estate expenditure on the improvement comprised several items, ranging from the complete financing of new schemes to the provision of part of the cost of draining and to the repair of existing work. The distinction between new and repair work was not regularly drawn in all estate accounts and as a result total estate outlay is likely to overestimate the intensity of the improvement.[58] On the other hand, where only part of the cost was met by the estate, multipliers have to be devised to accommodate the tenant element in the improvement so that an estimate of total expenditure may be produced for purposes of comparison with loan-financed draining. However, the representativeness of such multipliers of the tenant financial contribution can never be fully determined. The calculation of draining intensity on these estates also requires information on their acreage. Although the 1873 'Return of owners of land in England and Wales' and John Bateman's revision of the material provide a uniform source on estate acreage around 1875,[59] the data are not always accurate, record only total area and take no account of changes in size in the period 1845–99. In the present analysis, where possible, recourse has been made to estate maps and surveys to determine more precisely both the total and cultivated acreage of these properties.

Table 3.5 *Size and location of the Northumberland estate, Northumberland, 1827–1880*

Bailiwick	Total acreage 1850	Cultivated acreage in			
		1827–8	1850	1865	1878–80
Alnwick	12,644	–	9,284	9,685	9,579
Barrasford	9,494	8,920	8,022	8,098	6,935
Berwick	183	–	–	181	–
Chatton	20,265	18,332	19,456	19,391	17,869
Longhoughton	6,626	6,394	5,987	6,353	6,692
Lucker	9,785	8,817	9,038	9,330	8,880
Newburn	3,720	3,110	3,148	3,269	3,146
Prudhoe	6,057	4,992	5,042	4,998	4,251
Rothbury	16,507	10,127	15,624	15,524	14,835
Shilbottle	5,573	5,501	5,132	5,179	4,686
Tindale	59,108	52,541	58,710	60,748	59,793
Tynemouth	5,514	4,425	4,578	4,522	4,301
Warkworth	5,387	4,047	4,350	4,372	4,661
Total	160,863	127,206	148,371	151,650	145,628

–: No data available
Sources: Northumberland MSS: T. Bell and sons, 1850 survey and terrier; Annual returns of state of farms on bailiwicks, 1827–80; Business minutes, vol. 37, 16 February 1866

Estates in Northumberland

Draining data for the period were obtained for eight estates. The largest in the group and in the county belonged to the dukes of Northumberland. This grew in size over the century, rising from 134,462 acres in 1807 to 160,863 in 1850. Its maximum extent was reached in 1868 with 166,557 acres, when it formed 14 per cent of the county. From this peak it declined to 153,875 acres in 1899.[60] The estate was extensively distributed throughout the county and for administrative purposes was divided into bailiwicks, which by 1850 numbered thirteen, Berwick the most recent being purchased in 1842 (Fig. 3.14). These varied widely in acreage but, with the exception of Rothbury and Tindale to which much land was added in the first half of the nineteenth century, maintained a relatively stable size over the whole period (Table 3.5). Although five of the other seven estates in the sample could be classed as great estates with over 10,000 acres, none matched the extent of the duke of Northumberland's property.

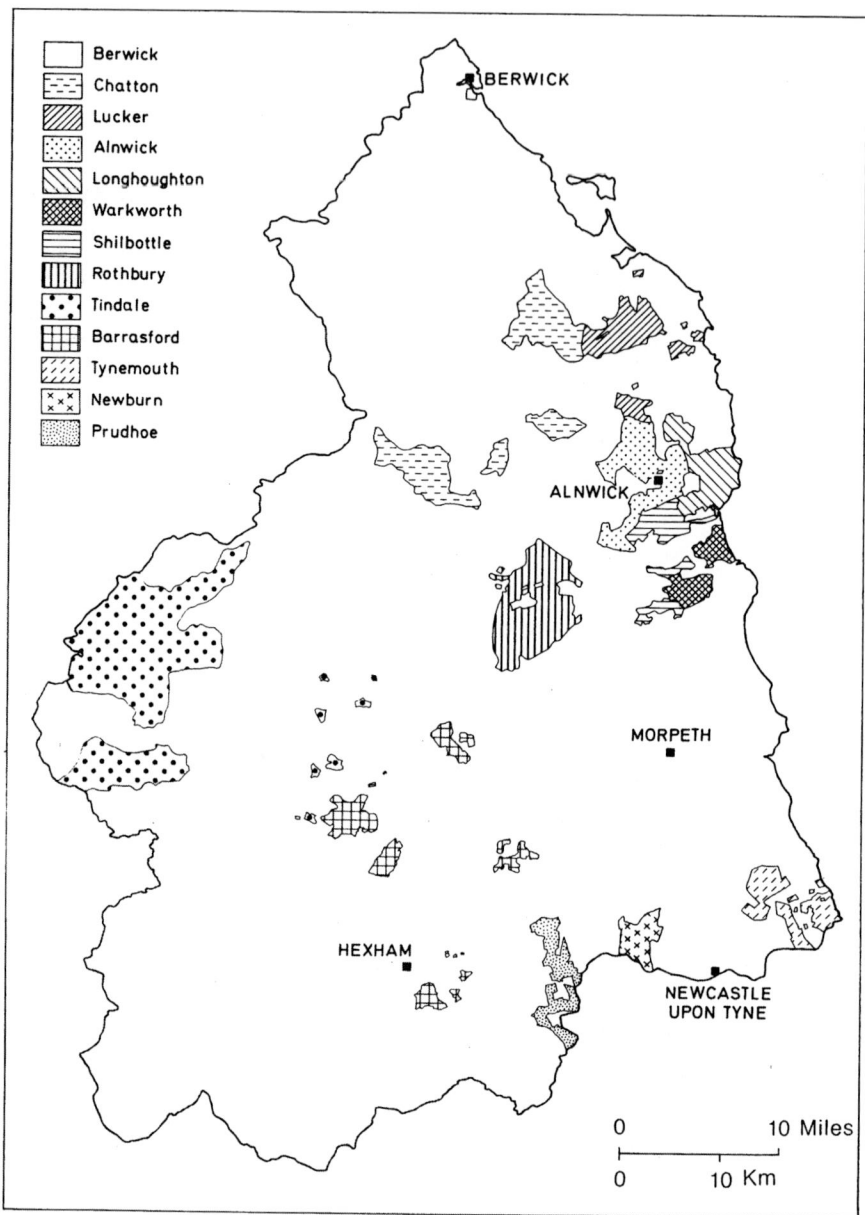

Figure 3.14 The bailiwicks of the duke of Northumberland's estate, 1850 (*Source:* Northumberland MSS: T. Bell and sons, Survey and terrier of the Northumberland estate, 1850: map indices to the bailiwicks)

Table 3.6 *Size and location of the Grey estate, Northumberland, 1803–1860*

Locale	Total acreage in		
	1803	1845	1860
Ancroft	2,692	2,541	2,494
Burton	1,734	2,193	2,189
Carham	4,490	4,751	4,745
Chevington	4,256	4,175	4,148
Grey's Forest	6,800	–	–
Horton	3,237	2,661	1,120
Howick	2,142	2,627	2,494
Total	25,351	18,948	17,190

–: Grey's Forest no longer recorded as part of estate
Sources: DPD, Grey MSS: Estate cropping book, 1845–78; M. Hughes, 'Lead, land and coal as sources of landlord income in Northumberland between 1700 and 1850', 1963, vol. 2, 60

Of the remaining five great estates, that of the earls Grey was next in size. In 1803, it covered 25,351 acres but by 1845 had fallen to 18,948 acres (Table 3.6). The Horton part of the estate was sold in 1849, reducing its total area to 17,438 acres, around which size the estate was recorded as still being in the 1870s.[61] Like that of the dukes of Northumberland, the Grey estate was widely distributed throughout the county (Fig. 1.3). The estates of the dukes of Portland, the earls of Carlisle and the Ridley family formed more compact and interlocking units in eastern Northumberland (Fig. 1.3). The first two extended over respectively 12,053 acres in 1861 and 12,916 acres in 1886.[62] Both had changed little from 1851 when they were mapped by J. T. W. Bell.[63] On rental evidence, the Portland estate retained this area until the 1890s, when some further land was added.[64] However, parts of the Carlisle property centred on Morpeth were sold from 1889, leaving from 1891 onwards only farms amounting to 1,127 acres paying rent.[65] The Ridley estate also fluctuated in size in the second half of the nineteenth century. In 1847 it amounted to 12,464 acres which, although concentrated at Blagdon to the south of the Portland property, included outliers at Hawkhope in the west of the county with 1,909 acres and at East Heddon, Byker and Heaton close to Newcastle with 1,481 acres. By 1868 the property at Hawkhope, Byker and Heaton had been sold, leaving 9,912 acres, to which 864 acres of the Carlisle estate at Netherton were added in 1889, creating a total area of 10,776 acres.[66] The smallest estate in the sample, that of the Baker-Baker family, also joined this block of great estates. Located at Stanton, it covered 2,067 acres in 1847, at which it was returned in 1873.[67]

The last two estates in the sample were located in the south of the county. That of the Blackett family expanded from 13,239 acres in 1810 to 15,192 acres in 1867, and rentals indicate little alteration of this area to the end of the century. It was administered in two parts: the West Water estate lay mainly to the west of Hexham, its acreage increasing from 6,749 in 1810 to 7,666 in 1869; and the Matfen estate to the east, its area growing from 6,490 acres in 1810 to 7,366 in 1867[68] (Fig. 1.3). The estate of the Middleton/ Monck family linked the Blackett Matfen and the Carlisle, Portland and Ridley properties (Fig. 1.3). Its total area varied little in the period, being recorded at 9,121 acres in 1853, 9,061 in 1872 and 8,973 in 1883. While the main body of the estate was at Belsay, there was a small detached portion at Shotley on the southern county boundary covering 1,185 acres in 1853.[69] In total, this sample of estates represented around 20 per cent of Northumberland from the 1850s to the 1870s, the cover being particularly dense in the east of the county.

On these estates, recorded draining expenditure related to new schemes and repair work, there being little evidence of partial outlay on the improvement. In general, all displayed high levels of draining investment. Absolute total outlay, including repairs, amounted to £252,878 on the Northumberland estate between 1845 and 1899,[70] a sum representing £1.52 per acre of the 1868 total area, or £1.67 per acre of the 1865 cultivated area. For this estate, areas drained were identified and this outlay at an average acreage cost of £5.47 accomplished the draining of 46,194 acres,[71] 28 per cent of the 1868 total area or 30 per cent of the cultivated land in 1865. Although the overall average per acre expenditure was low in comparison to others in the sample, this should not disguise the fact that large portions of the agricultural land of the Northumberland estate were drained in the second half of the nineteenth century, the intensity varying between bailiwicks. Expressed in terms of the 1865 cultivated area, draining outlay and area drained exceeded respectively £3.00 per acre and 50 per cent of the land in all bailiwicks save Barrasford, Chatton, Rothbury and Tindale, and at Alnwick, Longhoughton, Newburn, Shilbottle, Tynemouth and Warkworth around 75 per cent of the cultivated area was drained in the period (Table 3.7). Investment was much lower on the Barrasford, Chatton, Rothbury and Tindale bailiwicks, not more then £1.30 being spent per cultivated acre and the amount of land drained not exceeding 25 per cent. Tindale possessed the lowest levels on the estate with an outlay of £0.24 per acre and only 5 per cent of the cultivated area improved.

As assessments of land in need of draining were made on the estate, the extent to which the improvement had been adopted may be determined. In 1855 and 1866, the then chief commissioners of the estate, Hugh Taylor and J. Snowball, reported respectively to the fourth and fifth dukes the amount drained *and* to be drained on each farm, which together indicate the area in

Table 3.7 *Draining expenditure and acreage drained on the Northumberland estate, Northumberland, 1845–1899*

Bailiwick	Cultivated acreage in 1865	Draining outlay 1845–99 (£)	Outlay per 1865 cultivated acre (£)	Acreage drained 1845–99	Drained area as a percentage of 1865 area
Alnwick	9,685	38,302	3.95	7,068	73
Barrasford	8,098	10,504	1.30	1,909	24
Chatton	19,391	23,219	1.20	4,524	23
Longhoughton	6,353	25,906	4.08	5,129	81
Lucker	9,330	34,509	3.70	6,073	65
Newburn	3,269	13,860	4.24	2,340	72
Prudhoe	4,998	15,257	3.05	2,554	51
Rothbury	15,524	12,507	0.81	2,654	17
Shilbottle	5,179	22,202	4.29	3,989	77
Tindale	60,748	14,814	0.24	2,738	5
Tynemouth	4,522	20,599	4.56	3,574	77
Warkworth	4,372	21,199	4.85	3,642	83
Total	151,469	252,878	1.67	46,194	30

Source: Northumberland MSS: Draining volumes, 1–3, 1844–1903

need of the improvement.[72] The resulting area totalled 36,725 acres in 1855, 25 per cent of the 1850 cultivated land, which was increased to 44,770 acres in 1866, 30 per cent of the 1865 cultivated extent, the need varying greatly between bailiwicks (Table 3.8). Although Snowball admitted that precise calculation of areas to be drained was difficult, the amount of land drained on the whole estate exceeded both estimates, with no bailiwick possessing less than 80 per cent of the area in need being treated. Draining clearly was widely adopted on the estate in the period 1845–99 and, although the absolute area drained may have formed only 30 per cent of the cultivated acreage, virtually all land in need of the improvement had been dealt with.

Draining data are available for the Grey estate only to 1892, but from 1845 to 1892 total outlay amounted to £67,001, which represented a per acre expenditure of £3.53 of the 1845 estate area.[73] This average figure underrates the intensity of draining, for not all land on the estate required the improvement, at least 3,232 acres being described in 1847 as not in need of draining in an assessment of the drainage status of the cultivated area.[74] If such land is deducted from the 1845 estate area, average per acre draining expenditure rises to £4.25. Investment was not uniform on the different parts of the estate. Excluding the Horton part sold in 1849 and taking into account

Table 3.8 *Area in need of draining on the Northumberland estate, Northumberland, 1855 and 1866*

Bailiwick	Area in need of draining, 1855		Area in need of draining, 1866		Acreage drained, 1845–99, as percentage of area needing draining in	
	Absolute acreage	As percentage of 1850 cultivated area	Absolute acreage	As percentage of 1865 cultivated area	1855	1866
Alnwick	5,451	59	6,496	67	130	109
Barrasford	1,369	17	1,570	19	139	122
Chatton	3,806	20	5,054	26	119	90
Longhoughton	3,908	65	4,307	68	131	119
Lucker	4,288	47	6,093	65	142	100
Newburn	2,174	69	2,339	72	108	100
Prudhoe	2,454	49	3,196	64	104	80
Rothbury	1,741	11	2,296	15	152	116
Shilbottle	3,167	62	3,908	75	126	102
Tindale	1,417	2	2,383	4	193	115
Tynemouth	4,325	94	3,621	80	83	99
Warkworth	2,625	60	3,507	80	139	104
Total	36,725	25	44,770	30	126	103

Source: Northumberland MSS: Business minutes, vol. 17, 29 October 1855, and vol. 37, 16 February 1866

Table 3.9 *Draining expenditure on parts of the Grey estate, Northumberland, 1845–1892*

Locale	Draining outlay 1845–92 (£)	Outlay per acre of 1845 area (£)	Acreage in need of draining, 1847	Outlay per acre of 1847 area in need of draining (£)
Ancroft	13,499	5.31	2,418	5.58
Burton	8,379	3.82	1,440	5.82
Carham	10,907	2.30	3,718	2.93
Chevington	26,271	6.29	3,979	6.60
Howick	7,446	2.83	2,094	3.56
Total	66,502	4.08	13,649	4.87

Sources: DPD, Grey MSS: Uncatalogued ledger books, 1840–92; Draining volumes, 1841–86; Box 550, Draining reports for 1847

the areas described in 1847 as not requiring the improvement, expenditure was greatest on the Ancroft, Burton and Chevington sections of the estate, exceeding £5.50 per acre, and least at Carham and Howick, the relatively low level of the last being largely a product of the area of parkland and woodland held in hand and not drained (Table 3.9). However, the overall level of draining outlay, which approached the average acreage cost recorded on the estate of the dukes of Northumberland, would indicate a high proportion of the Grey estate being drained between 1845 and 1892.

Similar rates of draining investment were found on the adjoining Carlisle, Portland and Ridley properties. On the Morpeth estate of the earls of Carlisle, £41,960 was spent on draining in the period 1845–99, a sum equal to per acre outlays of £3.25 and £3.57 respectively of the total and cultivated area in 1886.[75] From 1856 acreage drained was recorded and by the end of the century 5,929 acres, 50 per cent of the 1886 cultivated area had been improved at an average cost of £5.06 per acre.[76] If the average cost of draining before 1856 was of the same order, total expenditure would have resulted in the draining of 8,290 acres, 70 per cent of the cultivated area of the estate. Draining expenditure over the whole period amounted to £43,843 on the Portland estate.[77] With per acre outlays of £3.64 of the total area and of £3.82 of the cultivated area of the estate in 1861, this sum would point to an intensity of draining little different from that on the Carlisle lands. Data are less complete for the Ridley estate, relating to the years 1847 to 1885. Over this period investment in draining was recorded at £32,520, £2.61 per acre of the 1847 estate area.[78] Although overall lower than the two neighbouring estates, the Ridley property was less compact and average expenditure concealed variations between the constituent parts. At Hawkhope in the west of the county, no draining outlay was made from 1847 until its sale in 1868. On the detached portions of the estate at East Heddon, Heaton and Byker, £4,137 was spent on the improvement at £2.79 per acre over the period, although Heaton and Byker, some 640 acres, were also sold in 1868. The remaining part of the Ridley lands experienced a draining outlay of £28,383 from 1847 to 1885, representing a per acre investment of £3.13 over its 1847 area of 9,074 acres, a level comparable in magnitude to that on the adjoining Portland estate.

Records of draining expenditure are available for the Middleton/Monck estate only for the period 1869–84. Nevertheless, in that time £39,993 was spent on the improvement, producing an acreage outlay of £4.41 of the 1872 estate area.[79] This level exceeded those on the Carlisle and Portland properties where expenditure related to the whole period 1845–99 and would indicate well over 70 per cent of the Middleton/Monck estate being drained. An unbroken run of draining accounts exist for the Blackett estates, revealing that £22,487 was laid out from 1845 to 1899, at £1.48 acre of the 1867 estate area.[80] This sum achieved the draining of 4,085 acres, 27 per cent of the estate at an average acreage cost of £5.50.[81] However, a marked distinction was evident between the two parts of the estate. Expenditure at West Water, £6,532 at £0.85 per acre of the 1867 area, resulted in only 16 per cent of the land being drained. At Matfen investment was higher, £15,955 being spent at £2.17 per acre, and land drained formed 39 per cent of the estate area, levels of activity that approached those of neighbouring estates. Draining data are least extensive for the smallest property in the sample, that of the Baker-Baker family. Yet from 1846 to 1873, outlay amounting to £4,179 was recorded on the improvement, producing in terms of its 1847 area a per-acre expenditure of £1.97.[82] Although this rate fell below those on the contiguous but larger Carlisle and Portland properties and which relate to longer time spans, fields covering 879 acres were drained at an average cost of £4.75, resulting in some 43 per cent of the estate being improved over the period.

Despite the generally high level of draining investment, variation in the intensity of activity, measured by either outlay per acre or area drained, is evident between and within estates in the sample. Part of this divergence may be related to the drainage needs of the soils on which the estates lay.

Unfortunately, where estate data are available, detailed examination of fields drained by soil type is precluded by the absence of a complete, large-scale series of soil maps for the country. In lieu of these, land drained on such estates in Northumberland, and also in Devon and Northamptonshire, has been plotted in relation to drift geology, an imperfect guide to the drainage needs of soils. However, the resulting maps indicate that, although much

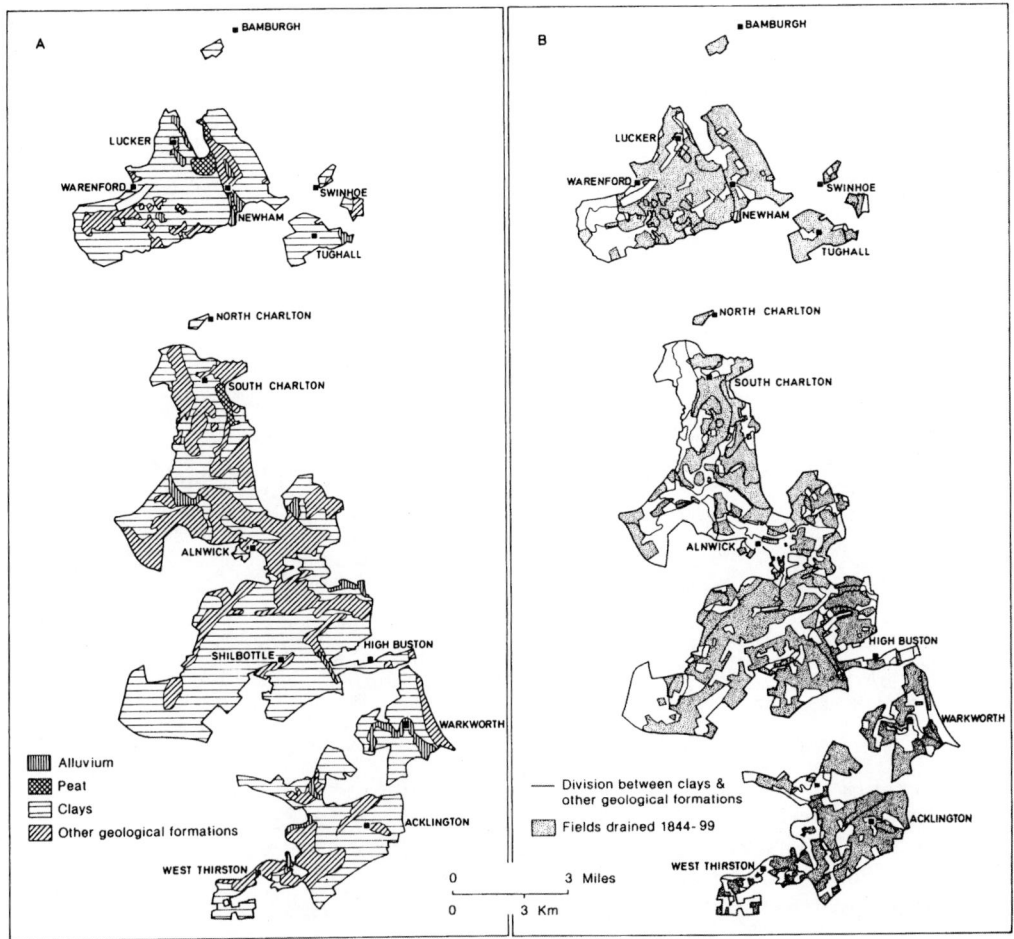

Figure 3.15 Land drained, 1844–1899, and drift geology, the duke of Northumberland's Alnwick, Lucker, Shilbottle and Warkworth bailiwicks (*Sources:* Northumberland MSS: Draining vols., 1–3; T. Bell and sons, Survey... of the Northumberland estate, 1850, map nos. 1–102; Geological Survey, 1 in drift sheets 4, 6 and 9)

draining was located on clay formations, the relationship was neither complete nor exclusive. Thus, on the Alnwick, Lucker, Shilbottle, Warkworth and Newburn bailiwicks of the duke of Northumberland's estate and on the Carlisle estate (Figs. 3.15a and b; 3.16a and b), the improvement was found on formations that geologically could not be classed as clays, while not all available clayland was drained, a situation more evident on the Barrasford bailiwick (Fig. 3.17). Further support of the value of draining to soils other than clays can be obtained from the Grey estate, where the nature of the land

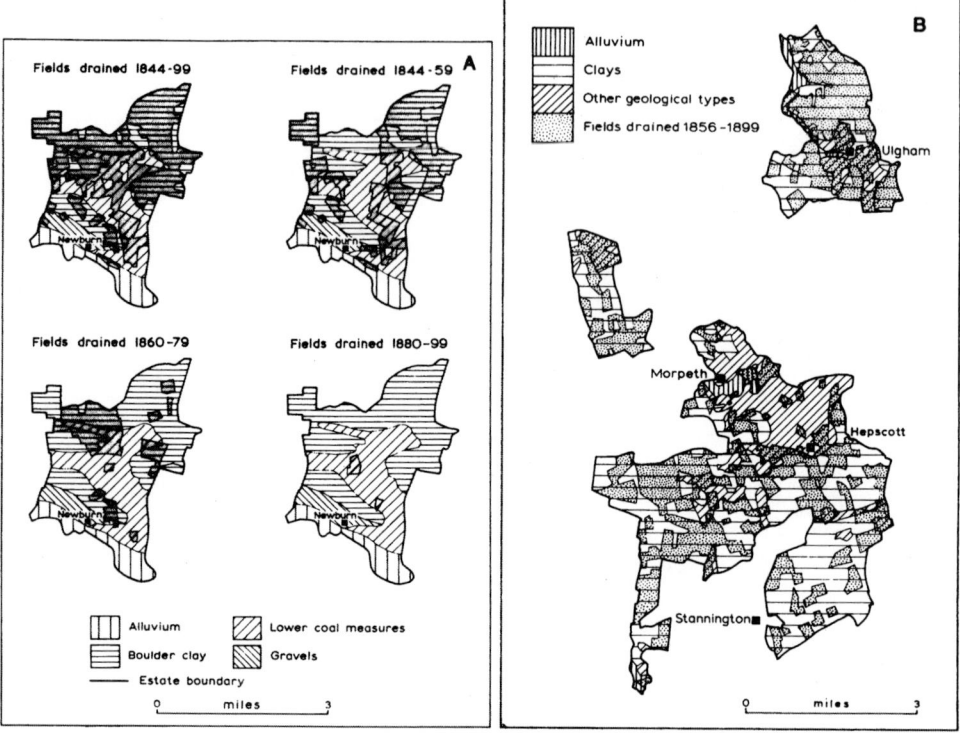

Figure 3.16 Land drained and drift geology on (A) the duke of Northumberland's Newburn bailiwick, 1844–1899, and (B) the earl of Carlisle's estate, 1856–1899 (*Sources:* Northumberland MSS: Draining vols., 1-3; T. Bell and sons, Survey... of the Northumberland estate, 1850, map nos. 129–37; DPD, Howard MSS: 99/2; Draining vols., 1–3; Geological Survey, 1 in drift sheets, 9, 14, 19–20)

drained between 1841 and 1847 was recorded. While clay and clay-loam soils dominated with 52 per cent of the 4,568 acres drained in the period, 35 per cent of the land was classed as clay and light soils, 9 per cent as turnip soils, sands and gravels, and 4 per cent as moor.[83] Although important on geologically defined clays, the evidence from these Northumberland estates, and indeed from Devon and Northamptonshire properties (Figs. 3.19–3.21, 3.23), serves to confirm that draining was not solely a clayland improvement.

In terms of the broad drainage needs of soils in Northumberland derived from the Soil Survey's 1983 *Soil Map of England and Wales* (Fig. 2.5), draining outlay was low on estates in the upland and moorland parts of the county. On the Tindale bailiwick of the duke of Northumberland's estate, lying mainly on open moorland over 1,000 ft, only 5 per cent of the 1865 cultivated area of 60,748 acres was drained by the end of the century, while Sir

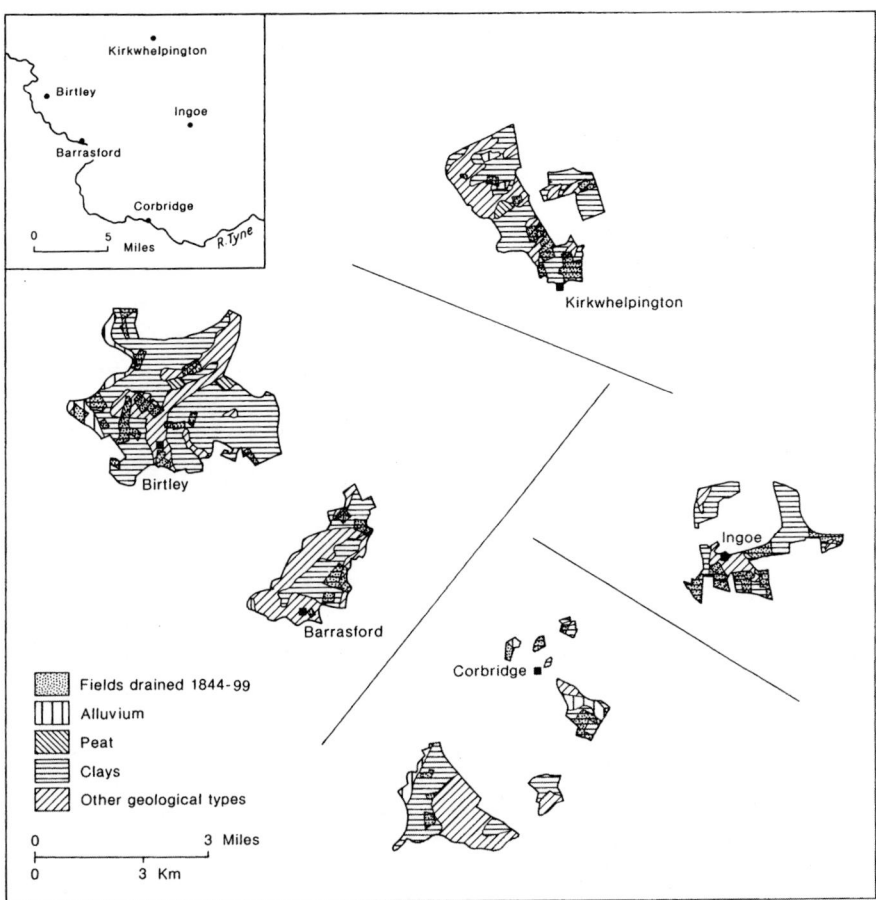

Figure 3.17 Land drained, 1844–1899, and drift geology, the duke of Northumberland's Barrasford bailiwick (*Sources:* Northumberland MSS: Draining vols., 1–3;
T. Bell and sons, Survey...of the Northumberland estate, 1850, map nos. 88–9,
152–63; Geological Survey, 1 in drift sheets, 13–14, 19–20)

Matthew Ridley made no draining expenditure between 1847 and 1868 on his
Hawkhope property, 95 per cent of which was moorland.[84] Such draining
that was undertaken on this land was in the form of open- rather than
underdrains. The other estates in the sample in upland locations, the
Barrasford, Chatton and Rothbury bailiwicks of the Northumberland estate
and the Blackett West Water estate, experienced slightly greater draining
outlay. Over the whole period, 20 per cent of the cultivated area of these
estates in the 1860s, some 50,679 acres, was drained with a per acre
expenditure of £1.04. These properties occupied lower land and contained
significant areas of clayey and loamy soils with impeded drainage.

Nevertheless, they formed largely poor-quality land, the rental value for example of Sir Edward Blackett's West Water estate in 1861 at £0.49 per acre being but 47 per cent of that of his Matfen estate, and quickly degenerated into moorland, which few landowners contemplated draining.[85] As a result, draining investment on these properties was limited.

With the exception of the Carham part of the Grey estate which contained a mixture of well-drained soils and loamy soils with impermeable clay subsoils and where draining expenditure averaged £2.30 per acre from 1845 to 1892, the remaining estates in the sample were predominantly situated on clayey and loamy soils with impeded drainage in lowland Northumberland.[86] Estates on these soils displayed the highest rates of draining investment in the county. Known draining expenditure on these properties amounted to £452,879 over the period 1845–99. At their recorded maximum extent in the second half of the nineteenth century and using where possible cultivated area, these estates covered 111,582 acres, with a draining outlay of £3.82 per acre. As average acreage draining costs over the whole period on those estates where the calculation can be made did not exceed £5.50, this level of expenditure would suggest that at least 70 per cent of the cultivated area of these Northumberland properties were drained from 1845 to 1899.

The location and intensity of draining on this sample of estates provide direct confirmation of the trends identified for Northumberland in the draining-loan data. However, the use of loans under the land-improvement legislation to finance draining ranged widely on these estates, irrespective of size. In all, loan capital formed 21 per cent of absolute draining expenditure recorded in the sample over the period 1845–99 (Table 3.10). This proportion is likely to overemphasize somewhat the importance of loan capital, for, whereas a near complete record of loans is extant, draining expenditure is not available for all estates for the whole period. Nevertheless, if this ratio were applicable to all draining-loan capital used in the county, the great proportion of the cultivated acreage and of soils with drainage difficulties in Northumberland would have been drained in the second half of the nineteenth century.

Estates in Northamptonshire

The full cost of the improvement was not met on all estates in Devon and Northamptonshire, a number of landowners in the period employing tenant capital either by providing allowances at half the cost of draining or by issuing draining materials to which the tenant had to add labour costs. To facilitate comparison of draining intensity, such limited estate involvement requires transformation into estimates of total expenditure. In those cases where allowances were made, the amounts may be doubled to arrive at a reasonable approximation of total draining outlay. The proportion that

Table 3.10 *Loan capital as a proportion of total draining expenditure on estates in Northumberland, 1845–1899*

Estate	Period for which data available	Total draining expenditure (£)	Amount of loan capital 1847–99 (£)	Loan capital as a percentage of total draining outlay
Baker-Baker	1846–73	4,179	4,079	98
Blackett	1845–99	22,487	19,636	87
Carlisle	1845–99	41,960	3,338	8
Grey	1845–92	67,001	31,453	47
Middleton/Monck	1869–84	39,993	39,993	100
Northumberland	1845–99	252,878	0	0
Portland	1845–99	43,843	4,168	10
Ridley	1847–85	32,520	4,947	15
Total		504,861	107,614	21

Sources: PRO, 1R3/6–38; MAF 66/1–6, 8–9, 11, 13–22, 25–39, 43–7; Northumberland MSS: Draining volumes, 1–3; DPD, Grey MSS: Ledger books, 1840–92; Draining volumes, 1841–86; Howard MSS: N. 101–6; N. 73/2; Estate accounts, 1875–1900; Draining volumes, 1–3; Baker-Baker MSS: 23/36–7, 98–118 *passim*; NdRO, ZSA/8/1–5; ZR1/44/4; ZBL/54/2 and 3; ZBL/277/6; ZBL/282/2; Belsay MSS: Box 12/IX

materials formed of the full cost of the improvement is less easy to calculate, much depending on their price and the draining system in use. However, the individual costs of draining were itemized on five estates in the two counties between 1847 and 1864. In this group, expenditure on drain-pipes and tiles amounted to £7,248, 30 per cent of the total draining outlay of £24,159.[87] As this proportion was derived from a range of draining systems and soil types, the assumption has been made that pipes and tiles formed about one-third of the full cost of the improvement, and the trebling of the amounts spent by estates in the samples solely on draining materials is likely to provide a broad indication of the total draining expenditure that was incurred from such partial outlay.

Of the eleven estates in Northamptonshire for which draining data exist for the period 1845–99, five contained areas in excess of 10,000 acres (Fig. 1.2). The lands of the dukes of Buccleuch were grouped into two compact estates centred on Boughton and Barnwell (Fig. 3.18). Both were stable in size throughout the nineteenth century, the acreage of the former being recorded at 11,423 and 12,110 in 1834 and 1896 respectively, that of the latter at 7,953 in 1813, 7,905 in 1860 and 7,813 in 1903.[88] Although part of the Barnwell estate lay in Huntingdonshire, 14 per cent in 1860, draining activity

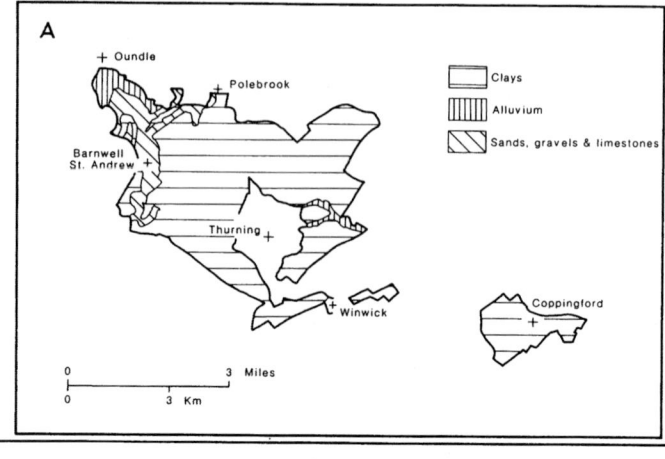

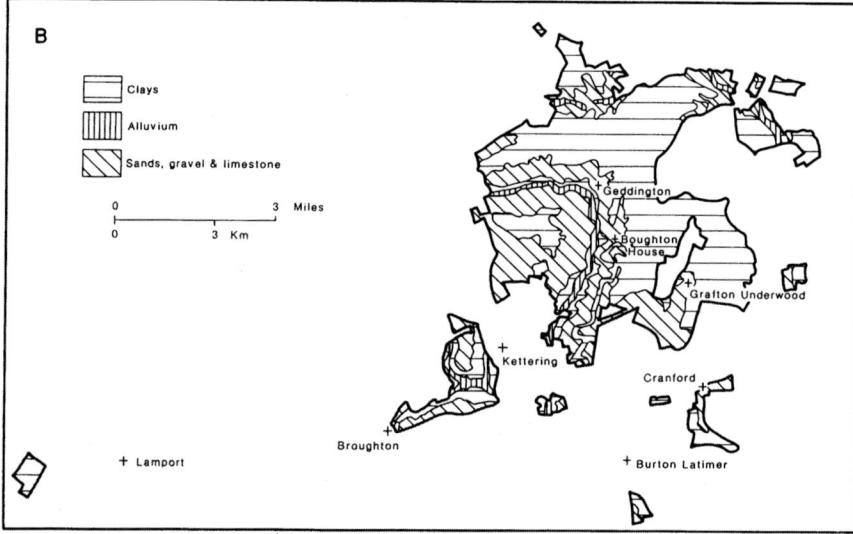

Figure 3.18 The Barnwell estate, 1860 (A), and the Boughton estate, 1895 (B), of the duke of Buccleuch (*Sources:* Buccleuch MSS: Numerical reference of the Boughton estate, 1895; NRO, Buccleuch MSS: Misc. Ledgers, 136 and 137; Geological Survey, 1 in drift sheets, 171, 186)

has been examined for the whole property. The estate of the earls Spencer fluctuated more in area, increasing from 13,955 acres in 1827 and 14,373 acres in 1859 to 16,073 in 1868 (Fig. 3.19). It remained at this size until 1890 when with land sales the area fell to 14,851 acres.[89] Throughout the period 1845–99, the main body of the estate was located around Althorp, with 61 per cent of the 1859 acreage, with outliers at Boddington, Brampton Ash, Elkington, Steane and Strixton, containing between them the rest of the

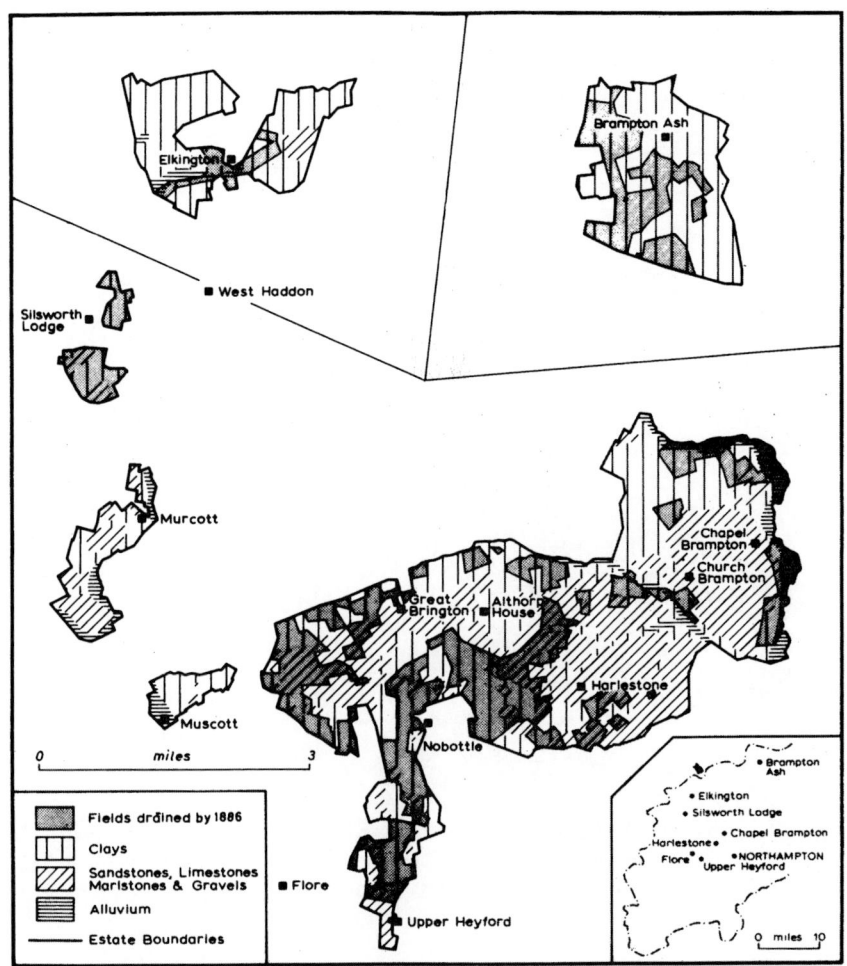

Figure 3.19 Fields drained by 1886 and drift geology on part of the earl Spencer's estate (*Sources:* Spencer MSS: Maps of estates at Althorp Park, Brampton Ash, Chapel and Church Brampton, Elkington, Heyford and Flore, Harlestone and Silsworth, 1838–1879; Estate cropping books, 1857–1899; Geological Survey, 1 in drift sheets, 185, 202)

estate area. Greater growth and dispersal characterized the estate of Lord Overstone. The size of the estate in 1832, when land was first bought in Northamptonshire, was 3,681 acres located at Overstone and Sywell. By 1850 through extensive purchase the property had expanded to 17,161 acres, which by 1877 had further increased to 18,816 acres.[90] As the estate was built up by purchase, it was well scattered throughout the county, the most

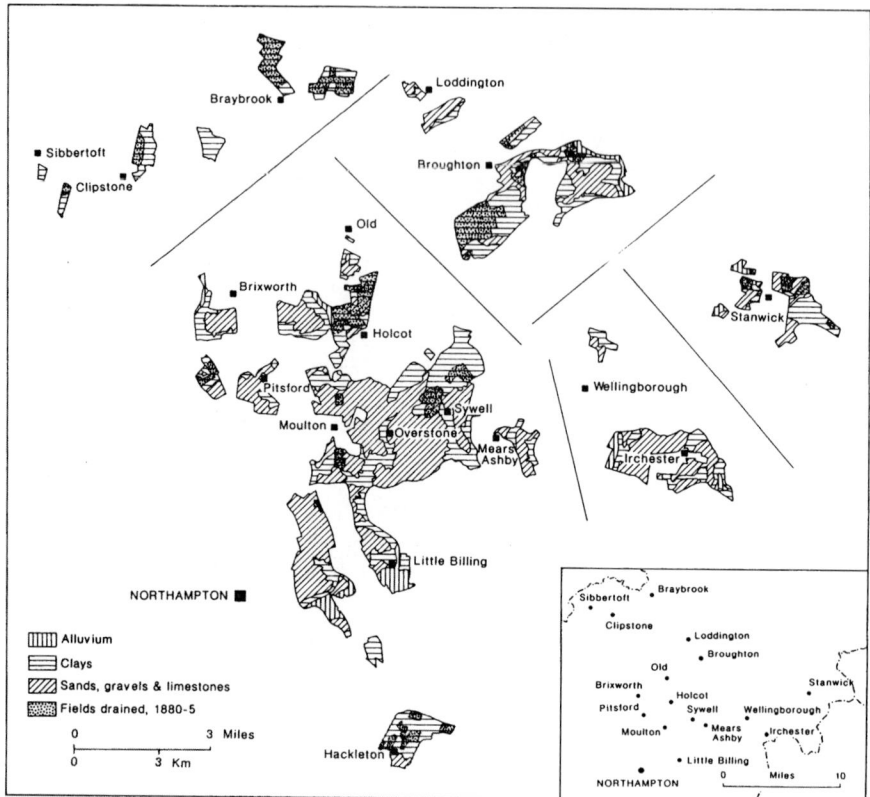

Figure 3.20 Fields drained, 1880–1885, and drift geology on part of Lord Overstone's estate (*Sources:* NRO, Overstone MSS: Ov. maps, 184–92, 194–5, 327, 329–31, 342, 354; Geological Survey, 1 in drift sheets, 170–1, 185–6, 202)

detached part with 1,970 acres in 1877 being located at Fotheringhay (Figs. 1.2; 3.20). Less precise information on size is available for the estates of the dukes of Grafton and of the Fitzwilliam family. The tenanted area of the former amounted to 12,718 acres in 1822, 65 per cent of which lay in the parishes of Ashton, Blisworth, Grafton Regis, Greens Norton, Paulerspury and Potterspury. The estate would seem to have changed little in area from that date to 1883 when Bateman recorded it at 14,507 acres in a corrected entry.[91] Prior to 1857, the Fitzwilliam lands in the county had belonged to the third earl, who on his death passed them to his second son, in whose family they remained for the rest of the century. They were administered in two parts. The smaller around Higham Ferrers contained 5,581 acres in 1857, of which 12 per cent was located in Huntingdonshire.[92] The acreage of the larger centred on Milton in the north of the county cannot be fully determined: land in parishes around Peterborough covered 13,373 acres in 1855, 83 per cent of which lay in Northamptonshire, but no data detail the area of the property in Warmington and Lutton in Northamptonshire and Great Gidding and Old Weston in Huntingdonshire.[93]

The sample also includes three estates in the smaller-size category of greater gentry, possessing between 3,000 and 9,999 acres. The land of the dukes of Cleveland at Brigstock remained constant in area throughout the century, with 3,627 acres in 1835, 3,620 in 1855 and 3,658 in 1902[94] (Fig. 3.21), as did the Wansford estate of the dukes of Bedford, containing respectively 4,128 and 4,233 acres in 1857 and 1895.[95] The Aynho estate of the Cartwright family, a small part of which lay in Oxfordshire, also displayed relative stability over the period, its area being recorded at 4,695 acres in 1859, 4,658 in 1877 and 4,939 in 1893.[96] To complete the sample, data have been obtained for three smaller properties, each covering less than 3,000 acres: the Brackley estate of the earls of Ellesmere amounted to 2,810 acres in 1837, an area little altered in Bateman's return of 1883;[97] the Dryden family's estate at Canons Ashby was recorded at 2,530 and 2,615 acres respectively in 1844 and 1873;[98] and the Naseby estate of the Ashby family remained around 1,050 acres from 1860 to 1903[99] (Fig. 1.2). In sum, these eleven properties throughout the period under consideration formed about 17 per cent of the total estate area of Northamptonshire.

Both the amount and intensity of recorded landlord investment in draining on these estates generally fell below the levels identified in the Northumberland sample. This situation may be partly the product of more incomplete data, draining records being extant for relatively shorter and occasionally intermittent runs for six of the estates examined. However, for those five properties with near-complete information, landlord expenditure never exceeded the average of £2.62 per acre on the Ashby estate (Table 3.11). The level also reflects the fact that not all landlords accepted full responsibility for the improvement and on five estates in the sample draining was only partially financed, recourse being made to allowances and the provision of draining materials. Yet, if the amounts allocated on these items are adjusted to represent total expenditure (Table 3.12), the corrected values, although greater, are not commensurate with intense draining activity. The conclusion is difficult to avoid that draining was less extensively adopted on these estates than on their Northumberland counterparts.

Such evidence that exists on acreages substantiates this view. The Cleveland estate exhibited one of the highest levels of draining expenditure in the sample at £2.24 per acre, but between 1848 and 1871, when 80 per cent of total outlay occurred, although fields covering some 66 per cent of the estate were affected, the acreage drained was limited to 1,577 acres, 44 per cent of the 1855 area[100] (Fig. 3.21). Fields drained by 1886 can be determined for part of the Spencer estate extending over 9,857 acres, 61 per cent of the

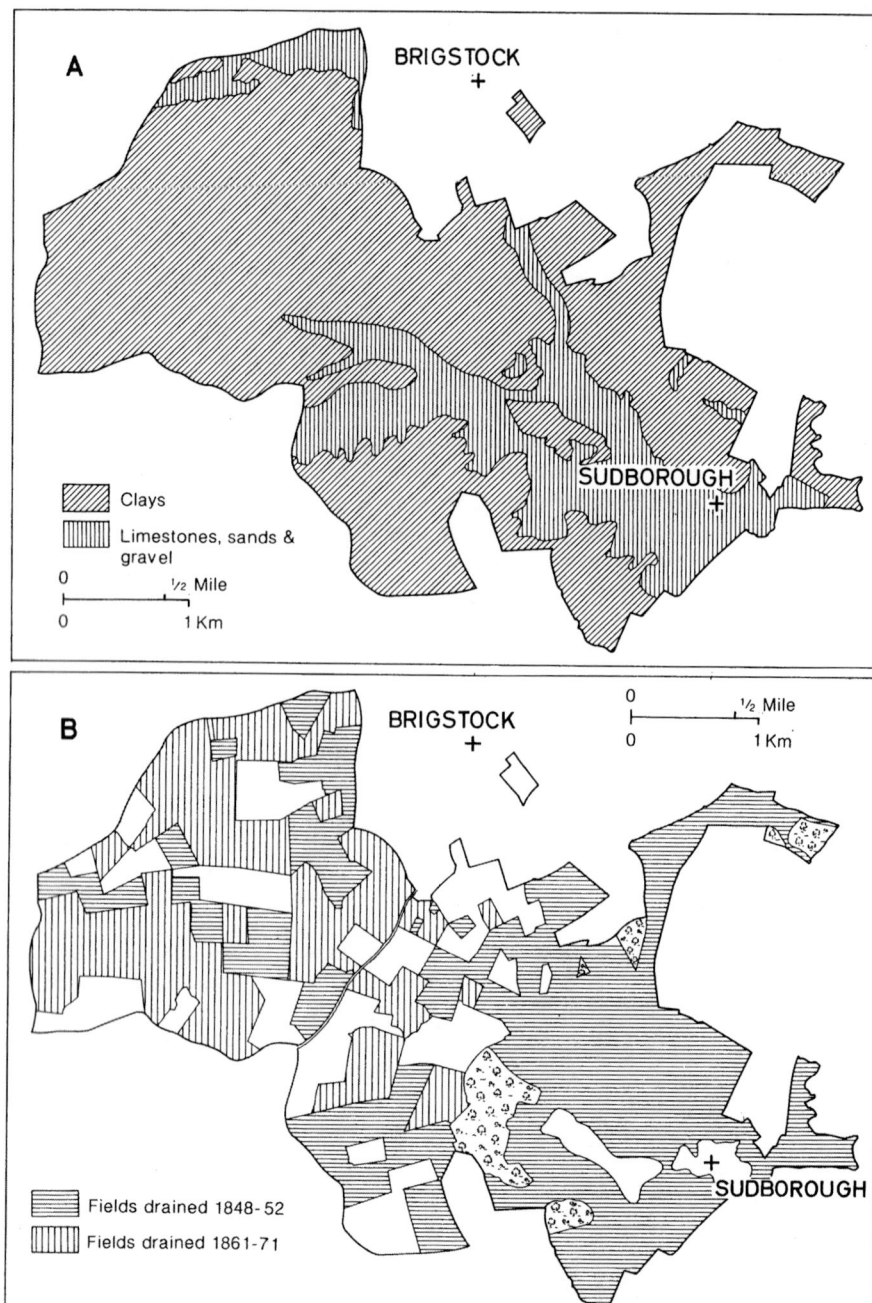

Figure 3.21 Fields drained, 1848–1871, and drift geology on the duke of Cleveland's estate (*Sources:* Raby MSS: Sudborough draining vol., 1848–53; Draining abstracts, 1848–53; Draining vols., 1861–1872; Plans of the Brigstock and Sudborough estates, 1855; Geological Survey, 1 in drift sheet, 171)

Table 3.11 *Recorded draining expenditure on estates in Northamptonshire, 1845–1899*

Estate	Period for which data available	Recorded estate acreage	Recorded draining expenditure (£)	Outlay per acre (£)
Ashby	1864–87	1,049	2,750	2.62
Bedford	1845–95	4,233	5,630	1.33
Buccleuch	1845–99	20,015	30,353	1.52
Cartwright	1851–70	4,695	8,391	1.79
Cleveland	1845–99	3,620	8,121	2.24
Dryden	1858–66	2,615	4,903	1.87
Ellesmere	1845–74	2,839	2,087	0.74
Fitzwilliam (Higham Ferrers only)	1845–99	5,581	985	0.18
Grafton	1845–86 *passim**	14,507	10,862	0.75
Overstone	1845–94 *passim***	18,816	18,357	0.98
Spencer	1845–99	16,073	7,562	0.47

* Data available for 33 years, 1845–6, 49–54, 59–71, 73–80, 82–3, 85–6
** Data available for 30 years, 1845–58, 72–4, 80–5, 88–94
Sources: NRO, Fisher-Sanders MSS: FS1/20, Ashby draining loans with the Lands Improvement Company; Cartwright MSS: C (A) 4740–57, Absolute orders under draining loans; C (A) 5242 and 5943, Draining sheets, 1856; C (A) 3844–71, Rentals, 1851–70; Dryden MSS: D (CA), 450, Draining schedules, 1858–66; Ellesmere MSS: X.461, Rentals, 1845–74; Fitzwilliam MSS: Estate accounts, 1845–99; Grafton MSS: G. 1810–2041, Estate vouchers, 1830–86; Overstone MSS: 0.906–38, Estate accounts, 1832–58; Ov. vol. 452, Estate accounts, 1883–4; Ov. Ledger vol. 1, Estate accounts, 1890–4; Bedford MSS: Annual reports, 1845–95; Buccleuch MSS: Barnwell and Boughton estate accounts and rentals, 1845–99; Raby MSS: Brigstock and Sudborough rentals, 1845–99; Sudborough draining volume, 1848–53; Draining volumes 1861–72; Spencer MSS: Estate accounts, 1845–99; BPP, 1894, XVI, part 1, 'Royal Commission on agricultural depression', appendix xvii

1868 estate, but they comprise only 24 per cent of the area[101] (Fig. 3.19). Data on fields drained also exist between 1880 and 1885 for 10,455 acres of the Overstone estate, 56 per cent of the 1877 total.[102] In those six years 17 per cent of the area was drained but the expenditure incurred formed 59 per cent of the adjusted total recorded for the whole estate over the period 1845–99 (Fig. 3.20).

The levels of draining investment found on estates in the sample vary in

Table 3.12 *Draining expenditure adjusted for partial outlay on five Northamptonshire estates, 1845–1899*

Estate	Total recorded outlay (£)	Outlay spent on			Total adjusted outlay (£)	Adjusted outlay per acre (£)
		Full cost of draining (£)	Draining allowances (£)	Draining materials (£)		
Buccleuch	30,353	18,070	9,696	2,587	45,223	2.26
Ellesmere	2,087	239	0	1,848	5,780	2.04
Fitzwilliam (Higham Ferrers)	985	308	33	644	2,306	0.41
Overstone	18,357	16,823	0	1,534	21,425	1.14
Spencer	7,562	3,583	0	3,979	15,520	0.97

Source: As for Table 3.11

response to soil conditions. The Fitzwilliam Higham Ferrers property with an adjusted per acre outlay of £0.41 from 1845 to 1899 was dominated by well-drained soils where the improvement was little needed, while significant areas of the Buccleuch Boughton and Spencer estates, with adjusted average per acre expenditures of £1.28 and £0.97 respectively in the period, were also occupied by such soils. Draining values were much higher on those estates which lay wholly or mainly on clayey and loamy soils with impeded drainage.[103] Thus, on the Ashby, Buccleuch Barnwell, Cleveland, Ellesmere and the Boddington, Brampton Ash and Elkington parts of the Spencer estates, containing 18,843 acres in the 1860s, draining investment, including adjustment for partial outlay, averaged £2.43 per acre between 1845 and 1899. Nevertheless, these rates were not of the order of those estates in Northumberland on similar soils, and on the evidence of the Cleveland estate would suggest 40 per cent of the land being drained.

Estates in the sample made varying use of loan capital to finance the improvement. Overall it formed 33 per cent of recorded landlord investment in draining (Table 3.13). However, as recorded landlord expenditure did not always represent total draining outlay, its importance has been exaggerated. Adopting figures that have compensated where necessary for partial financing of the improvement, the proportion of total expenditure provided by loan capital falls to 25 per cent. Although this percentage should be subject to further reduction, as not all estates in the sample possessed a complete run of data for the whole period, it is despite that of a level comparable to that derived from the Northumberland sample.

Table 3.13 Loan capital as a proportion of total draining expenditure on estates in Northamptonshire, 1845–1899

Estate	Period for which data available	Total recorded expenditure (£)	Total expenditure including adjustment for partial outlay (£)	Amount of loan capital 1847–99 (£)	Loan capital as a percentage of	
					Recorded expenditure	Adjusted expenditure
Ashby	1864–87	2,750	2,750	2,750	100	100
Bedford	1845–95	5,630	5,630	0	0	0
Buccleuch	1845–99	30,353	45,223	13,450	44	30
Cartwright	1851–70	8,391	8,391	7,847	94	94
Cleveland	1845–99	8,121	3,121	5,000	62	62
Dryden	1858–66	4,903	4,903	4,903	100	100
Ellesmere	1845–74	2,087	5,780	0	0	0
Fitzwilliam (both estates)	1845–99	4,797	9,372	760	16	8
Grafton	1845–86 passim	10,862	10,862	0	0	0
Overstone	1845–94 passim	18,357	21,425	0	0	0
Spencer	1845–99	7,562	15,520	0	0	0
Total		103,813	137,977	34,710	33	25

Sources: PRO, IR3/6–38, MAF 66/1–6, 8–9, 11, 13–22, 25–39, 43–7, and as for Table 3.11

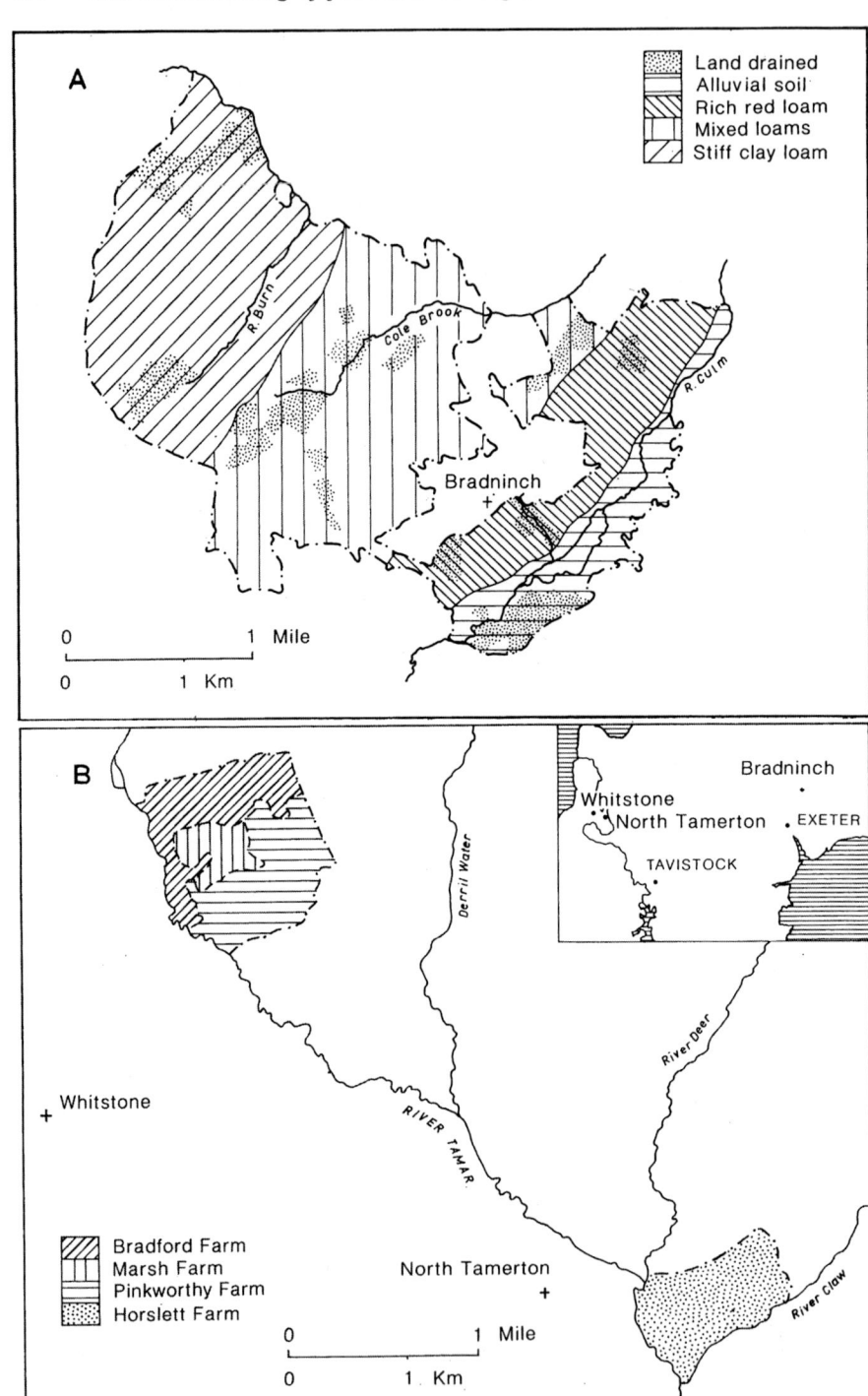

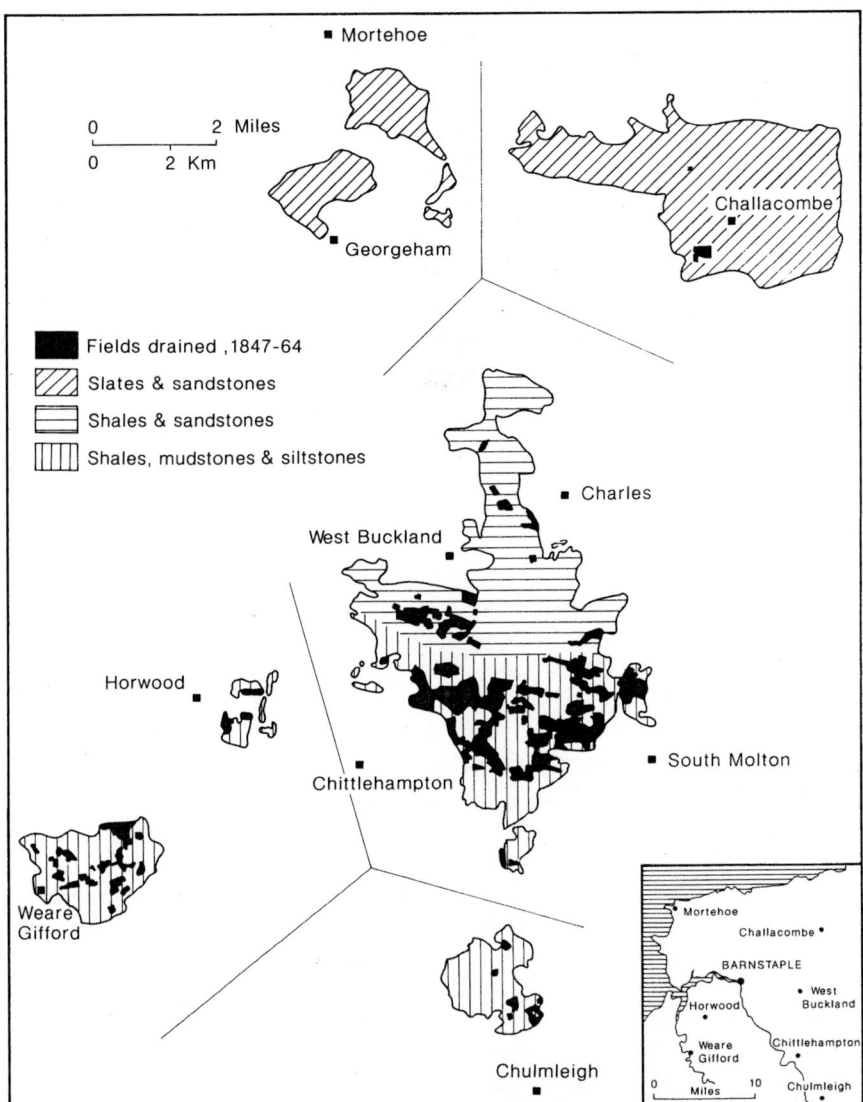

Mortehoe

0 ————— 2 Miles

0 ————— 2 Km

Georgeham

Challacombe

■ Fields drained ,1847-64

▨ Slates & sandstones

▤ Shales & sandstones

▥ Shales, mudstones & siltstones

■ Charles

West Buckland

Horwood

South Molton

Chittlehampton

Weare Gifford

Chulmleigh

Mortehoe
Challacombe
BARNSTAPLE
West Buckland
Horwood
Weare Gifford
Chittlehampton
Chulmleigh
0 ———— 10
Miles

Figure 3.23 Fields drained, 1847–1864, and drift geology on the earl Fortescue's north Devon estate (*Sources:* DRO, Fortescue MSS: 1262M/E1/103; 1262M/E22/ 10, 12, 16, 24, 51–7, Maps of the north Devon estate, 1880; Geological Survey, 1:50,000 sheets, 292, 293, 308, 309)

Figure 3.22 A. Land drained 1850–1899, and soils in 1855 after the agent, R. Watt, on the duchy of Cornwall's Bradninch estate; B. The Bradford estate of the duchy of Cornwall, 1869 (*Sources:* Cornwall MSS: Farm bundles, manors of Bradford and Bradninch; Inrolment books, 1862–1902; Map of... Bradninch, 1788; Valuation of ... Bradninch by R. Watt, 1855; Map of... Bradford, 1867)

Table 3.14 *Total and cultivated area of the Bedford estates, Devon, 1843–1895*

Estate	1843 Cultivated acreage	1857 Total acreage	1875		1895	
			Total acreage	Cultivated acreage	Total acreage	Cultivated acreage
Mid-Devon	2,885	2,994	4,702	4,254	4,675	4,618
North Devon	1,846	1,877	1,994	1,824	2,119	1,963
South Devon	502	510	528	372	532	366
Tavistock	10,889	15,237	16.155	10,418	16,887	11,091
Total	16,123	20,618	23,379	16,868	24,212	18,038

Source: Bedford MSS: Annual reports, 1843, 1857, 1875, 1895

Estates in Devon

The sample size for Devon is not as great as for the two other counties, data being examined for six estates. Three of these belonged to the category of greater gentry occupying between 3,000 and 9,999 acres and were widely distributed throughout the county (Fig. 1.1). The estate of the viscounts Sidmouth lay in east Devon, mainly in Upottery, and its total area increased throughout the period, being recorded at 3,919 acres in 1840–2, 4,103 in 1873 and 4,563 in 1894.[104] Larger in size was the estate of the dukes of Somerset, returned at 8,138 acres in 1873. It was administered as three separate units: the Wonwell estate with 724 acres in 1862; the Stover estate containing 2,758 acres at the same date; and the Berry Pomeroy estate, the acreage of which ranged from 4,498 in 1846 to 4,699 in 1872.[105] Draining data are available only for the last part, which lay almost entirely in Berry Pomeroy parish in south Devon. The final estate in this group was the property of the duchy of Cornwall. Although the duchy's holding in Devon was recorded at 48,457 acres in 1873, Dartmoor formed the greater part of the area and lowland agricultural property was restricted to two small estates included in the present analysis. In 1855, the Bradninch estate in east Devon contained 2,825 acres, an area it retained until the end of the century. The Bradford estate on the Cornish border in 1856 covered 530 acres in three farms. A fourth, Marsh farm, was added in 1869 enlarging the area to 597 acres, which remained unchanged for the rest of the period[106] (Fig. 3.22a and b).

The three other estates in the sample all exceeded 10,000 acres. The largest, that of the dukes of Bedford, grew from 20,618 acres in 1857 to 24,212 in 1895[107] and comprised four separate sub-estates, located in the Tavistock area, in mid-Devon predominantly in North Petherwin, in north Devon mainly in the parish of Swimbridge, and in Plymstock in south Devon (Table

Table 3.15 *Draining expenditure on the Devon estate, Devon, 1848–1899*

| Estate | Acreage in 1862 | Total recorded outlay (£) | Recorded outlay spent on | | Adjusted total outlay (£) | Adjusted outlay per 1862 acre (£) |
			Draining allowances (£)	Draining materials (£)		
Malborough	5,864	297	128	131	687	0.11
Moretonhampstead	6,116	833	418	316	1,882	0.31
Powderham	8,318	5,350	648	1,075	8,148	0.98
Tavistock	452	23	6	0	29	0.07
Wolborough	842	936	142	302	1,682	2.00
Total	21,592	7,439	1,392	1,824	12,428	0.58

Sources: DRO, Courtenay MSS, 1508M/Estate papers/14/A/111, Shelf III, 1862 valuation; 1508M/Estate vols./14/B/111/2–39

3.14). The estate of the earls of Devon was also scattered throughout the county. Its area amounted to 21,592 acres in 1862, having on rental evidence altered little from 1848. This size was broadly maintained until 1890, after which date the estate shrank, falling from a rental area of 17,913 acres in 1890 to 12,056 in 1891 and to 7,793 in 1892, an acreage that was retained until 1899.[108] The major part of the estate in 1862 was centred on Powderham, with large holdings around Moretonhampstead and Malborough in the south of the county. Smaller portions were found about Wolborough in Kingsteinton parish and in Tavistock parish (Table 3.15). From 1892 only lands at Powderham, Wolborough and Tavistock remained in estate control. Of similar extent was the estate of the earls Fortescue, in the early 1860s covering 21,260 acres. It consisted of two parts, the larger laying in north Devon containing 18,742 acres in 1864 (Fig. 3.23), the smaller known as the south Devon estate but mainly in Lamerton parish in the west with 2,518 acres in 1860.[109] The two estates adjoined and were intermixed with Bedford lands. Rentals suggest little change in estate size from 1850, and in 1880 total tenanted area amounted to 19,687 acres, 17,258 of which formed the northern property and 2,429 the southern.[110] Of the total estate area of the county noted by Bateman in 1883,[111] the recorded acreage of these six estates formed about 5 per cent from the 1850s to the 1880s, the proportion declining in the last decade of the century.

Although on most estates the full cost of the improvement was met by landowners, tenants sometimes contributed to part of the expenditure. Absolute draining outlay on the Sidmouth estate, available from 1850 to 1899, amounted to £13,853.[112] An assessment of the drainage status of the estate in 1850 indicated that the need for the improvement was extensive, being required on 2,200 acres, 64 per cent of the cultivated area.[113] As acreage drained was recorded from 1851 to 1866, some measure can be provided of the extent of adoption of the improvement. Over that 15 years, at an average cost of £6.10 per acre, 1,830 acres were drained, 83 per cent of the area described in need of draining in 1850.[114] Draining expenditure averaged £3.38 per acre of the 1873 estate area over the whole period, indicating a high investment rate in the improvement. Such levels were not evident on the duchy of Cornwall and Somerset estates. From 1850 to 1899, a total of £1,487 was spent at Bradninch achieving the draining of 254 acres, figures representing 9 per cent of the estate area and a per acre outlay of £0.53.[115] Draining activity was little different at Bradford and in the same period expenditure reached £532 at £0.89 per acre, resulting in the improvement of 22 per cent of the estate.[116] Although data relate only to the years 1850–75, a lower rate of investment was found on the Berry Pomeroy estate, outlay amounting to £421 at £0.09 per acre of the 1872 area.[117] Of this sum, £85 was spent on the provision of draining materials and, if this element is adjusted to represent the full cost of the improvement, total expenditure may be conjectured at £590. Even with this augmented sum, draining investment on the estate was still slight, being £0.13 per acre.

Expenditure from 1848 to 1899 was also low on the estate of the earls of Devon, averaging £0.34 per acre of the 1862 area. The recorded outlay of £7,439 included allowances at half the cost of the improvement and the provision of draining materials.[118] Transforming these partial costs to estimates of full cost, total draining investment on the estate may be conjectured at £12,430, raising the acreage rate slightly to £0.58 (Table 3.15). These modified figures reveal that draining intensity was greatest at Wolborough and Powderham, where much marshland was reclaimed, and insignificant on other parts of the estate. Draining was wholly landlord financed after 1845 on the remaining Fortescue and Bedford estates. However, overall investment levels on the Fortescue properties varied little from those on the Devon estate. Records cease for the Fortescue south Devon estate in 1880, but in the period 1845–80 draining involved the expenditure of £1,863 at £0.74 per acre of the 1860 area.[119] Outlay amounted to £10,726 between 1845 and 1899 on the north Devon estate, at a per acre rate of £0.57 of its 1864 area.[120] However, such a sum would seem insufficient to meet the need for the improvement on the property. In 1847, for the purpose of a loan, fields were selected for draining totalling 1,915 acres, 10 per cent of the estate area.[121] Josiah Parkes, who reported in 1848 on the draining undertaken by the loan to the Inclosure Commissioners, considered that the area to be drained had been underestimated: 'the number of wet areas...is so large as compared with his lordship's claim that the selection of the particular fields which would yield the greatest return from drainage would be extremely difficult'.[122] Between 1847 and 1864, the draining of 1,239 acres was accomplished at an average cost of £5.69 per acre, but which represented only 65 per cent of the area identified in need of improvement in 1847 and little more than 6 per cent of the estate.[123] As draining expenditure declined after 1864, much of the estate must have remained undrained.

A sum of £44,979 was spent on draining on the Bedford estates between 1845 and 1895, averaging on an acre basis £1.92 and £2.67 respectively of the total and cultivated area of the estate in 1875. The rate varied among the four properties, with no expenditure being made on the south Devon estate. Draining activity was modest on the Tavistock lands, with an outlay of £9,905 at £0.95 per acre of the 1875 cultivated area. However, on the mid- and north Devon estates the highest levels of draining expenditure in the sample were recorded, £26,512 on the former and £8,562 on the latter, representing in terms of the 1875 cultivated areas respective per acre investments of £6.23 and £4.69.[124]

Save for the Sidmouth and parts of the Bedford properties, draining expenditure was proportionally less in the sample of Devon estates than in those for Northamptonshire and Northumberland. Some of the low values relate to the varying physical need for the improvement.[125] The Moreton-hampstead estate of the earls of Devon lay on upland soils around 1,000 ft, virtual moorland, where the application of agricultural capital would have been slight (Fig. 2.6). Again, the rest of the Devon estate, the Bedford lands around Tavistock and in the south of the county, the Fortescue south Devon property and the Berry Pomeroy estate, which extended over 39,375 acres in the 1860s and 1870s, were predominantly located on well-drained soils where the area in need of draining was limited. Expenditure levels could be anticipated to reflect these soil conditions, and over the period 1845–99 the total draining outlay on these estates of £22,905, including adjusted sums for partial costs, represented an average of £0.58 per acre.

At the same time, part of the low rate of draining investment must be attributed to neglect of the improvement. The remaining estates in the sample, covering 32,963 acres in the 1860s, were dominated by clayey and loamy soils with impeded drainage.[126] Although overall draining activity was greater on these properties than others in the sample, there was no uniform response amongst them in dealing with the drainage needs of these soils. Thus, the duchy of Cornwall's Bradford and the Bedford mid-Devon estates both on the Cornish border, possessed soils described by their respective agents as poor, wet clayland,[127] but per acre expenditure on the former was 14 per cent of the latter. Similarly, the per acre rate on the Bedford north Devon estate was eight times greater than on the adjoining Fortescue land. The recorded draining outlay on these estates in the period 1845–99 amounted to £60,772, a per acre value of £1.84. At this level, with average acreage costs nearing £6 where data are available, perhaps 30 per cent

Table 3.16 Loan capital as a proportion of total draining expenditure on estates in Devon, 1845–1899

Estate	Period for which data available	Total recorded expenditure (£)	Total expenditure including adjustment for partial outlay (£)	Amount of loan capital 1847–99 (£)	Loan capital as a percentage of	
					Recorded expenditure	Adjusted expenditure
Bedford	1845–95	44,979	44,979	0	0	0
Cornwall	1850–99	2,019	2,019	0	0	0
Devon	1848–99	7,439	12,428	1,000	13	8
Fortescue	1845–99	12,589	12,589	6,676	53	53
Sidmouth	1850–99	13,853	13,853	12,422	90	90
Somerset	1850–75	421	590	0	0	0
Total		81,300	86,458	20,098	25	23

Sources: PRO 1R3/6–38; MAF 66/1–6, 8–9, 11, 13–22, 25–39, 43–7; DRO, Courtenay MSS: 1508M/Estate vols./14/B/111/2–39; Fortescue MSS: 1262M/E20/37–144; 1262M/E1/103; Sidmouth MSS: 152M/Memoranda, draining schedules, 1851–66; 152M/ Accounts, 1850–99; Seymour MSS: 1392M/Estate/Accounts/Bundle 4; Michelmore, Loveys and Carter MSS: 867/8; Bedford MSS: Annual reports, 1845–95; Cornwall MSS: Farm bundles, manors of Bradford and Bradnich; Inrolment books of patents and warrants, 1862–1902

of these estates would have been drained in the second half of the nineteenth century.

The evidence of the sample estates points to a general low intensity of draining investment in Devon, even on soils where the improvement was essential. Such an assessment corresponds to the picture of draining activity portrayed for the county from the analysis of the loan-capital data. As in Northamptonshire and Northumberland, varying use was made of draining loans by estates, but in total they formed 25 per cent of recorded expenditure in the sample (Table 3.16). If adjusted figures for partial outlay are included, the proportion falls to 23 per cent. As draining investment is not available for all estates over the whole period, the role of loan capital has to some extent been overemphasized but its importance in this sample is of the same order found in the other two counties.

The distribution and amount of land drained in England 1845–99

The analysis of draining on these sample estates in Devon, Northamptonshire and Northumberland provides corroboration of the patterns identified in the loan-capital data for the whole country. The evidence reinforces the importance of soil type in determining the broad location of the improvement, with draining being most intensive on estates with clayey and loamy soils with impeded drainage and least on properties with upland and free-draining soils. However, it also reveals that not all estates responded in a uniform fashion in supporting the improvement on similar soil types. At the same time, the estate material has endorsed not only the existence but more significantly the order of variation in the level of draining expenditure between counties in the loan capital data. Thus, the overall rates of draining investment on the sample estates in each county correspond with the relative intensities of loan-financed activity. Despite varying use made by individual estates, the relationship of loan capital in each of the three samples to total draining expenditure was remarkably constant, forming around 20 per cent of total outlay. The constancy of this proportion among estates in counties so widely divergent in the intensity of their adoption of loan capital would point to the broad reliability of the land-improvement data in indicating national trends in draining. On the evidence of these sample estates, loan-financed draining can be regarded as representative of the location and relative intensity of draining in the country as a whole in the second half of the nineteenth century, the loans being an index to general draining activity.

The spatial variations in loan-financed draining may be inferred to reflect real differences in the adoption of the improvement in the country. A clear contrast must be then recognized in the level of draining investment, both in terms of cultivated acreage and of area in need of drainage, between counties

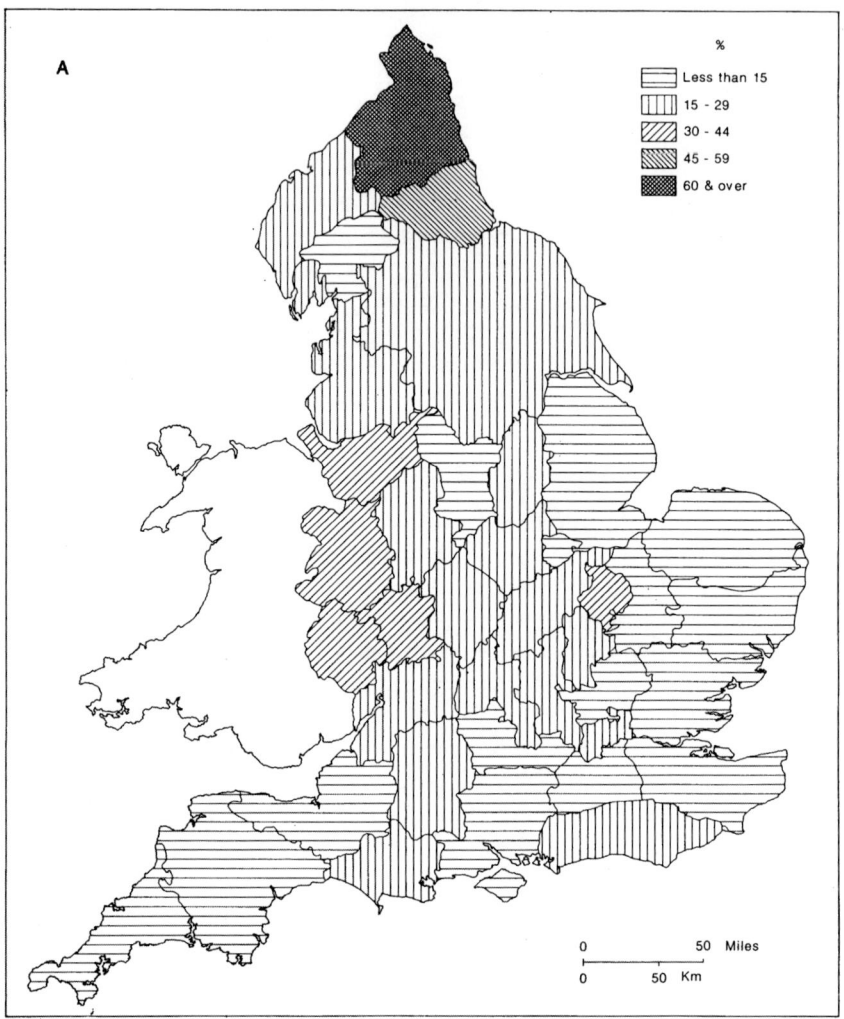

%

Less than 15

15 - 29

30 - 44

45 - 59

60 & over

0   50   Miles

0   50   Km

Figure 3.24 A. Conjectural proportion of cultivated land drained in the second half of the nineteenth century

in eastern, southeastern and extreme southwestern England and the rest of the country. Among the former, the relative lack of expenditure in Norfolk and Suffolk is particularly surprising in that systems of underdraining were reportedly widely practised in the two counties at the beginning of the century (Fig. 2.7) and that both possessed reputations for the early adoption of agricultural innovations. Nevertheless, the low rate of loan-financed draining corresponds to other evidence of a decline in landlord capital investment in

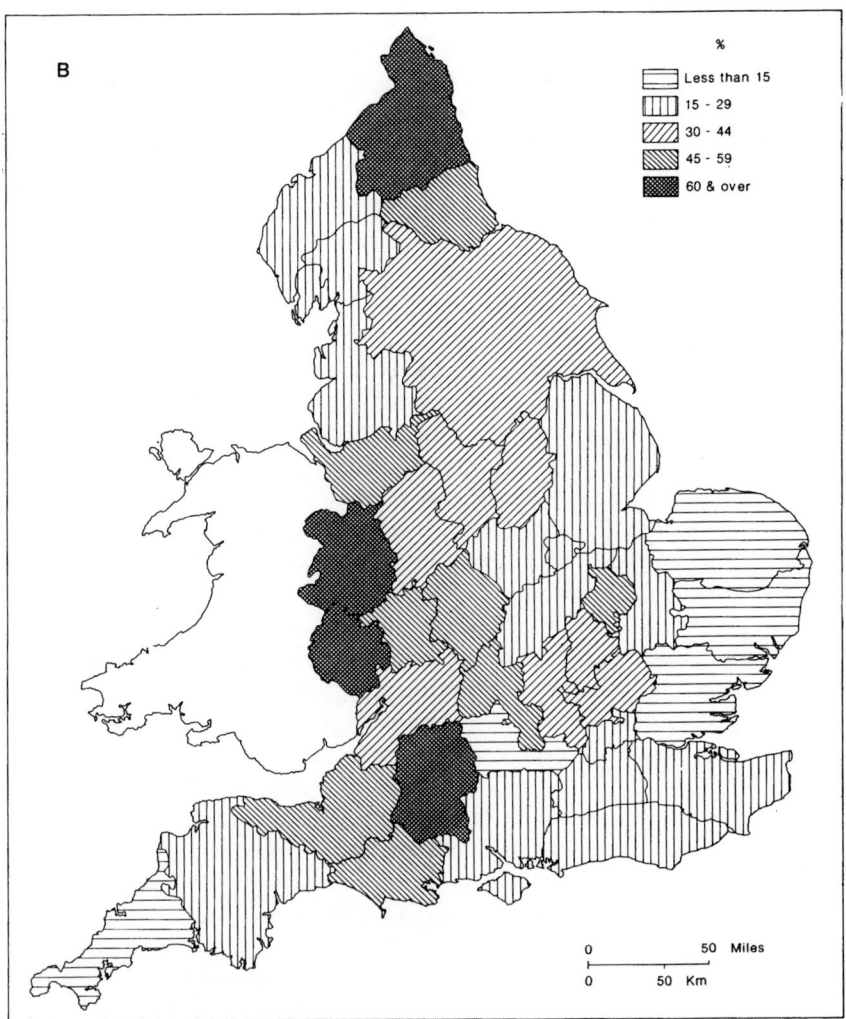

Figure 3.24 B. Conjectural proportion of clayey and loamy soils with impeded drainage drained in the second half of the nineteenth century

agricultural improvement in the area during the second half of the nineteenth century.[128] Throughout the rest of the country, draining expenditure was uniformly greater, with the highest levels being consistently recorded in the northeastern counties of Durham and Northumberland, and with the west midlands and the southwestern counties of Dorset, Somerset and Wiltshire being major centres of investment.

This broad pattern, based on actual investment in the improvement, diverges markedly from and demonstrates the limitations of existing distributions of draining derived from surrogate measures. Thus, M. Robinson suggested that the highest proportions of drained cultivated land in the period were located in the northwestern counties and Devon, but in terms of loan-financed draining these emerge respectively as areas of moderate and low investment.[129] Although fundamental to its location, the incidence of the improvement was not, as F. M. L. Thompson has argued,[130] purely a function of the occurrence of heavy, wet lands, for soils in need of draining were more intensively treated in some counties than in others. While investment levels were low in eastern and southeastern England, the pattern of loan-financed draining offers no support to R. W. Sturgess' view that draining was concentrated on claylands in counties to the north and west of Leicestershire.[131] It is evident that land in need of drainage in these counties was not treated uniformly, while counties outside the area displayed as high and sometimes greater rates of draining expenditure on such soils.

The land-improvement loans may reveal the variations in the intensity of draining throughout the country, but they do not tell how much land was drained in the second half of the nineteenth century. Although the precise acreage drained in this period is impossible to determine, the data emanating from the sample estates and the loans permit a tentative assessment of the extent of the improvement. If the proportion, around 20 per cent, that loan-financed draining formed of total draining expenditure on the sample estates in the three counties, is applied to draining loan capital in all counties, a figure of total draining outlay can be produced. If the average cost of draining of £6 per acre is again adopted, a rate which estate evidence indicates to be of a reasonable order, the amount of land drained in England may then be conjectured. On these bases, some £27,498,000 would have been spent on the improvement, involving the draining of 4.583 million acres. Such an acreage falls well below the high estimates of F. H. W. Green, B. D. Trafford and Robinson, who argue for virtually all land requiring draining being treated in the period.[132] However, the detailed estate analyses have indicated that the draining of all land physically in need of the improvement was rare. Even in Northumberland, where draining was intensively adopted, areas of wet land remained undrained on the most progressive estates. Although, in absolute form, the area approaches the figure suggested by E. J. T. Collins and E. L. Jones,[133] its contribution to nineteenth-century agriculture should not be minimized. The acreage would have resulted in 19 per cent of the cultivated land in the country being drained, its impact ranging from less than 10 per cent of the cultivated area in eastern England to around 35 per cent in the west midlands and to well over 60 per cent of Northumberland (Fig. 3.24a). In terms of land in need of draining, 35 per cent of the total area would have benefited, with Durham and Northumberland, a group of west-midland counties, and Dorset and Wiltshire having well over half such wet land drained (Fig. 3.24b). The conjectured absolute capital investment in the improvement is comparable to the total capital outlay of £29 million estimated by B. A. Holderness as having been spent on English parliamentary enclosure in public costs and subsequent expenditure on ditching, hedging and fencing.[134] Such a sum confirms the importance of underdraining as a capital improvement in English agriculture in the second half of the nineteenth century. It serves further to emphasize the scale of investment in the period on the wet and heavy lands of the country.

4

The temporal pattern of underdraining in the nineteenth century

The spatial pattern of draining in England has been examined for the period 1845–99 as a whole. Yet the sequence of maps of parishes with draining loans (Figs. 3.4–3.6) indicates that the improvement was not adopted uniformly over those years, there being considerable temporal variation in its occurrence. To understand the timing and rate of spread of the improvement, a chronology of draining activity over the nineteenth century is an essential starting point. Such a time series allows not only the identification throughout the country of those periods of the century when draining was an important element in agricultural investment but also an assessment of the impact of changing physical, economic and technical factors on the adoption of the improvement. A guide to this temporal pattern may be obtained for the whole country in the second half of the nineteenth century from the provision of draining-loan capital, while trends in landlord expenditure on draining in the three groups of sample estates permit a more detailed analysis of the sequence over a longer time period.

The chronology of draining loans

The supply of loan capital for draining under the land improvement acts of the middle of the nineteenth century was not uniform from 1847 to 1899 and two peaks of activity may be recognized: in the 1850s and 1860s, and on a much smaller scale in the early 1880s (Table 4.1). From the granting of the first loan in 1847, the amount lent for draining rose quickly to £216,500 in 1852 (Fig. 4.1). This level was maintained until 1864, when on average £214,500 had been annually borrowed. From 1865 the quantity of draining-loan capital declined almost continuously to 1878, when just over £51,000 was made available. An upward movement in the use of loans occurred over the years 1879–85, averaging £91,000 per annum. After 1886, the sums borrowed

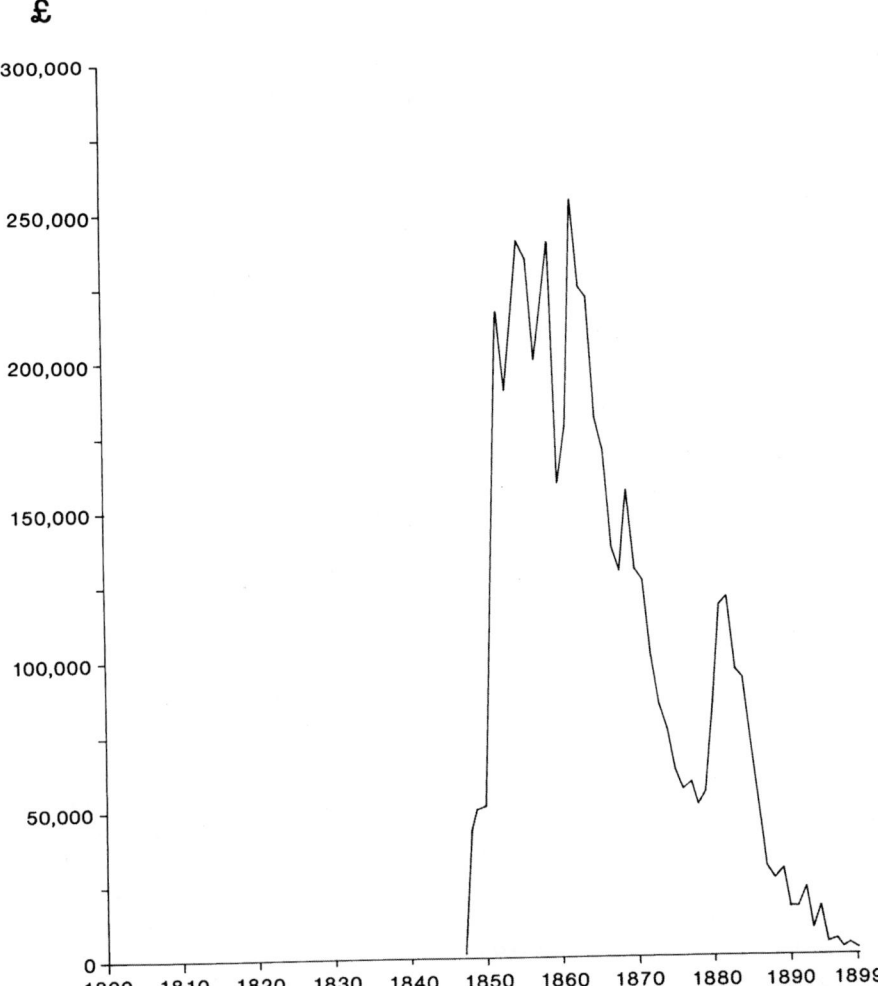

Figure 4.1 The supply of total draining-loan capital in England, 1847–1899 (*Sources:* PRO, 1R3/6–38; MAF 66/1–6, 8–9, 13–22, 25–39, 43–7)

for draining fell sharply to £2,300 in 1899. The overall pattern in the supply of draining-loan capital approximates closely to the theoretical curve of the rate of adoption of agricultural innovations,[1] the main difference being that the use of such capital is skewed towards early adoption. This trend may be demonstrated by examining the proportion of draining-loan capital adopted in each decade over the whole period. Thus, by 1849 only 2 per cent of the total had been used, but in the next two decades the respective percentages

Table 4.1 *Amount of loan capital used for draining by quinquennium,*
1847–1899

Period	Amount (£)	Percentage of total
1847–9	96,769	1.7
1850–4	813,040	14.8
1855–9	1,136,898	20.7
1860–4	1,039,753	18.9
1865–9	779,651	14.2
1870–4	526,076	9.6
1875–9	286,740	5.2
1880–4	513,565	9.3
1885–9	207,992	3.8
1890–4	81,049	1.5
1895–9	18,115	0.3
Total	5,499,648	

Sources: PRO, 1R3/6–38; MAF 66/1–6, 8, 9, 11, 13–22, 25–39, 43–7

had grown to 35 and 33, so that 70 per cent of the loan capital available for
the improvement had been employed by 1869 (Table 4.2). In the decades
1870–9 and 1880–9, the proportions of total loan capital used were lower,
respectively 15 and 13 per cent, leaving only 2 per cent of the sum to be spent
in the last ten years of the century.

This temporal sequence in the supply of draining-loan capital was not
replicated uniformly throughout the country. Nevertheless, the county distribution of amounts lent for draining by decade emphasizes the pre-eminence
of the 1850s and 1860s in the adoption of loan capital by landowners in the
greater part of the country, all but five counties absorbing 50 per cent or
more of total expenditure by 1869 (Table 4.2). From that date, two differing
temporal trends can be identified in the country as measured by the supply
of loans. In much of England the adoption of loan capital was heavily concentrated in the earlier part of the half century, the speed with which it was
used varying widely between individual counties, with little investment being
recorded after 1880. Thus, in 23 counties, 85 per cent or more of total loan
capital had been employed by 1879, a rate at or above the national average.
However, at the same time there were eleven counties 8 percentage points or
more below the national average. Although in most cases larger sums had
been borrowed before 1879 than after, these counties exhibited a marked
resurgence in the use of loan capital in the last two decades of the century at
a rate over 50 per cent greater than the national average. These counties
formed distinct spatial groupings: Cambridgeshire, Essex and Norfolk in the

Table 4.2 *County supply of draining-loan capital by decade, 1847–1899*

County	Total loan capital (£)	Percentage of total borrowed in					
		1847–9	1850–9	1860–9	1870–9	1880–9	1890–9
Bedfordshire	73,324	0	32	44	19	3	2
Berkshire	28,840	0	32	50	7	10	1
Buckinghamshire	72,374	3	30	37	18	12	0
Cambridgeshire	37,261	0	41	29	1	21	8
Cheshire	231,224	1	31	34	19	14	1
Cornwall	4,000	35	24	39	2	0	0
Cumberland	136,526	8	36	24	8	19	5
Derbyshire	57,517	0	23	61	12	3	1
Devon	100,009	11	20	48	15	6	0
Dorset	132,148	0	25	37	23	14	1
Durham	260,095	4	45	35	7	8	1
Essex	53,907	1	30	25	18	20	6
Gloucestershire	150,058	2	6	27	7	17	1
Hampshire	92,341	0	25	51	18	6	0
Herefordshire	177,001	0	20	50	21	9	0
Hertfordshire	37,186	0	57	24	3	16	0
Huntingdonshire	99,996	2	43	32	9	13	1
Kent	91,329	0	26	44	25	5	0
Lancashire	191,354	1	34	27	15	21	2
Leicestershire	127,736	0	23	9	13	42	13
Lincolnshire	147,765	0	32	19	15	30	4
Middlesex	29,000	0	26	21	40	12	1
Norfolk	13,905	3	71	0	0	16	10
Northamptonshire	124,064	1	18	36	9	29	7
Northumberland	771,609	1	44	32	13	9	1
Nottinghamshire	84,044	1	33	37	9	19	1
Oxfordshire	91,189	0	55	33	7	5	0
Rutland	12,056	0	0	0	53	47	0
Shropshire	307,759	1	31	34	23	10	1
Somerset	117,839	1	35	19	23	14	8
Staffordshire	156,434	0	35	44	9	12	0
Suffolk	24,744	0	35	16	30	17	2
Surrey	62,145	6	30	22	27	13	2
Sussex	106,762	4	30	37	19	10	0
Warwickshire	144,610	2	15	29	22	30	2
Westmorland	40,479	17	12	15	19	35	2
Wiltshire	185,913	0	17	37	36	10	0
Worcestershire	167,028	0	43	39	9	8	1
Yorkshire	758,077	2	47	31	8	10	2
England	5,499,648	2	35	33	15	13	2

Source: As for Table 4.1

east, where between 26 and 29 per cent of all loan capital was borrowed in the period 1880–99; the midland counties of Leicestershire, Lincolnshire, Northamptonshire, Rutland and Warwickshire, where the proportion of loan capital used in the last two decades of the century ranged from 32 to 55 per cent of the total; and Cumberland, Lancashire and Westmorland in the northwest, where loans from 1880 to 1899 accounted for between 23 and 27 per cent of the total. These may be regarded as laggard areas in the adoption of loan capital, and towards the end of the century landowners attempted to remedy the relative lack of earlier investment in the improvement.

These temporal sequences relate only to the adoption of loan capital and may not be typical of the chronology of all draining activity in the second half of the nineteenth century. For example, D. Spring has suggested that the decline in the amount of loans from the middle of the 1880s was a reflection not of a diminution in draining but of the introduction by means of the Settled Land Act, 1882, of alternative sources for landowners to finance agricultural improvement.[2] By this act, a tenant for life was allowed to sell land subject to settlement and, if he chose, to apply the capital money to the improvement of his estate.[3] The Board of Agriculture estimated that £1,360,000 had been laid out on the wide range of schemes permitted under the act between 1893 and 1899.[4] Yet T. H. Elliott, permanent secretary to the Board of Agriculture, was doubtful whether such capital monies would have been used extensively for an improvement like underdraining, which by that time was recognized to possess a limited life. Both he and E. P. Squarey, chairman of the Land Loan Company and a noted land agent, were of the opinion in 1894 that the overall decline in draining loans from the middle of the 1880s was a product of a general falling off in the implementation of the improvment in the country rather than the effect of the Settled Land Act.[5]

The supply of draining capital on estates

More direct support that the time series obtained from the supply of loan capital typified general trends in draining activity in the second half of the nineteenth century is provided in the temporal patterns of landlord expenditure on the improvement on the sample estates in Devon, Northamptonshire and Northumberland. Runs of annual draining expenditure do not exist for all estates sampled and where they do survive do not always cover the whole century. However, where data are available, the chronology of landlord investment in draining was not uniform, and varied not only between counties but also between estates in the same county. In Northamptonshire, landlord expenditure on the duke of Cleveland's estate peaked in the 1850s, some forty years earlier than on the neighbouring duke of Buccleuch's properties (Figs. 4.8e, 4.9b and c). Both contrasted to the relative stability in draining outlay recorded on the estate of the earls Spencer. Again, the main landlord supply of draining capital was provided a good decade earlier on the Grey estate than on the Carlisle property in Northumberland (Fig. 4.10a and d). Even on land belonging to the same estate in a county varying draining sequences can be identified, reflecting differential adoption of the improvement. Thus, the major period of draining expenditure on the mid-Devon estate of the dukes of Bedford was about ten years later than on the Tavistock and north Devon properties (Fig. 4.7b, c and d). A similar range in the timing of draining expenditure was evident on the Northumberland bailiwicks of the dukes of Northumberland (Fig. 4.11b–g).

To reduce these individual variations between estates to general trends, the relevant estate material has been aggregated for each county. For Devon, runs of draining expenditure have been used from the Bedford, Cornwall, Devon, Fortescue and Sidmouth estates; for Northamptonshire from the Bedford, Buccleuch, Cartwright, Cleveland, Ellesmere, Fitzwilliam Higham Ferrers, Spencer and Cardigan estates, the last, as yet undiscussed, a property containing 4,550 acres in 1812 in the north of the county[6] (Fig. 1.2); and for Northumberland from the Blackett, Carlisle, Grey, Northumberland and Portland estates. Most of the landlord investment in draining relates to estates sized over 3,000 acres – in the case of Northumberland over 10,000 acres – and this emphasis on larger properties, a product of the availability of source material, may create some bias in the resulting chronologies. To facilitate recognition of changes in the level of draining expenditure through the nineteenth century, landlord outlay on the sample estates has been expressed as quinquennial investment per acre. Save for a few estates among those sampled, acreages were not recorded annually. However, for most properties, sufficient surveys detailing area are extant during the century to permit identification of the chief alterations to estate size, and where possible cultivated acreages have been used. Although not precise, the results should not be regarded as too artificial (Table 4.3).

The chronology of draining activity on these estates has been based solely on landlord expenditure on the improvement. However, as already seen, certain estates in the Devon and Northamptonshire samples provided only part of the draining capital, requiring some financial co-operation from the tenantry. Although after 1845 a reasonable assessment of the tenant contribution can be made as draining systems, materials and costs became more standardized and were detailed in estate accounts, before that date when the practice of draining was characterized with less order and greater variety such an exercise becomes more problematical. As a result, no attempt has been made to incorporate for those estates affected an estimate of tenant financial involvement over the century. Yet as such tenant capital was directly linked to landlord expenditure on the improvement, the exclusion of such sums should not unduly influence the chronology of draining investment.

The resulting analysis indicates distinct trends in the provision of landlord

Table 4.3 Landlord expenditure per acre on draining on sample estates in Devon, Northamptonshire and Northumberland, 1800–1899

Period	Devon			Northamptonshire			Northumberland		
	Aggregate expenditure (£)	Estate acreage	Quinquennial expenditure per acre (£)	Aggregate expenditure (£)	Estate acreage	Quinquennial expenditure per acre (£)	Aggregate expenditure (£)	Estate acreage	Quinquennial expenditure per acre (£)
1800–4	–	–	–	569	48,153	0.01	–	–	–
1805–9	–	–	–	32	48,153	0	155	26,150	0
1810–14	–	–	–	2,177	51,763	0.04	5,066	51,510	0.10
1815–19	–	–	–	3,174	46,183	0.07	1,822	51,510	0.04
1820–4	–	–	–	2,242	43,810	0.05	858	51,510	0.02
1825–9	–	–	–	621	27,840	0.02	708	51,510	0.01
1830–4	601	21,260	0.03	2,505	47,749	0.05	573	51,510	0.01
1835–9	295	21,260	0.01	1,228	47,749	0.03	1,900	51,510	0.04
1840–4	679	21,260	0.03	2,414	50,559	0.05	21,155	51,510	0.41
1845–9	5,390	36,880	0.15	5,382	50,049	0.11	60,040	205,970	0.29
1850–4	13,576	65,750	0.21	8,244	49,890	0.17	79,986	204,460	0.39
1855–9	9,376	69,980	0.13	9,121	48,950	0.19	80,314	204,460	0.39
1860–4	14,447	69,980	0.21	4,906	54,750	0.09	63,464	216,510	0.29
1865–9	15,703	69,980	0.22	5,704	54,750	0.10	43,729	224,160	0.20
1870–4	9,044	73,230	0.12	2,457	51,570	0.05	17,930	224,160	0.08
1875–9	6,706	73,230	0.09	3,881	48,710	0.08	18,333	224,320	0.08
1880–4	3,523	69,230	0.05	1,782	21,210	0.08	28,495	217,310	0.13
1885–9	1,406	69,230	0.02	10,704	48,800	0.22	10,931	217,310	0.05
1890–4	1,038	42,340	0.02	6,517	47,650	0.14	6,515	182,250	0.04
1895–9	1,208	33,040	0.04	403	44,100	0.01	5,416	182,250	0.03

-: No data available

The county figures were made up of the following estate runs:

Devon: Bedford estates, 1845–94; Cornwall estates, 1850–99; Devon estate, 1850–99; Fortescue north Devon estate, 1830–99; Fortescue south Devon estate, 1830–79; Sidmouth estate, 1850–99;

Northamptonshire: Bedford estate, 1845–94; Buccleuch Barnwell estate, 1800–24, 1830–99; Buccleuch Boughton estate, 1800–24, 1830–79, 1885–99; Cardigan estate, 1800–24; Cartwright estate, 1800–46, 1855–69; Cleveland estate, 1810–99; Ellesmere estate, 1840–74; Fitzwilliam Higham Ferrers estate, 1800–54, 1860–99; Spencer estate, 1800–79, 1885–99;

Northumberland: Blackett estate, 1805–99; Carlisle estate, 1805–99; Grey estate, 1810–89; Northumberland estate, 1845–99; Portland estate, 1860–99

Sources: Bedford MSS: Annual reports, 1843–95; Cornwall MSS: Valuation of the manor of Bradninch, 1855; Report on the manor of Bradford, 1862; Farm bundles, manors of Bradford and Bradninch; Inrolment books of patents and warrants, 1862–1902; DRO, 1508M/Estate papers/14/A/III/Shelf III, 1862 valuation; 1508M/14/B/III/2–39; 152M/Memoranda on draining schedules, accounts, 1850–99 and draft terrier, 1894; 1262M/E1/102, 103; 1262M/E20/16–144; 1262M/E22/45; 1262M/E29/58; Buccleuch MSS: Barnwell and Boughton estate accounts, 1800–99; Particular...of Boughton estate, 1834; Numerical reference of Boughton estate, 1895; Raby MSS: Rentals, 1806–99; Draining volumes, 1848–72; Particulars of Brigstock and Sudborough rents, 1835, 1846, 1849, 1855; Fieldbook 1871–8; Northamptonshire estate reference book, 1901–2; Spencer MSS: Estate accounts, 1800–26; Estate accounts and rentals 1827–99; Survey of parishes of Strixton and Bozeat, 1827; Survey of Harlestone, 1831; Reference to estate of earl Spencer, 1859; Abstracts of rents and acreage of Lord Spencer's tenants, 1882; NRO, Brudenell MSS: ASR 95 and 96; ASR 256–303, 490–513; Cartwright MSS: C(A) 3502, 3503, 3574–81, 3784, 3787–9, 4740–57, 5914; Ellesemere MSS: Box X.461, 471–2, 3695; Fitzwilliam MSS: Estate accounts, 1800–99; Misc. vols., 656, 738; Buccleuch MSS: Misc. vols., 136, 137, 140, 144; Northumberland MSS: Annual return of state of farms, 1827–80; Survey and terrier of the duke of Northumberland's estate, 1850; Business minutes, vol. 119, 21 November 1906; General accounts, 1808–48; Ledger balances, 1847–65; Abstract of accounts, 1865–1907; Variation in rental vols., 1827–1912; Draining vols., 1844–1903; Nd RO, Blackett MSS: ZBL/4/10; ZBL/54/1–3; ZBL/274/6–10; ZBL/275/5 and 6; ZBL/282/2; Sample MSS: ZSA/8/1–5; ZSA/18/2/1 and 2; DPD, Grey MSS: Rentals, 1803–42; Cash and ledger books, 1806–92; Building and improvement vol., 1841–58; Draining vols., 1841–86; Estate cropping book, 1845–78; Howard MSS: N. 73/2; N. 80/14–19; N. 99/ 2; N. 101–6; Estate accounts, 1875–1900; Rental vols., 1879–1912; Draining vols., 1856–1901; Newcastle Central Library, L. 622.33, J. T. W. Bell, Map of the Blyth and Warkworth Coal District, 1851; F. M. L. Thompson, 'The economic and social background of the English landed interest, 1840–70...', 1956, appendix xi; M. Hughes, 'Lead, land and coal as sources of landed income in Northumberland between 1700 and 1850', 1963, vol. 2, 60, 101–2

A

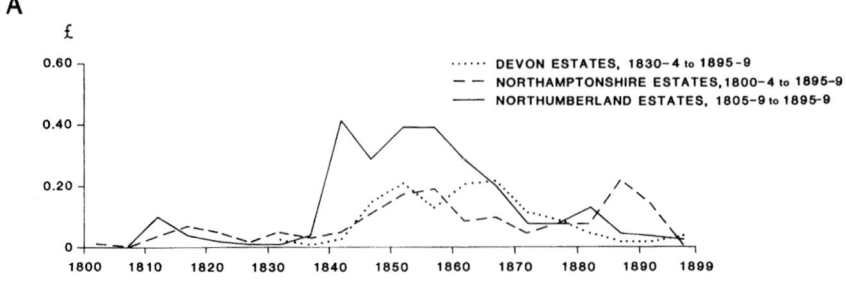

B

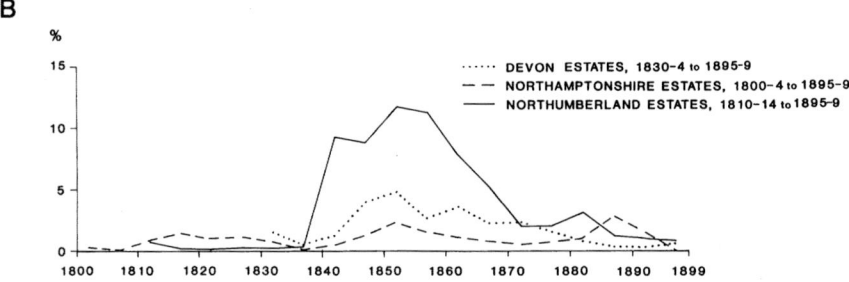

C

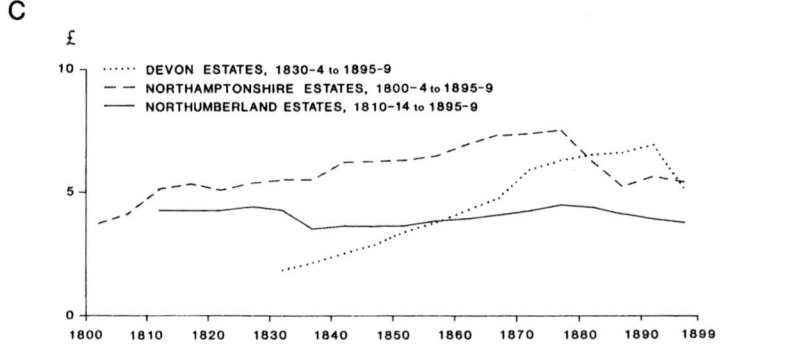

Figure 4.2 A. Quinquennial draining expenditure per acre on sample estates, 1800–1804 to 1895–1899; B. Draining expenditure as a percentage of gross agricultural rent due on sample estates, 1800–1804 to 1895–1899; C. Quinquennial gross agricultural rent due per acre on sample estates, 1800–1804 to 1895–1899 (*Source*: As for Tables 4.3–4.5)

capital for draining over the nineteenth century which, although per acre levels of investment vary, are common to all three groups of sample estates (Fig. 4.2a). Landlord outlay was greatest in all three counties between 1840 and 1869, with the expansion in the supply of capital being earliest on the Northumberland estates and being more laggard on properties in Devon and Northamptonshire. Investment per acre was less on the sample estates in the

30 years after 1870, although there was considerable variation in the pattern of provision between the three counties. Landlord expenditure on the Devon estates was characterized by an almost continuous decline, reaching levels after 1880 little higher than in the 1830s. On the Northumberland estates an overall fall in draining outlay was also evident, being interrupted by a slight rise in activity in 1880–4. Landlord involvement on the Northamptonshire estates presented a different trend, being marked by an increase in the provision of capital over the decade 1885–94 comparable to the 1850s and revealing a late revival of landlord commitment to draining in the county.

The time series of landlord outlay on the improvement on the sample estates from 1850 relate closely to those recorded for their respective counties in the draining-loan data. That material disclosed an upswing in the use of loans in Northamptonshire after 1880 following a relatively high level of activity in the 1850s and 1860s, while there was a rapid decline in draining-loan capital applied to estates in Devon and Northumberland from 1870. Such correspondence would suggest a real basis for regarding the supply of draining capital under the land-improvement legislation as representative of the temporal pattern of landlord investment in the improvement throughout England in the second half of the nineteenth century. Indeed, on the evidence of the sample estates the issuing of loans from 1847 comes in the wake of the initial expansion of landlord expenditure on draining in the early 1840s.

The level of landlord outlay on draining before 1840 contrasts markedly with that recorded between 1840 and 1869. Although data for the Devon estates are not available earlier than 1830, the overall provision of landlord capital for the improvement was slight in all three samples. Thus, for those estates with nearly complete data runs, the proportion that landlord expenditure on draining before 1840 formed of the total never exceeded the 22 and 24 per cent recorded respectively on the Buccleuch estates in Northamptonshire and the Blackett estates in Northumberland (Figs. 4.9b and c; 4.10b and c). On most estates the proportion was much lower, being 2 and 6 per cent respectively on the Grey and Carlisle estates in Northumberland (Fig. 4.10a and d) and 5 and 14 per cent on the Cleveland and Spencer properties in Northamptonshire (Figs. 4.8e and 4.9a). In that 40-year period, recorded rates of outlay were at their highest for estates in Northamptonshire and Northumberland from 1810 to 1824. However, such increased landlord involvement in those fifteen years was limited to a few estates, on the Grey property being devoted to the improvement of land in hand.

Translation of these rates of expenditure into amounts of land drained presents difficulties before 1840, as little direct evidence specifies the areas improved. Such that is available would indicate that the acreages affected were small. On the estate of the dukes of Northumberland, the number of rods of drains cut in each bailiwick, save Alnwick, were recorded from 1834 to 1839. In the early 1840s the dominant system of draining on the estate was to lay drains 18 to 24 ft apart. If the median interval of 21 ft is taken, 100 rods of drains were required to drain one acre. Between 1834 and 1839, 23,113 rods of drains were cut, effectively draining 231 acres, an area representing 0.2 per cent of the 1827–8 cultivated acreage of the estate exclusive of Alnwick, or 23 per cent of the acreage drained in 1844 and 1845 when the improvement was first widely introduced on the estate.[7] For other properties in the samples, the descriptive accounts of agents around the middle of the century reveal a similar lack of draining before 1840. F. Thynne, in his report on the Sidmouth estate in Devon in 1850, noted that the property was 'charged with water injuring the proper cultivation and starving the produce of the land', being surprised that it should have been allowed 'to continue in so sadly neglected a state'.[8] On the duchy of Cornwall estates at Bradford and Bradninch, the respective agents E. C. Marriott and R. Watt identified no landlord expenditure on draining before 1850, while in spite of continuous outlay on the improvement on the Fortescue north Devon estate from 1830 (Fig. 4.7f) the excessive wetness of the land was very evident to Josiah Parkes in 1848.[9] Similarly, in Northamptonshire on the Ellesmere property, James Loch admitted to Lord Francis Egerton in 1837 that draining 'would benefit the estate wonderfully', a view repeated in 1850 by Tycho Wing of the duke of Bedford's Wansford estate.[10] In general, the relative lack of landlord capital bestowed on draining before 1840 may be regarded as an indication that the improvement had been but little adopted on the sample estates in that period.

The limited adoption of the improvement by 1840 that characterized these estates was also reported from many parts of the country in contemporary agricultural literature. A number of witnesses before the select committees on agriculture in 1833 and 1836–7 spoke of the need for draining and described the introduction and some expansion of the improvement in individual midland and southern counties during the 1830s.[11] However, the rate of growth would seem to have been low. The report of the 1833 Select Committee recorded that heavy claylands had lacked capital investment, while C. S. Lefevre, commenting on the evidence before the 1836 Committee, noted the relative absence of draining on such soils throughout the country.[12] Contributors to the first numbers of the *Journal of the Royal Agricultural Society* wrote both of recent indifference towards the adoption of draining and the extensive need for land to be drained, sentiments reiterated in more detail by the authors of the early prize essays on the agriculture of individual counties also published in the *Journal*.[13]

No single source provides statistical data that allow the determination of the extent of underdraining in England by 1840. However, this omission may be partly remedied by recourse to local tithe agents' and assistant tithe commissioners' references to the improvement in their reports on tithe agreements arising from the Tithe Commutation Act, 1836, contained in the general body of tithe files. This evidence is abstracted in R. J. P. Kain's *Atlas and Index of the Tithe Files*. The availability of these reports is neither complete nor uniform over the country. Thus, their incidence is greatest in the East Anglian counties of Essex, Norfolk and Suffolk and the western counties of Dorset, Herefordshire and Somerset with over 60 per cent cover, and least in the midlands, where tithes in many places had been commuted at the time of parliamentary enclosure, the cover for Leicestershire, Northamptonshire and Nottinghamshire being under 20 per cent. In addition, discussion of the improvement was not a mandatory requirement, mention of draining in the reports depending on both the enthusiasm and awareness of individual assistant tithe commissioners.[14] Although the data in the tithe files on the improvement cannot be regarded as comprehensive, the mapping of those tithe districts where the practice was observed is likely to provide an indication of the relative importance of draining in England around 1840.

The results endorse the general trends indicated on the sample estates and in the literary evidence of the small scale of adoption of the improvement by that date. Thus, many tithe districts throughout the country were described as possessing land in need of draining (Fig. 4.3). At the same time, underdraining with tiles or with more traditional materials, such as bushes, straw or stones, was recorded as being practised in much fewer tithe districts (Fig. 4.4). Only Norfolk and Suffolk deviated from this pattern with numerous tithe-district references to the improvement. These counties had formed part of the area reported with above-average draining activity at the beginning of the nineteenth century (Fig. 2.8), a situation seemingly unchanged by 1840. The draining systems employed were distinct, being based on the use of bushes or straw as fill, recognized to enjoy a limited life, perhaps at most fifteen years, resulting in the need for land to be treated periodically, and as a consequence were regarded as little more than an element in tenants' cultivation practices.[15] Yet, although the extent of tenant involvement in draining in these counties by 1840 was considered by contemporaries as exceptional, much of their clayland could still be described by agents working in the area as in need of the improvement at the middle of the century.[16] And accounts of the level of landlord capital provision for draining were more in accord with conditions found in other parts of the country. Thus, James Caird noted that landowners in Suffolk before the middle of the 1840s had supplied little capital for the improvement, a view substantiated by the Norfolk and Suffolk witnesses, particularly H. Keary, agent to the Tollemache estate, before the 1848 Select Committee on agricultural customs.[17]

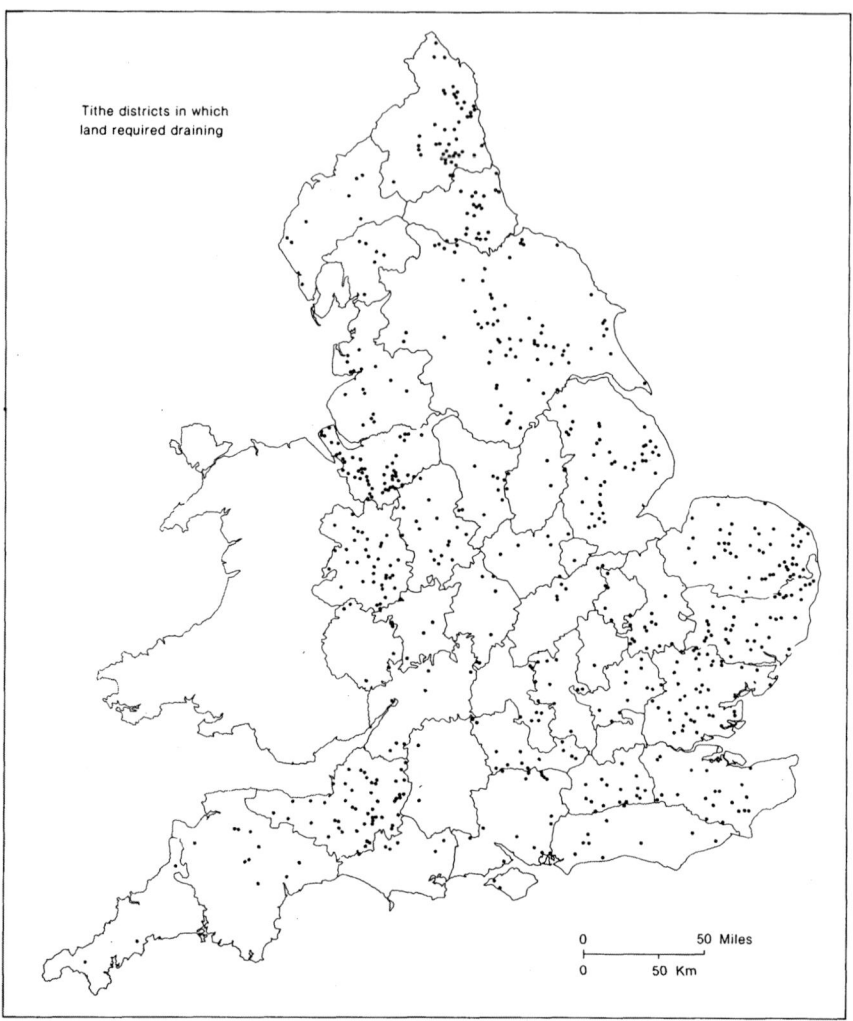

Tithe districts in which
land required draining

0 ——————— 50 Miles
0 ——————— 50 Km

Figure 4.3 Tithe districts in which land was described in need of draining around 1840
(*Source:* R. J. P. Kain, *An Atlas and Index of the Tithe Files*, 1986, 562–631)

General determinants of the temporal sequence of draining

A clear chronology can be distinguished in the supply of landlord capital for
draining over the nineteenth century, with expenditure expanding rapidly
between 1840 and 1869 from an earlier period of limited outlay and
subsequently undergoing severe contraction. Variation in the temporal
patterns of draining investment attracted the attention of contemporary

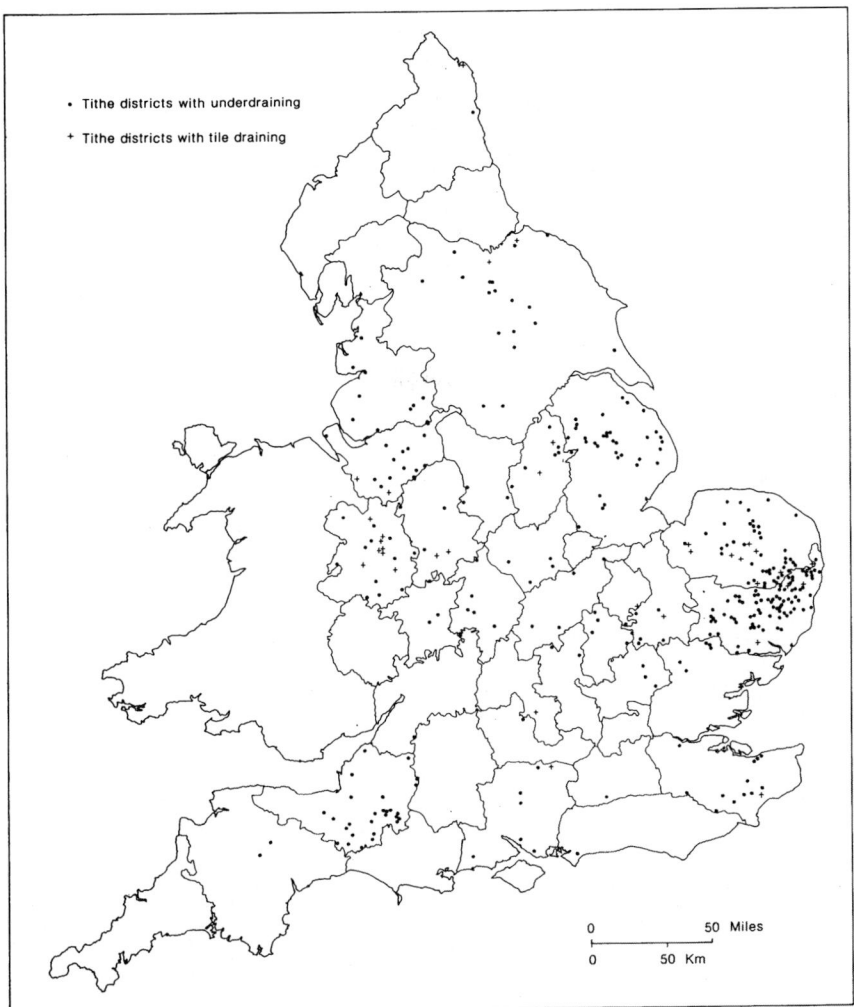

Figure 4.4 Tithe districts in which underdraining and tile draining were reported around 1840 (*Source:* As for Fig. 4.3)

agriculturalists and several factors were suggested that were thought to relate to trends in draining activity.

The incidence of draining in relation to variation in rainfall

A number of commentators considered that the incidence of draining was a function of change in the quantity of rainfall, wetter years encouraging adoption, drier seasons reducing interest.[18] Illustrating this linkage, J. Bailey

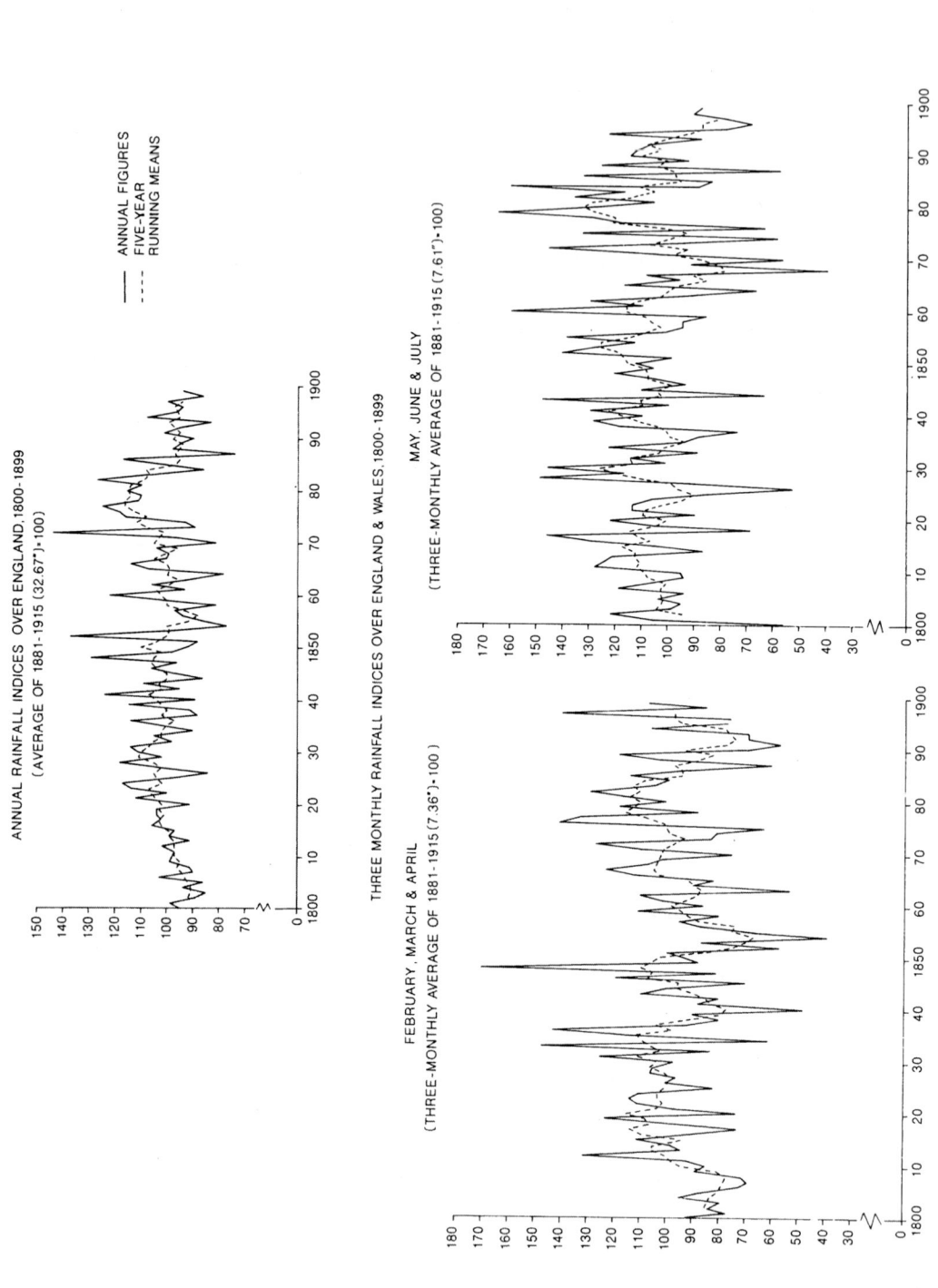

ANNUAL RAINFALL INDICES OVER ENGLAND, 1800-1899
(AVERAGE OF 1881-1915 (32.67″)=100)

———— ANNUAL FIGURES
------- FIVE-YEAR
RUNNING MEANS

THREE MONTHLY RAINFALL INDICES OVER ENGLAND & WALES, 1800-1899

FEBRUARY, MARCH & APRIL
(THREE-MONTHLY AVERAGE OF 1881-1915 (7.36″) =100)

MAY, JUNE & JULY
(THREE-MONTHLY AVERAGE OF 1881-1915 (7.61″)=100)

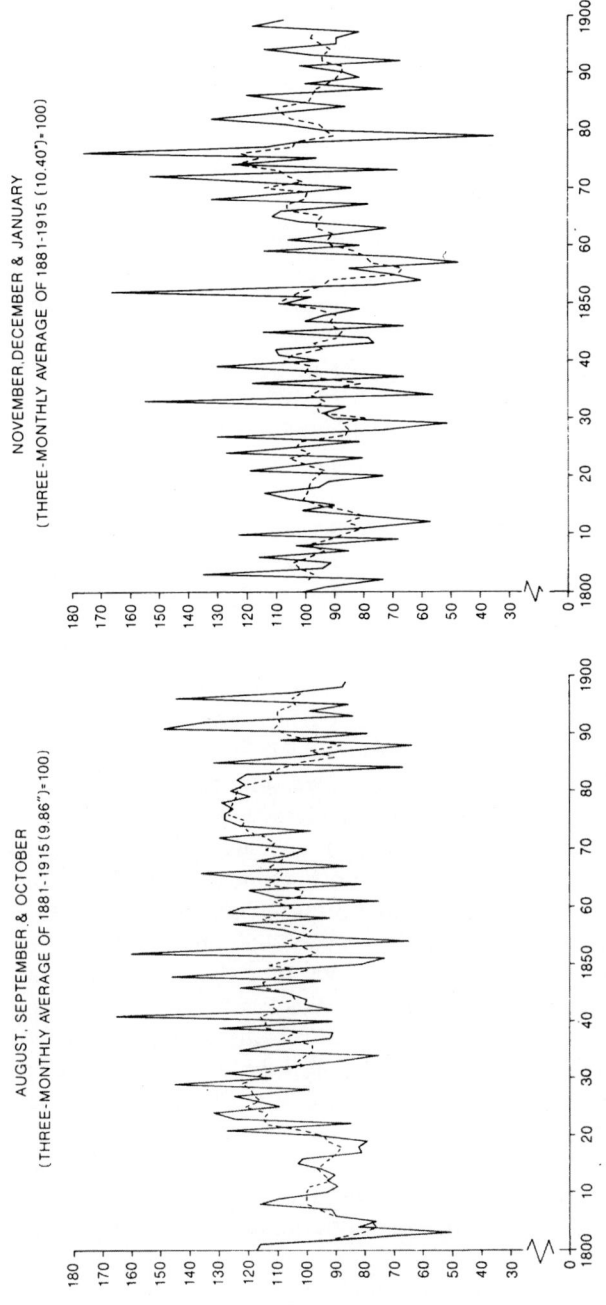

Figure 4.5 Annual and seasonal rainfall indices over England, 1800–1899 (*Sources*: J. Glasspoole, 'Two centuries of rain', 1928, 1–6; F. J. Nicholas and J. Glasspoole, 'General monthly rainfall over England and Wales, 1727–1931', 1931, 299–306)

Denton drew the distinction between 1859 with below average rainfall and 1860 with much wetter conditions, which he regarded as an inducement to the spread of the improvement.[19] The relationship was more frequently commented on in the early 1880s. In 1883, the Land Commissioners, successors to the Inclosure Commissioners in the administration of draining loans, reported that 'in the last five years during the pressure of wet and unfavourable seasons, the annual expenditure for drainage increased considerably'.[20] This view was repeated by several witnesses and assistant commissioners to the Royal Commission on the depressed condition of the agricultural interest, 1880–2, and by a number of agricultural writers. Thus, J. H. Tiffen, a Holderness land agent, asserted in 1884 that the large rainfall of recent times had 'rendered the evils of insufficient drainage so evident that many landlords have expended large sums of money in draining their estates'.[21]

However, detailed examination of annual and seasonal rainfall trends over the nineteenth century would suggest that they had little direct influence on the course of landlord investment in draining (Fig. 4.5). Thus, the generally increasing annual rainfall up to 1830 was not mirrored in a constant growth in landlord outlay on the sample estates in Northamptonshire and Northumberland. The decade 1850–9 was marked in the main by declining levels of annual rainfall but represented a major period in the use of draining loans in the country. From the early 1860s to the late 1870s, annual rainfall amounts gradually increased, a trend accompanied by a slackening in the sums borrowed under the land-improvement acts, even in the exceptionally wet year of 1872.[22] A degree of correspondence between the provision of loan capital and rainfall levels can be identified only in the late 1870s and early 1880s, as the Land Commissioners suggested, but the county distributions of such funds indicate that an increase in their use was not found universally or uniformly throughout the country at this time (Table 4.2). Of the seasonal trends in rainfall, that for the summer months offers the best fit to the overall supply of loan capital in England between 1850 and 1899, but there is little statistical relationship between the two time series, the Spearman rank correlation coefficient being 0.19.[23]

At the scale of individual estates, there is also little evidence of a direct connection between rainfall amounts and landlord investment in draining. No mention can be found on the sample estates of rainfall as a factor behind the expansion of the improvement in the period 1840–69. The series of wet seasons around 1880 were reported from many of the estates and were recognized as bringing attention to land requiring draining and more frequently to the need to repair existing systems. As was noted on the Spencer estate in Northamptonshire in 1879, 'the wet seasons have washed out the goodness from the land and burst up a large portion of drains', conditions also described by tenants on the Northumberland estate of Sir Edward

Blackett between 1880 and 1884.[24] Although the wet seasons may have pointed to land requiring the improvement, reaction to the problem ranged widely between the three estate samples (Fig. 4.2a). Even where estates accepted that excess rainfall had created a need for draining, outlay on the improvement was dependent on other factors. Thus, G. Herriot, agent of the duchy of Cornwall's Bradninch property in Devon, in recommending the draining of 42 acres on Park and Fordishaies farms between 1881 and 1884, considered that although 'the wet seasons has caused a necessity for the drainage' the expenditure should be made only to 'maintain the present rents...and avert...the consequent loss of tenant', especially as the holdings were highly rated.[25] While draining may have had the general effect of mitigating the impact of wet seasons on heavy-land cultivation,[26] the provision of landlord capital for the improvement cannot be seen as a straightforward response to changing physical conditions.

Agricultural prices, rents and draining expenditure

As the improvement necessitated the application of capital, the temporal sequence of the supply of landlord investment in draining was frequently related by contemporaries to the changing patterns of agricultural prosperity as determined by levels of agricultural prices or rent. The relationship was most commented on in periods of agricultural depression and was expressed in inverse terms, landlords investing in draining to offset falls in rent being explicitly described by Parkes in 1845: 'If rents have a tendency to fall, the drainage of an estate which is waterlogged may keep them up to what they now are and even increase them.'[27] Such a link had been identified by William Blamire, a future Inclosure Commissioner, before the 1833 Select Committee on agriculture; it was repeated by Squarey to the Royal Commission on the depressed condition of the agricultural interest in 1881 and was incorporated into the final report of the Royal Commission on agricultural depression in 1897.[28] A similar inverse association was recognized between draining outlay and agricultural prices, Denton for example suggesting that the capital landowners were prepared to provide for the improvement in periods of low wheat prices dried up when these increased.[29]

The time series of draining loans makes possible some examination of this connection for England as a whole in the second half of the nineteenth century. At this level, the data provide little support for contemporary opinion. In terms of agricultural prices, investment in draining as represented by loan capital was greatest when, although subject to annual fluctuation, they were broadly stable or gradually increasing, as occurred from 1850 to 1879 (Fig. 4.6). The decline in the price of both cereal and livestock products after 1880 did not coincide with a large-scale adoption of loan capital for

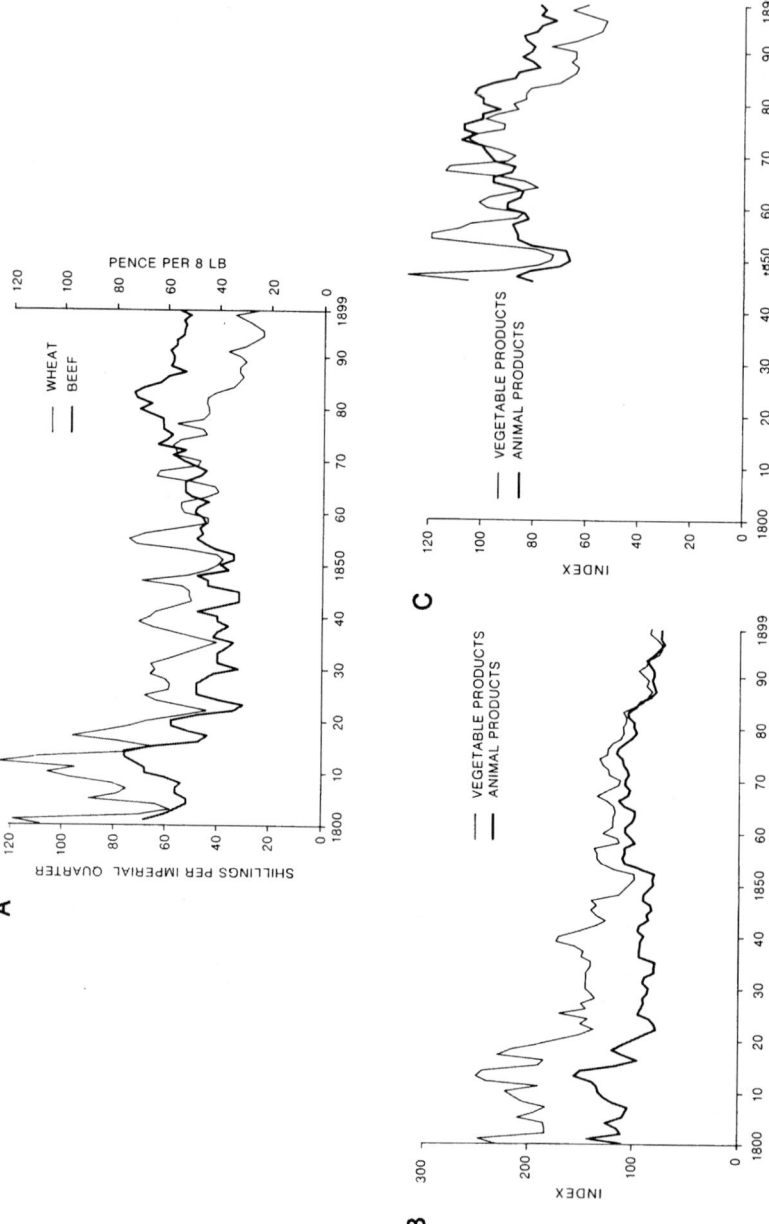

Figure 4.6 A. Wheat prices per imperial quarter, 1800–1899, and London beef prices per 8 lb, 1800–1899. (*Sources*: 1800–1899 (average of 1865 and 1885 = 100); B. Rousseaux price indices, G. R. Porter, *Progress of the Nation*, 3rd edn, 1851, 589; BPP, 1878–9, LXV, 'Return of the average prices of butchers' meat at the metropolitan cattle market, 1828–78'; B. R. Mitchell and P. Deane, *Abstract of British Historical Statistics*, 1962, 471–5 and 488–9; P. J. Perry, *British Farming in the Great Depression, 1870–1914*, 1974, 46; R. Edwards and R. Perren, 'A note on regional differences in British meat prices, 1828–1865', 1979, 123–34)

draining, the amount borrowed in the period 1880–99 forming only 22 per cent of that used between 1850 and 1869 (Table 4.1). The supply of loan capital after 1859 should be seen as responding positively to periods of prosperity in agricultural prices. Of the various agricultural products, use of draining loans was most closely related to the movement of wheat prices. Thus, the expansion in loan capital in the 1850s corresponds to a distinct upward movement in wheat prices, while rapid falls characterize the price of wheat and the quantities of draining loans towards the end of the century. Indeed, the relationship was statistically significant, the Spearman rank correlation coefficient producing a value of 0.68 between the annual amount of draining-loan capital and the annual price of wheat over the period 1850–99. With a sample size of 50, this value is significant at the 0.001 level, implying a real correlation between the two factors.[30] Expenditure in draining would seem to be attractive to landlords at times of relatively high prices, so that tenants might secure greatest financial return on the improvement.

At the same country-wide level, the pattern of supply of draining loans would also suggest that in general landowners were prepared to invest more on the improvement in periods when rentals were rising, and therefore their overall prosperity, and were inclined to be less expansive with capital for draining when they were decreasing. From 1850 to 1869, when 68 per cent of draining loan capital was employed, total agricultural rent in England derived from assessments of land rent for Schedule A of the income tax rose consistently and overall by 13 per cent.[31] The income tax assessments indicate that rent fell almost continuously after 1879, a decline of 27 per cent between 1880 and 1899. Although there was some response between 1880 and 1885 to the initial downward movement in rent by an increase in the use of loan capital, the amount was on a smaller scale and represented a smaller proportion of rental income than at any time during the 1850s and 1860s. From 1885 with continued falls in rent, the level of loan capital dwindled. Overall, draining investment in the second half of the nineteenth century as measured by the supply of loan capital reflected trends in agricultural prosperity, and it is clear that the relationship perceived by many contemporaries was inaccurate. With the exception of the early 1880s, when the beginnings of rental decline were accompanied by some growth in expenditure, by far the greatest amount of loan capital was absorbed in a period of rising farming prosperity, being linked to upward movements in agricultural prices, especially of wheat, and to expanding rental incomes.

The appropriateness of this general relationship, particularly between landlord draining outlay and rental movement, may be assessed in more detail and over a longer time span on the sample estates (Figs. 4.7–4.11). To generalize from these individual estates, draining expenditure has been expressed quinquennially as a percentage of gross agricultural rent due,

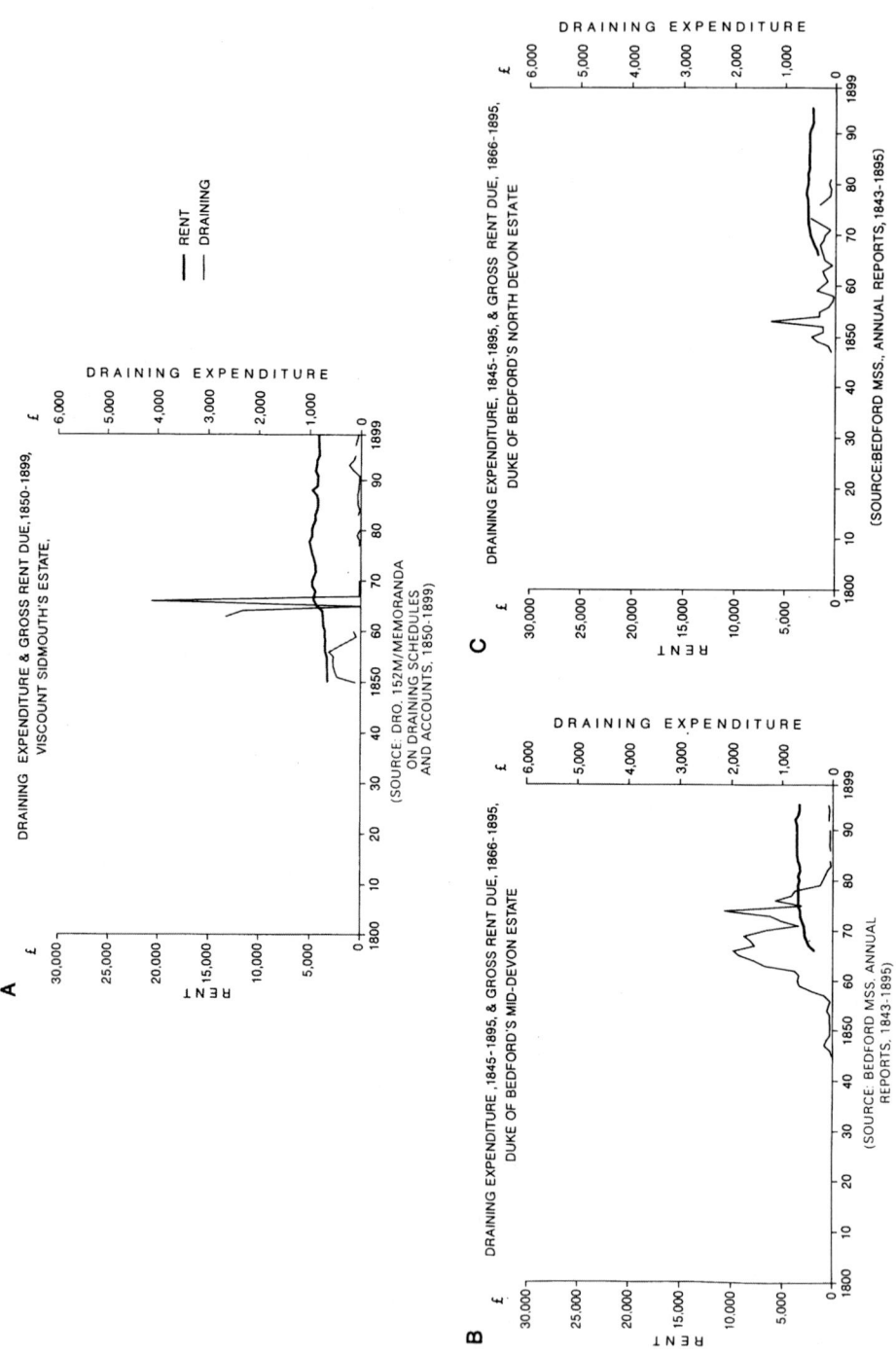

A

DRAINING EXPENDITURE & GROSS RENT DUE, 1850-1899,
VISCOUNT SIDMOUTH'S ESTATE,

(SOURCE: DRO. 152M/MEMORANDA
ON DRAINING SCHEDULES
AND ACCOUNTS. 1850-1899)

— RENT
— DRAINING

B

DRAINING EXPENDITURE , 1845-1895, & GROSS RENT DUE, 1866-1895,
DUKE OF BEDFORD'S MID-DEVON ESTATE

(SOURCE: BEDFORD MSS. ANNUAL
REPORTS. 1843-1895)

C

DRAINING EXPENDITURE, 1845-1895, & GROSS RENT DUE, 1866-1895,
DUKE OF BEDFORD'S NORTH DEVON ESTATE

(SOURCE:BEDFORD MSS., ANNUAL REPORTS, 1843-1895)

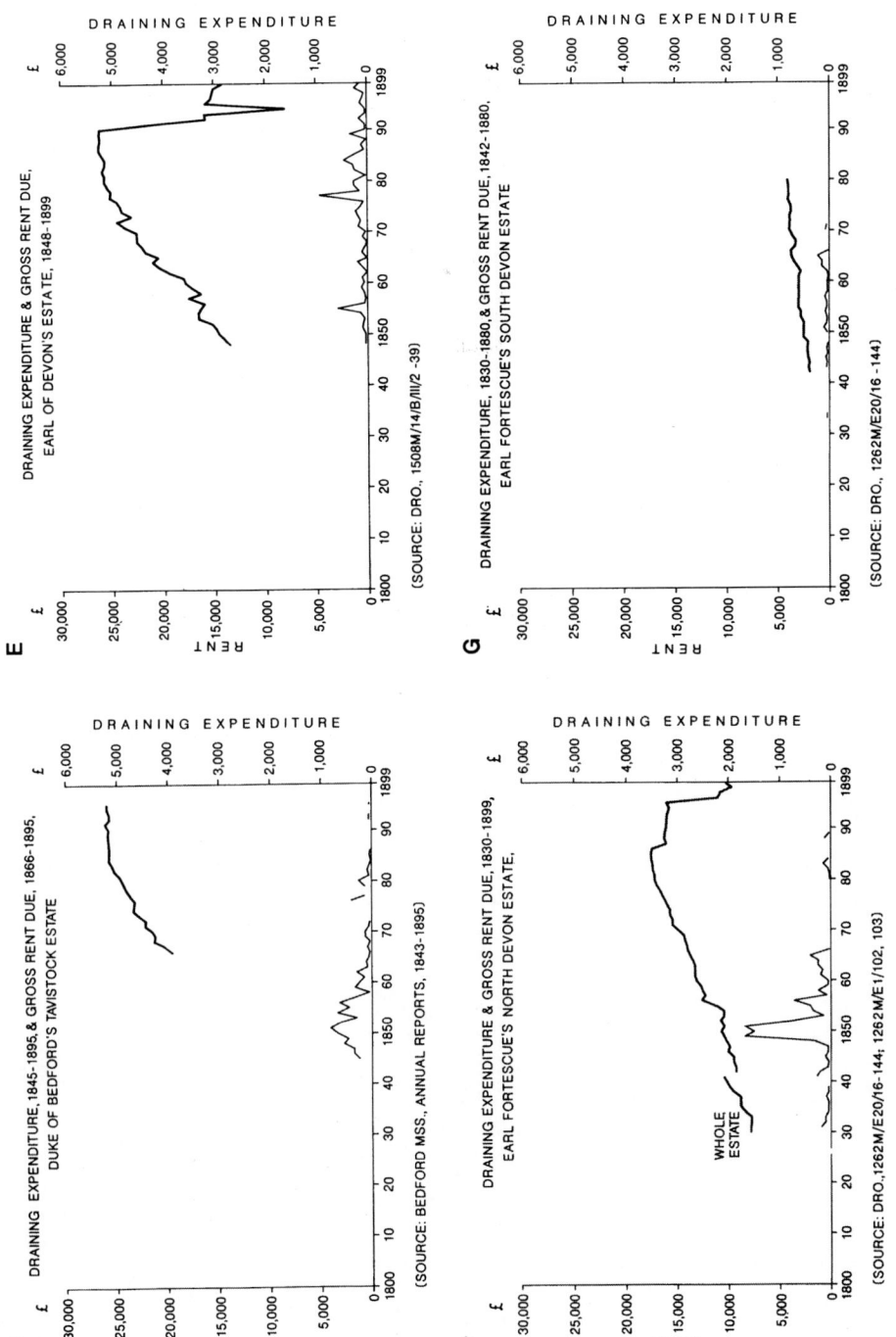

Figure 4.7 Draining expenditure and gross agricultural rent due on sample Devon estates, 1830–1899

143

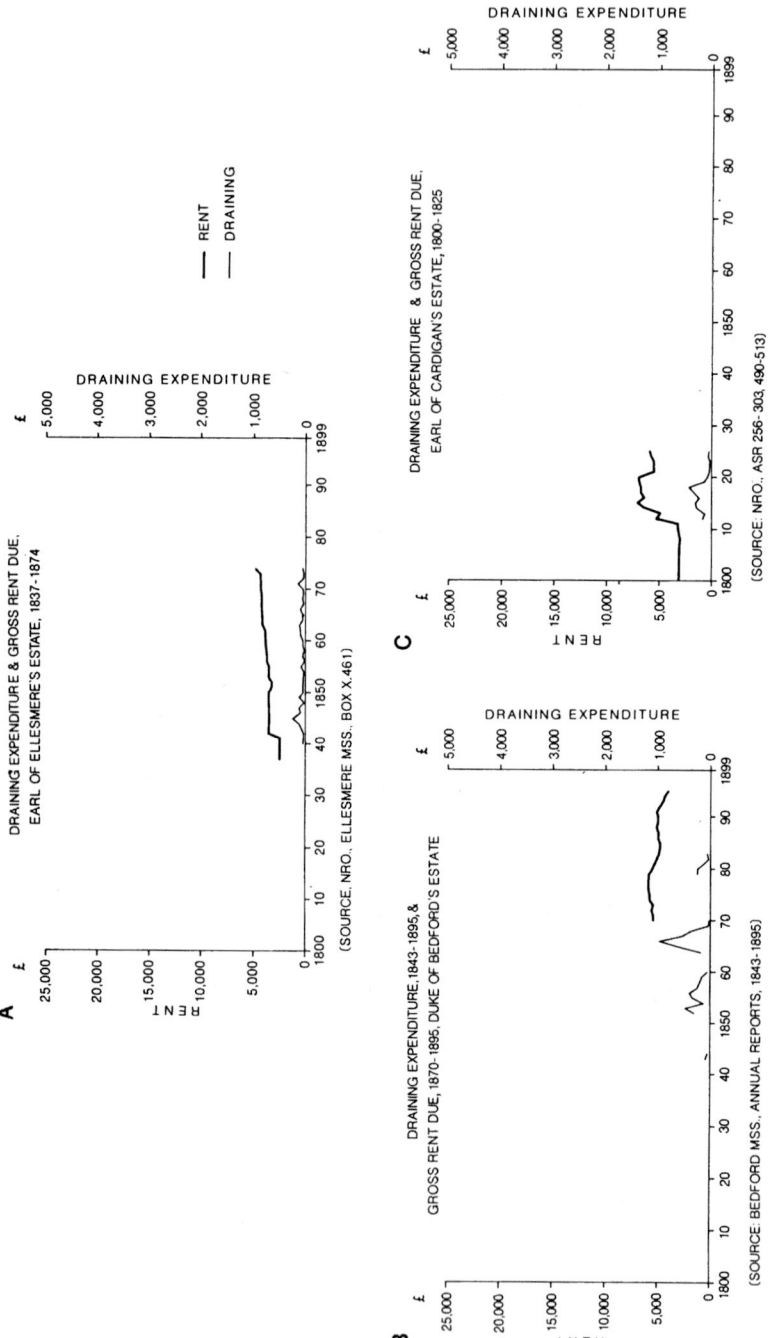

A

DRAINING EXPENDITURE & GROSS RENT DUE,
EARL OF ELLESMERE'S ESTATE, 1837-1874

(SOURCE, NRO. ELLESMERE MSS., BOX X.461)

B

DRAINING EXPENDITURE, 1843-1895, &
GROSS RENT DUE, 1870-1895, DUKE OF BEDFORD'S ESTATE

(SOURCE: BEDFORD MSS., ANNUAL REPORTS, 1843-1895)

C

DRAINING EXPENDITURE & GROSS RENT DUE,
EARL OF CARDIGAN'S ESTATE, 1800-1825

(SOURCE: NRO., ASR 256: 303, 490-513)

—— RENT
—— DRAINING

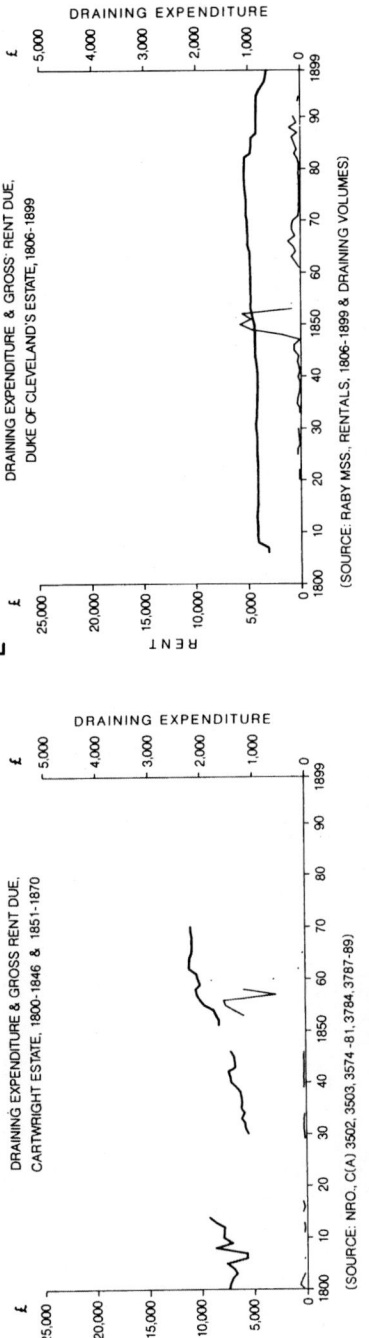

Figure 4.8 Draining expenditure and gross agricultural rent due on sample Northamptonshire estates, 1800–1899

145

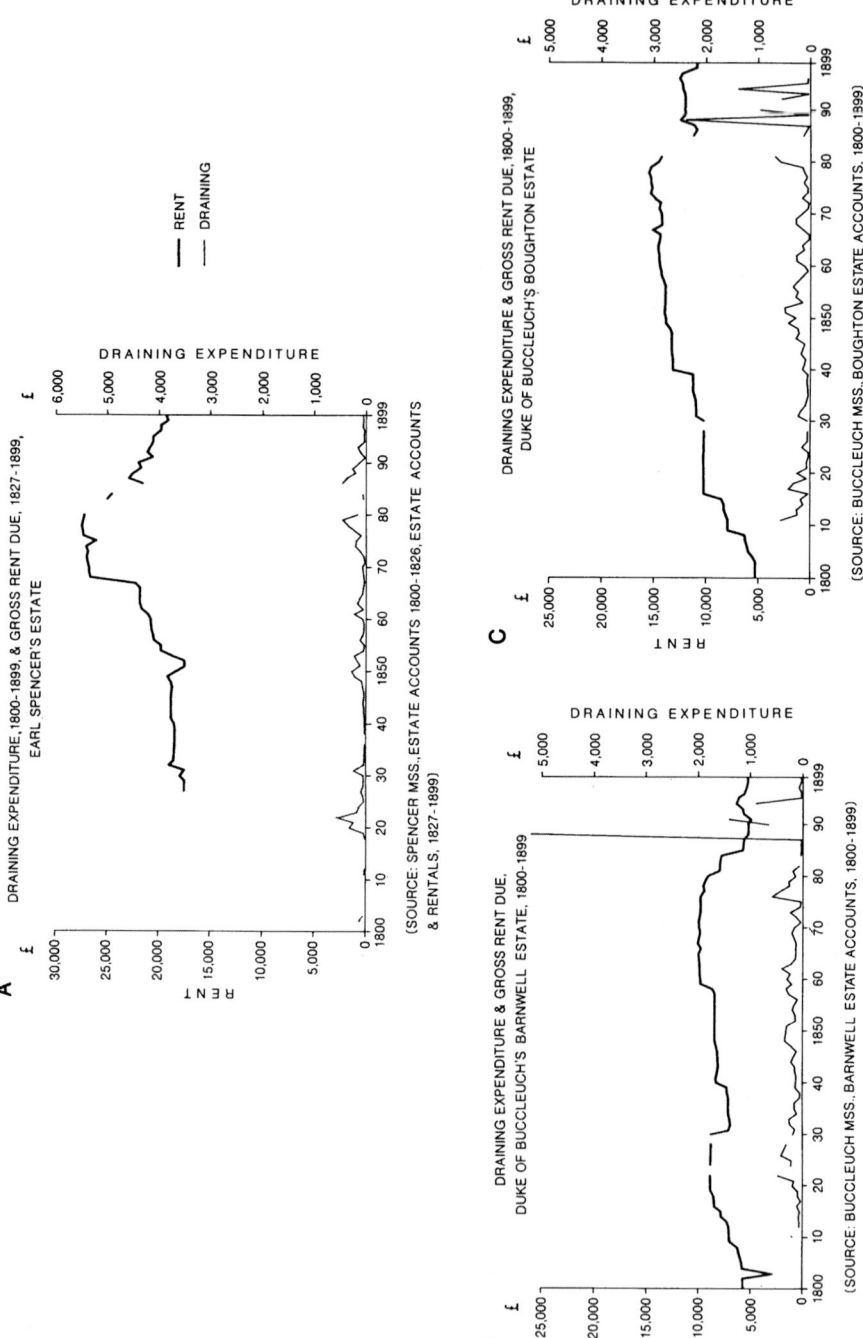

A

DRAINING EXPENDITURE, 1800-1899, & GROSS RENT DUE, 1827-1899,
EARL SPENCER'S ESTATE

RENT
DRAINING

(SOURCE: SPENCER MSS., ESTATE ACCOUNTS 1800-1826, ESTATE ACCOUNTS
& RENTALS, 1827-1899)

B

DRAINING EXPENDITURE & GROSS RENT DUE,
DUKE OF BUCCLEUCH'S BARNWELL ESTATE, 1800-1899

(SOURCE: BUCCLEUCH MSS. BARNWELL ESTATE ACCOUNTS, 1800-1899)

C

DRAINING EXPENDITURE & GROSS RENT DUE, 1800-1899,
DUKE OF BUCCLEUCH'S BOUGHTON ESTATE

(SOURCE: BUCCLEUCH MSS., BOUGHTON ESTATE ACCOUNTS, 1800-1899)

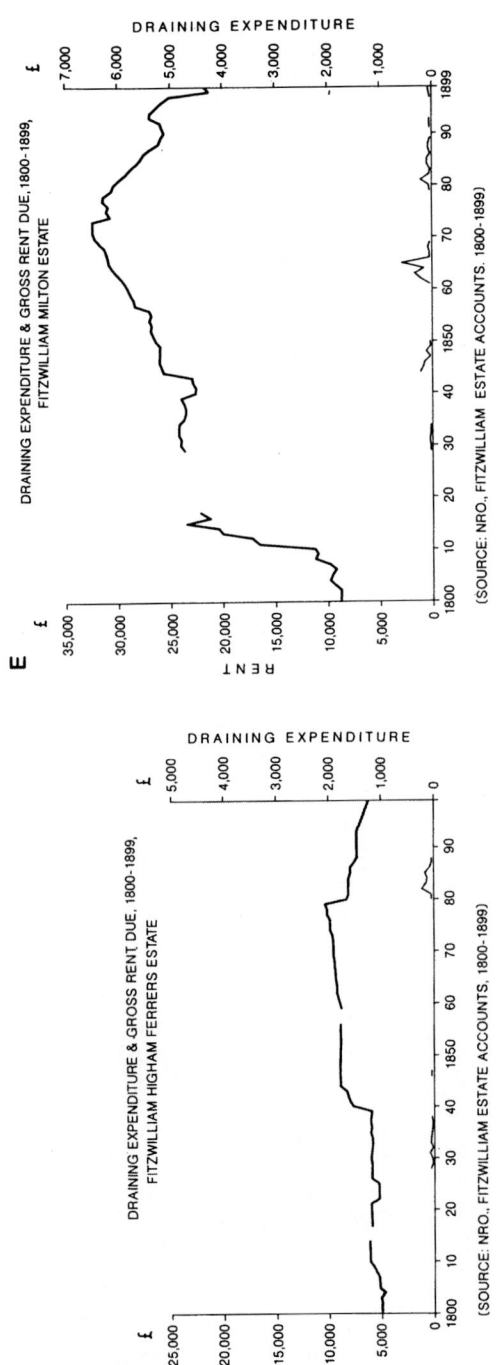

Figure 4.9 Draining expenditure and gross agricultural rent due on sample Northamptonshire estates, 1800–1899

147

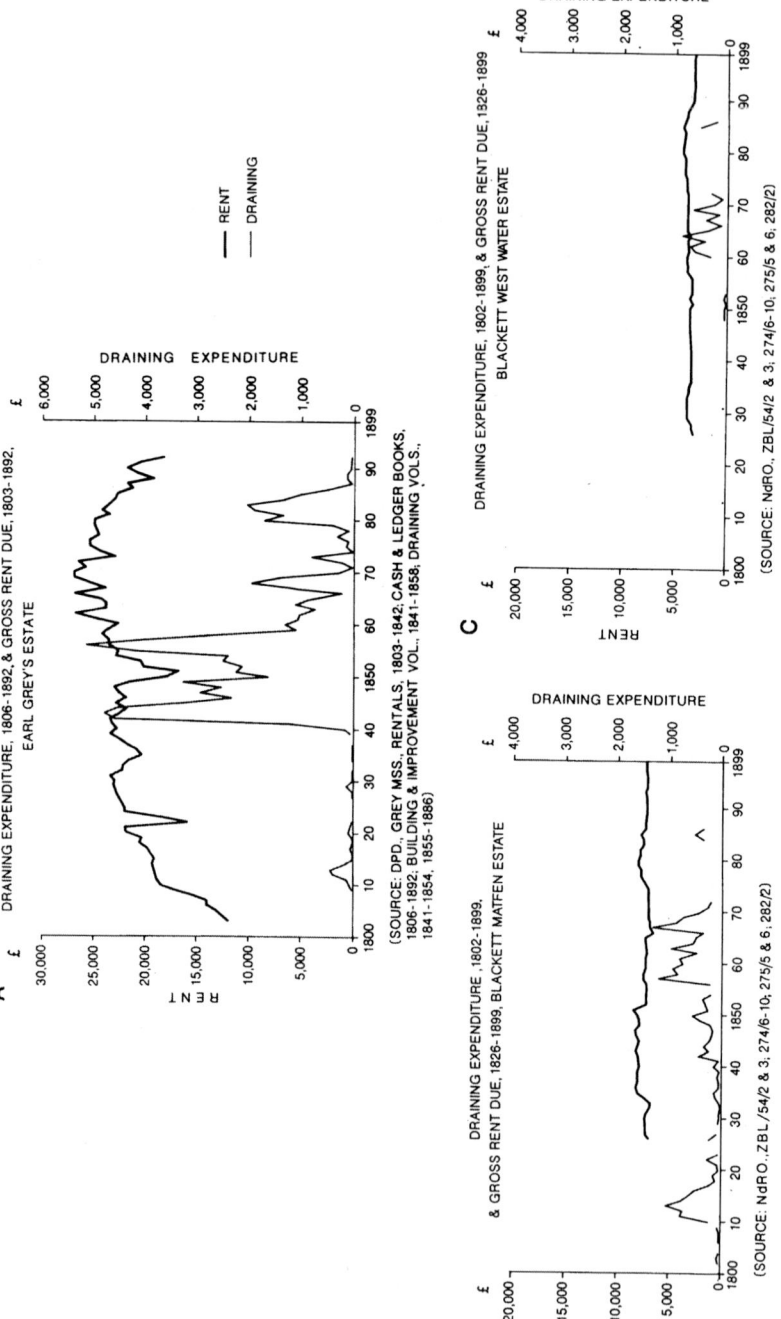

A £

DRAINING EXPENDITURE, 1806-1892, & GROSS RENT DUE, 1803-1892,
EARL GREY'S ESTATE

DRAINING EXPENDITURE

RENT

DRAINING

(SOURCE: DPD., GREY MSS., RENTALS, 1803-1842; CASH & LEDGER BOOKS,
1806-1892; BUILDING & IMPROVEMENT VOL., 1841-1858, DRAINING VOLS.,
1841-1854, 1855-1886)

B £

DRAINING EXPENDITURE, 1802-1899,
& GROSS RENT DUE, 1826-1899, BLACKETT MATFEN ESTATE

DRAINING EXPENDITURE

RENT

(SOURCE: NdRO., ZBL/54/2 & 3, 274/6-10, 275/5 & 6, 282/2)

C

DRAINING EXPENDITURE, 1802-1899, & GROSS RENT DUE, 1826-1899,
BLACKETT WEST WATER ESTATE

DRAINING EXPENDITURE

RENT

(SOURCE: NdRO., ZBL/54/2 & 3, 274/6-10, 275/5 & 6, 282/2)

148

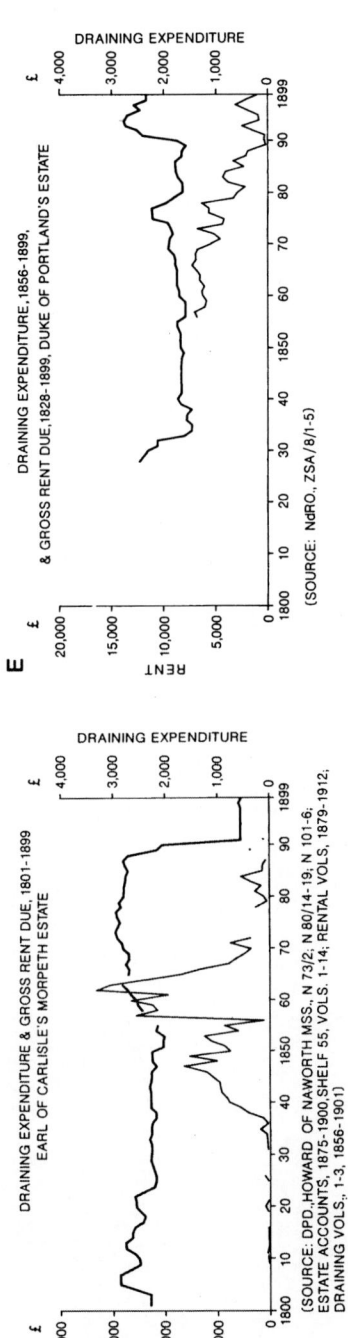

Figure 4.10 Draining expenditure and gross agricultural rent due on sample Northumberland estates, 1801–1899

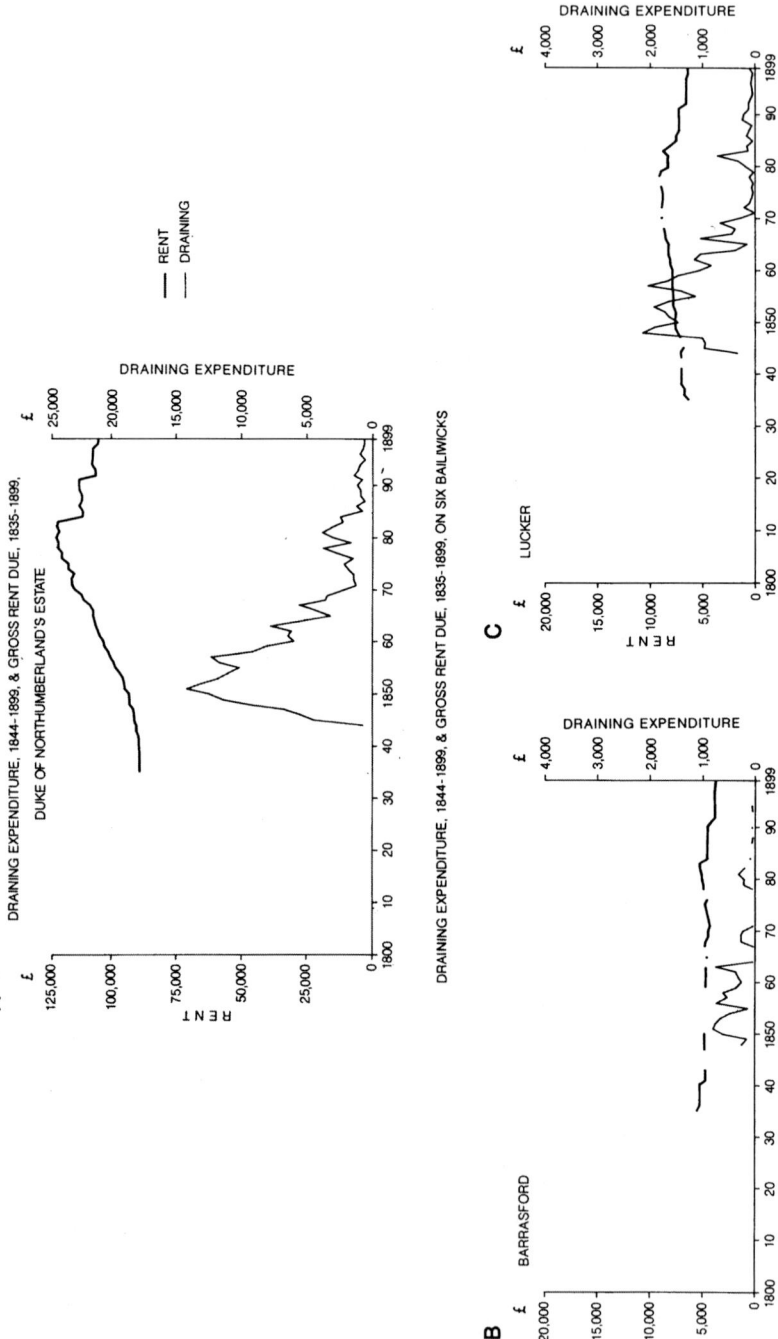

A
DRAINING EXPENDITURE, 1844-1899, & GROSS RENT DUE, 1835-1899, DUKE OF NORTHUMBERLAND'S ESTATE

—— RENT
—— DRAINING

B
BARRASFORD

C
LUCKER

DRAINING EXPENDITURE, 1844-1899, & GROSS RENT DUE, 1835-1899, ON SIX BAILIWICKS

150

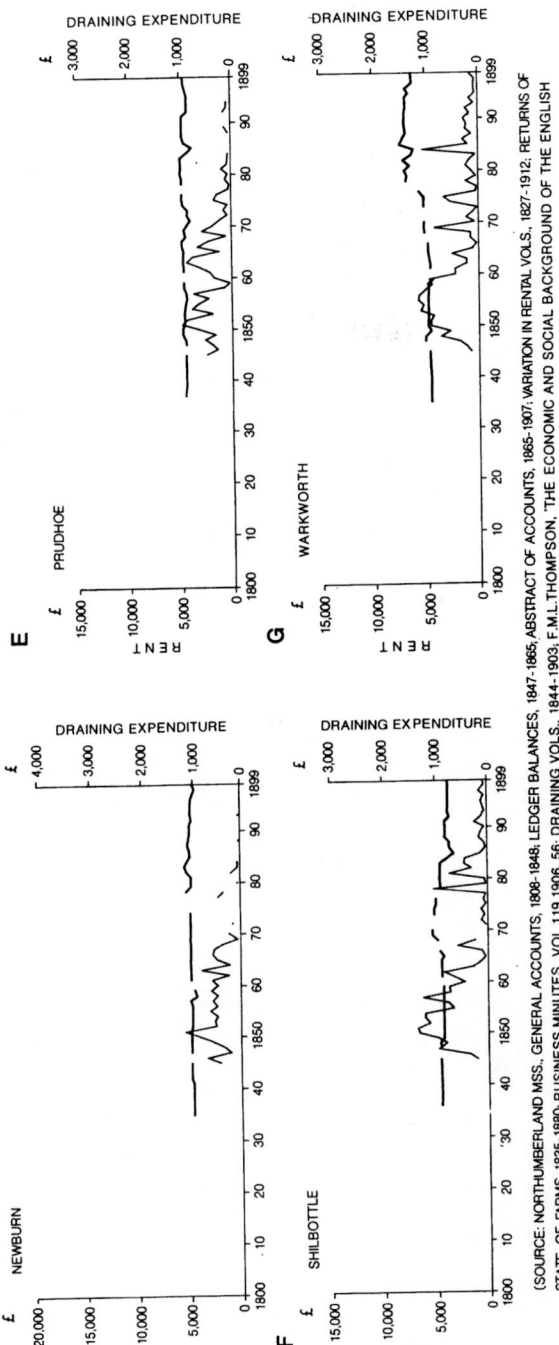

Figure 4.11 Draining expenditure and gross agricultural rent due on the duke of Northumberland's estate, 1835–1899

(SOURCE: NORTHUMBERLAND MSS, GENERAL ACCOUNTS, 1808-1848; LEDGER BALANCES, 1847-1865, ABSTRACT OF ACCOUNTS, 1865-1907; VARIATION IN RENTAL VOLS, 1827-1912; RETURNS OF STATE OF FARMS, 1835-1880; BUSINESS MINUTES, VOL. 119, 1906, 56; DRAINING VOLS, 1844-1903; F.M.L.THOMPSON, 'THE ECONOMIC AND SOCIAL BACKGROUND OF THE ENGLISH LANDED INTEREST, 1840-70 ...', 1956, APPENDIX XI)

Table 4.4 *Landlord outlay on draining as a percentage of gross agricultural rent due on sample estates in Devon, Northamptonshire and Northumberland, 1800–1899*

Period	Devon			Northamptonshire			Northumberland		
	Aggregate gross rent (£)	Aggregate draining outlay (£)	Percentage	Aggregate gross rent (£)	Aggregate draining outlay (£)	Percentage	Aggregate gross rent (£)	Aggregate draining outlay (£)	Percentage
1800–4	–	–	–	175,333	400	0.2	–	–	–
1805–9	–	–	–	207,402	32	0.0	–	–	–
1810–14	–	–	–	279,136	2,139	0.8	161,011	1,369	0.9
1815–19	–	–	–	231,450	2,948	1.3	161,758	215	0.1
1820–4	–	–	–	164,789	1,877	1.1	161,999	357	0.2
1825–9	–	–	–	179,159	2,150	1.2	210,983	708	0.3
1830–4	39,172	601	1.5	384,092	2,685	0.7	223,542	573	0.3
1835–9	45,820	295	0.6	393,324	1,231	0.3	222,799	1,900	0.9
1840–4	54,609	679	1.2	432,095	2,644	0.6	228,734	21,155	9.2
1845–9	61,072	2,386	3.9	417,892	5,907	1.4	687,310	60,040	8.7
1850–4	159,862	7,560	4.7	456,888	9,941	2.2	682,872	79,986	11.7

1855–9	177,253	4,744	2.7	478,056	7,734	1.6	742,114	82,811	11.2
1860–4	197,883	7,026	3.6	501,186	5,371	1.1	800,926	61,076	7.6
1865–9	222,011	5,081	2.3	546,190	3,903	0.7	840,263	43,729	5.2
1870–4	377,873	9,006	2.4	543,428	2,466	0.5	878,637	17,930	2.0
1874–9	402,813	6,706	1.7	523,021	3,914	0.7	905,562	18,333	2.0
1880–4	395,043	2,920	0.7	385,917	3,856	1.0	886,602	28,495	3.2
1885–9	396,426	1,285	0.3	395,564	10,535	2.7	831,392	10,931	1.3
1890–4	349,006	1,038	0.3	408,040	6,574	1.6	681,650	6,515	1.0
1895–9	154,027	988	0.6	356,661	510	0.1	661,623	5,416	0.8

–: No data available

The county figures were made up of the following estate runs:

Devon: Bedford estates, 1870–94; Devon estate, 1850–99; Fortescue north Devon estate, 1830–99; Fortescue south Devon estate, 1830–79; Sidmouth estate, 1850–99;

Northamptonshire: Bedford estate, 1870–94; Buccleuch estates, 1800–99; Cardigan estate, 1800–24; Cartwright estate, 1800–14, 1830–44, 1850–69; Cleveland estate, 1805–99; Ellesmere estate, 1835–74; Fitzwilliam Higham Ferrers estate, 1800–99; Fitzwilliam Milton estate, 1800–19, 1830–99; Spencer estate, 1825–99;

Northumberland: Blackett estate, 1825–99; Carlisle estate, 1810–99; Grey estate, 1810–89; Northumberland estate, 1845–99; Portland estate, 1855–99

Source: As for Table 4.3

153

providing an indication of landlord financial commitment to the improvement (Table 4.4). Given the range in the intensity of investment, the proportion that draining outlay formed of rent varied widely between the three groups of estates. Thus, in any quinquennium the proportion on the Northamptonshire properties never exceeded 2.7 per cent of gross agricultural rent, peaked at 4.7 per cent in 1850–4 in the Devon sample, and was greater than 11 per cent over the decade 1850–9 on the Northumberland estates. The high values in Northumberland should not be ascribed to the large absolute draining outlay made on the lands of the dukes of Northumberland: indeed the inclusion of data from the estate depresses the overall proportion and in 1855–9 and 1860–4, for example, landlord investment on the other properties in the sample amount respectively to 13.0 and 10.7 per cent of rent. Landlord involvement in draining was far more developed from 1840 on the sample estates in Northumberland than on those in Devon and Northamptonshire.

The temporal pattern of landlord outlay on draining as a percentage of agricultural rent conforms to that of expenditure per acre (Fig. 4.2b). Before 1840, the proportion that draining represented of rent was insignificant, not exceeding the 1.5 per cent recorded in 1830–4 in Devon. Landlord investment as a percentage of rent grew from 1840, and for the period 1840–69 a higher proportion of rent was devoted to draining than at any other time in the nineteenth century, especially on estates in Devon and Northumberland. From 1870 there was a marked decline in the proportion of rent allocated to draining expenditure in the Devon and Northumberland samples, although some reverse of this trend is evident on estates in the latter county in 1880–4. On estates in Northamptonshire, however, draining outlay as a percentage of rent rose from 1870, peaking in 1885–9 at 2.7 per cent.

Rents on these sample estates have also been presented quinquennially and on an acreage basis (Table 4.5). Gross agricultural rent due has been employed, industrial and urban revenues, particularly significant in Northumberland but evident on properties in Devon and Northamptonshire, having been excluded, and has been expressed where possible in terms of tenanted acreage. Rental movements, however, may be affected by the changing composition of the samples. Thus, the inclusion for the first time in 1835–9 of data from the estate of the dukes of Northumberland, which encompassed much low-valued land, intensified the per acre decline in rent from 1830 to 1834 found on other properties in the Northumberland sample. Although care should be taken in being too specific in their interpretation, the patterns do reflect real trends in rent (Fig. 4.2c).

No consistent relationship between landlord outlay on draining and rental trends can be determined on the sample estates before 1840 (Fig. 4.2a and c). Irrespective of rental movement in this period, landlord reaction on the Northumberland properties was to provide little funding for the improvement. However, on the Northamptonshire estates, declining and stable rents between 1815 and 1829 did result in some increased landlord provision for draining, but of a limited and restricted nature. For example, Robert Edmonds, agent to the Buccleuch Barnwell estate, noted in 1822, when draining allowances had amounted to £422, that the sums had been provided 'with a view to the improvement of the estate, also to assist the tenants in finding employment for the labourers'. Yet the reduction of net receipts through such expenditure had not been to the landlord's liking and Edmonds admitted that even in those times of depression outlay had gone beyond both what he intended and should have done, and acknowledged the necessity to guard against such 'extraordinary expenses' in future.[32]

However, the major landlord financial commitment to draining from 1840 to 1869 took place against a background of rising rental receipts, although the rate of growth was less rapid on the Northumberland estates than on those in Devon and Northamptonshire. Decline in gross agricultural rent was common to all three estate samples after 1880 but occurred at different times and intensities. Rental fall was delayed on the Devon estates until 1896–9 and did not correspond with any distinct resurgence of landlord investment in the improvement. On the Northumberland estates, reduction in rent dated from 1880–4 and was accompanied by some expansion in draining outlay. The decline in agricultural prosperity was recognized on most estates in the sample, G. A. Grey, agent to the Carlisle property, noting in 1880 that they had 'had an awful time...with tenants giving up farms, or requiring abatement of rent and more failing',[33] as was the need for additional landlord financial assistance in such conditions. J. Snowball, chief comissioner to the duke of Northumberland, considered that some attempt should be made 'to relieve...tenantry to some extent of the serious losses they have sustained' and to prevent, as Sir Edward Blackett feared, farms falling in hand.[34] Increased draining outlay may be seen as an element in a general effort to stem rental decrease, as T. Sample wrote on handing over the agency of the Blackett estate to his son, C. H. Sample, in 1888: 'draining...will always help the future letting of the farm'.[35] Nevertheless, levels of landlord expenditure on draining were well below those recorded in the 1850s and 1860s. On the sample estates in Northamptonshire, which experienced the greatest percentage reduction in rent after 1880, there was growth in landlord outlay on the improvement, reaching proportions of agricultural rent comparable to those at the middle of the century. Yet although the increase, particularly on the Buccleuch and Spencer estates, involved a degree of compensation for the relative lack of landlord investment and considerable use of tenant capital on the improvement before 1880, the overall proportion that draining outlay formed of rent in this period can only be described as slight.

For the whole of the nineteenth century, it is not possible to distinguish a

Table 4.5 Gross agricultural rent due per acre on sample estates in Devon, Northamptonshire and Northumberland, 1800–1899

Period	Devon			Northamptonshire			Northumberland		
	Aggregate rent (£)	Estate acreage	Quinquennial rent per acre (£)	Aggregate rent (£)	Estate acreage	Quinquennial rent per acre (£)	Aggregate rent (£)	Estate acreage	Quinquennial rent per acre (£)
1800–4	–	–	–	129,796	34,202	3.79	–	–	–
1805–9	–	–	–	142,281	34,202	4.16	–	–	–
1810–14	–	–	–	193,607	37,810	5.12	161,011	38,267	4.21
1815–19	–	–	–	146,406	27,534	5.32	161,768	38,267	4.23
1820–4	–	–	–	129,273	25,162	5.14	161,999	38,267	4.23
1825–9	–	–	–	50,053	9,189	5.45	168,432	38,267	4.40
1830–4	39,172	21,260	1.84	263,435	47,749	5.52	271,572	63,559	4.27
1835–9	45,820	21,260	2.16	267,097	47,749	5.59	706,699	200,099	3.53
1840–4	54,609	21,260	2.57	315,167	50,559	6.23	720,444	200,049	3.60
1845–9	61,072	21,260	2.87	287,635	45,923	6.26	727,899	205,527	3.55
1850–4	159,862	46,771	3.42	288,983	45,759	6.32	724,841	204,017	3.55

1855–9	177,253	46,771	3.79	260,962	40,131	6.50	715,164	191,101	3.74
1860–4	197,883	46,771	4.23	329,569	45,927	6.98	746,948	194,380	3.84
1865–9	222,011	46,771	4.75	335,881	45,927	7.31	840,263	209,249	4.02
1870–4	377,873	63,451	5.96	383,172	51,569	7.43	878,637	209,249	4.20
1875–9	402,813	63,451	6.35	367,550	48,708	7.54	905,562	203,227	4.46
1880–4	395,043	59,449	6.65	134,168	21,212	6.33	886,602	203,227	4.36
1885–9	396,426	59,449	6.67	172,476	32,726	5.27	831,392	203,227	4.09
1890–4	349,006	50,163	6.96	276,575	47,653	5.80	681,650	174,003	3.92
1895–9	154,027	29,614	5.20	236,464	43,411	5.45	661,623	174,003	3.80

–: No data available

The county figures were made up of the following estate runs:

Devon: Bedford estates, 1870–94; Devon estate, 1850–99; Fortescue north Devon estate, 1830–99; Fortescue south Devon estate, 1830–79; Sidmouth estate, 1850–99;

Northamptonshire: Bedford estate, 1870–94; Buccleuch Barnwell estate, 1800–19, 1830–99; Buccleuch Boughton estate, 1800–24, 1830–79, 1885–99; Cardigan estate, 1800–24; Cartwright estate, 1800–14, 1830–44; Cleveland estate, 1810–99; Ellesmere estate, 1840–74; Fitzwilliam Higham Ferrers estate, 1800–14, 1820–54, 1860–99; Spencer estate, 1830–79, 1890–9;

Northumberland: Blackett estate, 1830–99; Carlisle estate, 1810–54, 1865–99; Grey estate, 1810–89; Northumberland estate, 1835–99; Portland estate, 1830–99

Source: As for Table 4.3

constant or precise relationship between levels of landlord expenditure on draining and of agricultural rent on these sample estates. However, on most of them after 1840, large-scale investment in the improvement occurred in a period of advancing rents and, although some increase attended rental decline before 1840 and after 1880, the amounts involved both absolutely and proportionally were small. The estate evidence must be seen as corroborating the country-wide trends between investment in draining and agricultural prosperity derived from the loan-capital data.

Advances in draining techniques

Investment, however, was also dependent on the effectiveness of the improvement and many contemporary agriculturalists saw the temporal adoption of draining largely as a response to its technical development. Technical change in the improvement was most rapid in the late 1830s and 1840s, and during that period more space was devoted in the agricultural literature to the discussion of draining than at any other time in the nineteenth century. A revaluation of the improvement was described in the literature at that time which made it potentially attractive of landlord capital, involving a reappraisal of draining systems, materials, costs, permanence and productivity.

During the 1820s and much of the 1830s, specific draining and general agricultural accounts maintained the view prevalent at the beginning of the century that wetness in land was capable of being treated by a variety of draining systems. Emphasis was still laid on the removal of water arising from springs, using the method devised by Joseph Elkington of a few deep cuts near the eye of the spring. For excess wetness resulting from rainfall on impermeable soils, a choice existed between surface draining, the system of shallow, frequent underdrains as practised in parts of East Anglia, and furrow draining, the laying of shallow drains in the furrows of ridged land, as reported from the Cumberland estates of Sir James Graham in 1829.[36] From 1831 and particularly from 1836 in evidence before the Select Committee into the state of agriculture, James Smith presented a systematic and unified approach to the problem.[37] For the first time, he identified the removal of surface water as the prime objective of the improvement, but his system of thorough draining, which comprised throughout a field a series of frequent, parallel drains at a minimum depth of $2\frac{1}{2}$ ft and at intervals dependent on soil texture, could also deal with any spring water that might arise. Modification of the system was suggested from the early 1840s, chiefly by Parkes, who, in a series of papers and in evidence before the 1845 Select Committee of the House of Lords to enable possessors of entailed estates to charge such estates for the purpose of draining, argued that drain depth should be increased to 4 ft.[38] Such additional depth not only allowed the laying of drains at greater intervals, so reducing labour and material costs, but was also considered to achieve a more effective drainage of soil water. Although there was debate over the merits of respective depths,[39] a coherent system of draining was recognized that was applicable to most conditions of soil wetness.

Associated with this new approach, the literature disclosed the beginnings of the mechanization of the production of the most reliable of drain fills, a process reported both to reduce total costs and to increase permanence. Excluding spring draining, by the middle of the 1830s a range of drain types was known and from the examples cited by witnesses before select committees on agriculture and in the agricultural press some measure of their relative costs may be obtained. Mole draining was regarded as the cheapest form, being recorded in 1821 at £0.90 per acre.[40] Bush and straw drains varied between £2 and £3 per acre over the 1820s and 1830s in East Anglia, while plug and turf drains were priced at £1.50 to £2.50 per acre.[41] Stone and tile drains were more expensive although they lasted longer. With stones readily available and drains laid at 20 ft intervals, stone draining was estimated at £6 per acre under Smith's system, a figure which rose to £8 and more where stones were not at hand and required carting.[42] Variously shaped semi-circular tiles laid on soles were recognized as a more effective form of drain, not only limiting the tendency to siltation common to all other drain fills but also possessing a comparative ease of handling. Despite the removal of brick-and-tile duties from tiles used for draining purposes,[43] the costs were high. In Cumberland in 1821 tiles ranged from £2.10 to £2.60 per 1,000, excluding soles, while Philip Pusey reported in 1841 that tiles and soles cost between £2 and £3 per 1,000 in Berkshire, Gloucestershire and Norfolk.[44] As few tiles were more than 12 in long, to drain an acre of land at 20 ft intervals would require at least 2,000 tiles, involving an outlay for materials alone in excess of £4.[45]

The high price of tiles stemmed from their being moulded and made by hand. However, with the invention of machines which mass-produced tiles and soles, the costs were greatly reduced. In 1835, Robert Beart of Godmanchester, Hungtingdonshire, introduced such a machine, to be followed quickly by the marquis of Tweeddale and F. W. Etheredge of Southampton.[46] On average, these machines were claimed to produce tiles and soles from £1 to £1.25 per 1,000, and their value was rapidly appreciated.[47] By 1843, Parkes noted that 19 Tweeddale and 50 Etheredge machines had been purchased.[48]

The use of tiles and soles as fill was superseded by the development of cylindrical drain-pipes. Hand-made drain-pipes had been reported from various parts of the country from early in the nineteenth century.[49] The existence of several simple machines that produced limited numbers of drain-pipes was noted by 1840 in Essex, Kent, Suffolk and Sussex.[50] The design of these was improved to provide greater output and standardization of form, involving an outlay that varied from £23 to £50.[51] The first of these new drain-pipe machines was accredited to T. Scragg of Calveley, Cheshire, in 1842, but the growth in their number was swift, 34 being exhibited at the Royal Agricultural Society's show at York in 1848, while R. Boyle writing in 1853 was aware of 45 pipe machines on the market.[52]

While the drain-pipe was acknowledged to represent a technical advance over the tile and sole in that it reduced the likelihood of blockage, the new machines were regarded as contributing to lowering the overall cost of the improvement. In 1845, pipes with a diameter of 1 in were described as costing £0.60 per 1,000 from machines devised by H. Clayton and T. Hatcher.[53] At this price, both Parkes and Pusey could claim that heavy clayland could be drained using the new systems for no more than £3 per acre.[54] Although doubts were raised at the time of the ability of 1 in pipes to maintain an open channel underground, 2 in pipes which were priced around £1 to £1.25 per 1,000 and raised acreage costs to nearer £5 being considered safer,[55] effective underdraining was reliably reported as being obtainable at the same cost as traditional methods.

Added to these cost reductions, the new draining systems were attributed with a high degree of permanence. Although views of their duration varied, mole, bush, straw, turf and plug drains were accepted as temporary improvements with the need for regular renewal.[56] However, many agriculturalists, impressed by the technical efficiency with which tiles and especially pipes were produced, saw the new systems as possessing a much longer, if not perpetual, life. While recognizing a finite existence to the improvement, the fifty years suggested by Parkes represented a life-expectancy at least twice that of traditional draining methods.[57] Pusey and Hewitt Davis predicted an even longer life for the new systems of draining, doubting if pipes and tiles properly laid would ever be stopped.[58]

Technical efficiency, permanence and low cost of drains were linked in the agricultural literature to increased productivity, which was largely assessed in terms of crop yields, especially those of wheat. Although a considerable number of estimates were made, landowners were led to believe that a growth in wheat output of between 10 and 30 per cent per acre could be reasonably expected on the adoption of these new draining methods, with authorities such as Denton, Pusey and Smith placing the amount firmly in the range 15 to 25 per cent.[59] Such productivity was further seen as providing landowners with an opportunity of obtaining a clear return on the capital investment in the improvement in the form of increased rent. Estimates of this return varied in the 1840s but they were rarely less than 10 per cent and often as high as 25, averaging about 15 per cent, rates quoted amongst others in answers to questions put by the earl of Carlisle in 1848 to a group of seven agents and draining experts, including Parkes and Smith, as part of the minutes of information collected by the General Board of Health in respect of draining land.[60] In sum, the change in draining systems that occurred in the late 1830s and 1840s was presented in the agricultural literature not simply as a technical advance resulting in the more effective removal of excess water in soils but also as an economic improvement that had real implications for enhancing land values. As J. H. Charnock noted in 1844, draining was 'a permanent improvement of the fee', a view that received endorsement in 1845 in the report of the Select Committee of the House of Lords to enable possessors of entailed estates to charge such estates for the purpose of draining.[61]

Many of the claims made for the improvement were questionable, being based on enthusiasm rather than experience. Thus, to credit pipe drains with permanence after being laid for only a few years was precipitate. The low costs of pipe drains involved the use of prices at which pipes were produced rather than sold and of generous calculations for labour. The estimates of increased yields and landlord return reflected anticipation rather than achievement. Nevertheless, that these views were retailed by leading and respected agriculturalists of the standing of Pusey and Parkes and were replicated throughout the agricultural press[62] must have raised contemporary perceptions of the value of draining. And the developments in tile and pipe production and in drain design were real.

The expansion of landlord investment on the sample estates after 1840 involved a response to this technical revaluation of the improvement. Most of the sample estates displayed an awareness of technical developments in underdraining, and the degree of landlord confidence in their effectiveness had a major influence on the provision of capital. The importance of technical improvement may be demonstrated from the Northumberland sample, where advances in draining methods were readily diffused between estates.

The first estate in the sample to adopt thorough draining was that of the duke of Portland in 1834. From 1829 spring draining with a few deep cuts over 4 ft had been employed on the estate[63] (Fig. 4.12). Stones had been used to fill such drains until 1830, when a tilery was built at Bothal to produce tiles and soles.[64] The introduction of thorough draining was on the insistence of the duke who had seen the technique on his Scottish lands and considered it more valuable for strong clay soils than spring draining.[65] William Sample, his Northumberland agent, was unsure both of the new technique and of tenant response, writing in 1833 that he had no person in his employment who was conversant with the system. To overcome this difficulty, an estate employee, George Taylor, was sent in 1833 to the duke's Ayrshire estate to learn the art of thorough draining.[66] In 1834 a second tilery was constructed to cater for the increased demand for tiles and soles created by the new system, and in the autumn of that year thorough draining began on the estate.[67] In January 1836, Sample wrote that, since its introduction 'and its effects seen on the few specimens that have been done, many of the tenants are becoming anxious to try the system'. Indeed, so successful was the new system that Sample was 'at a loss to [know] what Your Grace may be disposed to pursue...annually', especially as he noted in 1836 'I am strongly inclined to believe from what I have already seen of its effects that [thorough] draining will be found to be increasingly beneficial to the land in general on the Bothal estate and will also greatly tend to promote an increase of produce.'[68] From 1834, expenditure on spring draining declined markedly on the estate.

Other estates were quick to adopt thorough draining, replacing the existing emphasis on spring draining and involving an increase in landlord investment. On the adjoining Carlisle lands, thorough draining was first recorded in 1837, and between 1839 and 1842 a tilery was built at Hepscott to produce tiles and soles.[69] Lord Grey had been gathering information during 1837 and 1838 on methods of thorough draining and on the establishment of tileries. In 1840 an agreement was signed with the Tweeddale Patent Tile Company to erect two tileries at Ancroft and Broomhill to provide tiles and soles for the thorough draining of the estate.[70] The system was first reported on Sir Edward Blackett's lands in 1840, where Sample was also agent. His experiences on the Portland estate no doubt contributed both to its adoption and to the building of a tilery at East Matfen in 1839 for the necessary tiles and soles[71] (Fig. 4.10a and b). The use of thorough draining came later on the estate of the dukes of Northumberland. Although the system was noted in 1840, the estate embarked on a large-scale draining programme based on thorough draining only from 1843[72] (Fig. 4.11a). In that year, Etheredge was brought from Southampton to Alnwick to set up tileries on the estate at which with his machine tiles and soles could be produced.[73] The first tilery was built at Shilbottle in 1844 and the estate spent £547 in 1844-5 on the purchase of Etheredge machines from J. R. and A. Ransome.[74] An estate drainer, John Loraine, who had come from the marquis of Tweeddale's estate, was appointed to apply a system of drains $2\frac{1}{2}$ to 3 ft deep and 18 to 24 ft apart.[75]

Further advances in the technique of draining, as exemplified by drain-pipe machines and deep-draining systems, were readily adopted on the estates and necessitated the maintenance of landlord investment in the improvement. In 1846, the duke of Portland expressed enthusiasm for the purchase of a pipe-making machine and insisted that Sample should obtain one made by Clayton. In reply to Sample's inquiry as to size of pipe to be used, the duke sent a copy of Parkes' paper on deep draining that appeared in the *Journal of the Royal Agricultural Society*, a system which he wanted implemented on the estate, adding that 2 in diameter pipes should be employed instead of the 1 in pipes advocated by Parkes.[76] It is interesting to

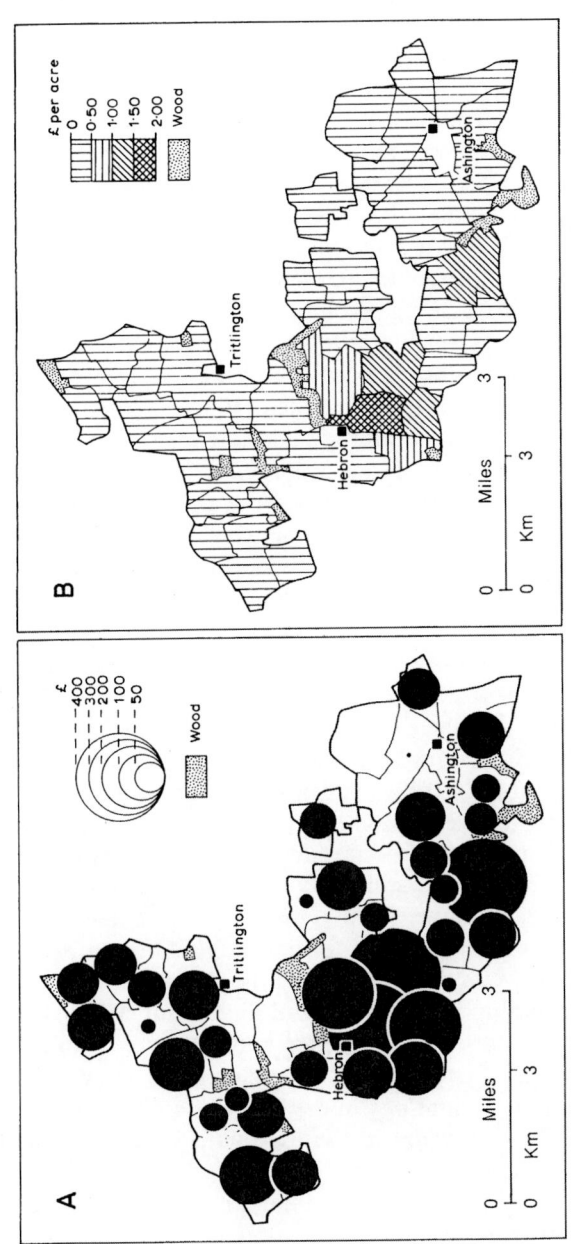

Figure 4.12 A. Absolute expenditure on spring draining, 1829–1834, on the duke of Portland's estate by 1861 farm size; B. Per acre expenditure on spring draining, 1829–1834 (*Sources*: NdRO, Sample MSS: ZSA/51/29; ZSA/18/2/1 and 2; ZSA/8/2–5)

note that on the other estate in Sample's control, a pipe-making machine was also purchased in 1846 for the Blackett Matfen tilery, although of different make, being produced by J. H. Charnock.[77]

Some estates went further in introducing the most recent technical developments. On the succession of the fourth duke of Northumberland in 1847, the estate commitment to major draining investment was continued. However, the duke identified that 'the general object in draining is that the work be effectual', and was convinced that the new system of deep draining would achieve 'the permanency of the drains'.[78] In 1848 he informed his commissioners that Parkes should be engaged to advise on the draining of the estate.[79] In his report Parkes recommended the adoption of a 4 ft deep draining system at 30 to 40 ft intervals, using 1 in diameter pipes, and the purchase of two pipe-making machines by either Scragg or J. Whitehead of Preston for the tileries at Shilbottle and Longhoughton. The Shilbottle tile-maker, R. Hall, was to be sent to Whitehaven to examine the tileries that Parkes had established for Lord Lonsdale. Finally, draining on the estate was to be carried out by two draining superintendents under the guidance of Parkes. The greater part of the report was accepted, the duke however insisting that 1 in pipes were unsatisfactory and nothing smaller than $1\frac{1}{2}$ in pipes should be used.[80] The new techniques were readily assimilated and in 1855 Hugh Taylor, the chief commissioner, suggested that 'considering the progress made and the information and experience obtained by Messrs Loraine and Locking [the draining superintendents], perhaps Your Grace might be pleased to direct that Mr. Parkes' services be dispensed with after the present year', to which the duke agreed.[81] On similar lines, the earl of Carlisle, who had obtained first-hand knowledge of Parkes' deep-draining system in collecting material on draining for the General Board of Health, also employed him to introduce a draining programme on his Northumberland estate, and between 1848 and 1850 three pipe-making machines were purchased and an additional tilery constructed at Hepscott.[82]

Not all estates responded uniformly to the revaluation of the improvement in the county. For example, John Aynsley, agent to the Middleton/Monck estate, identified in 1849 the need for draining on the property.[83] Sir Charles Monck, however, was indifferent to the improvement, and large-scale investment in draining had to wait until Sir Arthur Middleton inherited the estate in 1867.[84] Clearly, the incidence and intensity of draining on estates were dependent on factors besides technical efficiency. Nevertheless, the progress in draining technology in the late 1830s and 1840s was an essential prerequisite to the expansion in the use of the improvement that occurred after 1840.

Together, the supply of loan capital and landlord investment on the sample estates provide specific data by which to assess the timing and rate of adoption of draining throughout the nineteenth century. Although some regional divergence is apparent, the temporal pattern so revealed indicates that, whilst outlay on the improvement occurred over the whole century, draining activity was at its most intense in the thirty years between 1840 and 1969. The time series derived from this material is at variance with those that have been based either on technological change in draining or on trends in agricultural investment. Thus, landlord expenditure did not increase steadily throughout the nineteenth century as improved draining techniques became available, but exploded as draining methods and materials were revalued around 1840. Yet, the attainment of technical mastery did not result, as M. Robinson and B. D. Trafford have argued,[85] in an unchanging level of draining activity for the greater part of the century, and outlay fell away in varying degrees after 1870.

Although it represented only one of the improvements supported by landlord funds, the pattern of draining expenditure also deviated from the sequence of capital investment in agriculture described by F. M. L. Thompson.[86] Rather than running counter to the course of boom and slump, draining outlay, both absolutely and as a proportion of rent, tended to mirror agricultural prosperity.[87] Thus, although some increase in draining investment can be detected on the Northamptonshire sample estates in the 1820s and 1830s, the scale of adoption was well short of that suggested by D. C. Moore as a response to declining agricultural rents and prices.[88] The highest levels of draining provision in the period 1840–69 took place at a time when rents and farm income were growing. Draining outlay approached Thompson's general pattern of agricultural investment only in the agricultural depression from the late 1870s. After an initial growth in expenditure to maintain rents and to aid tenants, which was aggravated by a series of wet seasons and which tended to be greater on those estates and in those parts of the country where the improvement had been relatively neglected before 1870, landlord financial involvement in draining became less pressing.

Although failing to correspond fully to either of the proposed chronologies, the overall temporal pattern of draining should be seen as incorporating aspects of both. Landlord investent in draining was initially dependent on its technical effectiveness. Reliable draining systems and materials were not available until the late 1830s and only from that time could the improvement be expected to receive widespread landlord financial attention. However, the provision of landlord capital for these improved techniques closely reflected trends in agricultural prosperity. The introduction of efficient draining methods coincided with a period of rising agricultural rents which made the commitment of financial resources to the improvement more acceptable to landowners. At the same time, investment in draining on heavy lands promised financial return to both landlords and tenants by means of higher yields, and increases in product prices, especially of cereals, in a period of prosperity encouraged greater expenditure on the improvement, as can be seen from the correlation between the use of loan capital and wheat prices. The downward movement of prices and rents from the late 1870s made outlay on the improvement less attractive to landlords and less rewarding for tenants, despite the technical capabilities of draining systems. The rapid growth in the provision of landlord capital for draining from around 1840 in response to technical proficiency and hence increased output, and the concentration of such outlay from 1840 to 1870 against a background of agricultural prosperity establish the improvement as essentially a product of the 'high farming' era.[89]

5

Capital provision and the management of the improvement

Within the overall spatial and temporal pattern of draining, the decision to adopt the improvement was individual to each estate. The undertaking of agricultural improvement in the nineteenth century has been widely recognized as a joint enterprise between landlords and tenants, although the precise relationship for particular improvements remains unclear.[1] The adoption of draining and the intensity of its use on specific estates were largely a function of the amount of capital that was provided. However, the availability and use of capital depended on the financial resources of estates and on the respective attitudes of landlords and tenants to the agricultural value of the improvement. The roles of landlords and tenants in the capital provision and management of draining are therefore crucial factors in understanding the spread of the improvement.

The development of landlord investment in draining

The landlord represented the major source of capital for agricultural improvement in the century. The technical advances in draining in the 1840s convinced many contemporary agriculturalists that, because of the improvement's permanence, cost and potential for providing a financial return through increased productivity, it should be both financed and undertaken by landlords.[2] The sample estates reveal that individual arrangements between landowners and tenants varied considerably, resulting in differing levels of draining activity.

Landlord responsibility for the improvement was least developed in Northamptonshire, particularly on the largest properties. On the Buccleuch estates the capital provision and management of draining did not come completely under landlord control until the early 1880s. When the improvement was first recorded as an item of estate expenditure in 1804,

167

allowances amounting to half the cost were given to tenants. From 1822 a second system of financing was introduced, that of providing the cost of draining materials, leaving labour to be found by the tenants. Both systems were employed until the early 1880s (Table 5.1). The full expense of the improvement was met only occasionally by the estate before 1880 and occurred in exceptional circumstances such as at a change in tenancy or when a tenant was in difficulty. Partial financing of the improvement restricted estate control over the methods of draining adopted. For draining covered by allowances, an estate drainer, J. West, superintended work up to 1839 but thereafter no evidence of estate supervision can be detected. Where only materials were provided, the method of draining would seem to have been in the hands of the tenants. These approaches to draining were abandoned in 1880–1, from which time the estate supplied the total cost of materials and labour, and undertook the improvement, processes intensified by the use of a loan from the General Land Drainage Company in 1888.[3]

Similar systems characterized the Ellesmere, Fitzwilliam, Overstone and Spencer estates in the country (Table 5.1). John Beasley, agent to the last two properties from 1832 and 1826 respectively to 1873, actively restrained landlord investment in draining. He considered that 'a landlord may put an estate into perfect order but it is the tenants only who can keep it in order', and insisted on the use of tenant resources in draining.[4] On the Spencer estate, the practice of draining allowances was abandoned in 1831 and replaced by the provision of draining materials to tenants, which was also employed on the Overstone estate from 1832. Despite limited landlord outlay, Beasley attempted to manage the extent and method of draining. Tenants applied to him for materials, which he distributed where he thought necessary, and he recommended drain depths and materials for different types of draining.[5] Nevertheless, the estate could not be sure of the draining systems used by tenants when digging in the materials. Both estates accepted full responsibility for the cost, layout and implementation of the improvement only from the early 1880s.[6]

Draining expenditure was first recorded on the Ellesmere estate in 1840 and took the form of providing draining materials. By 1874, when accounts cease, 90 per cent of landlord outlay on the improvement had been spent on materials. Although the agent, James Loch, was responsible for introducing the improvement and insisted that the materials should be laid by tenants under the guidance of the sub-agent, neither the capital provision nor the management of the improvement was completely in landlord control.[7] Again, on the Fitzwilliam estates, despite an expansion in the supply of landlord capital in the 1860s, largely the result of a loan of £760 from the Lands Improvement Company in 1866 for the Milton property, reliance was firmly placed on the provision of draining materials throughout the second half of the nineteenth century.[8]

Table 5.1 *Forms of draining capital on the Buccleuch, Fitzwilliam and Spencer estates, Northamptonshire, 1800–1899*

Decade	Landlord outlay (£)	Percentage on full cost	allowances	materials
Buccleuch estate				
1800–9	15	0	100	0
1810–19	2,841	20	80	0
1820–9	3,194	3	96	1
1830–9	2,524	0	97	3
1840–9	4,708	0	92	8
1850–9	5,029	25	66	9
1860–9	3,959	21	75	4
1870–9	3,289	43	35	22
1880–9	9,740	90	0	10
1890–9	5,988	98	0	2
Fitzwilliam estate				
1800–9	0			
1810–19	0			
1820–9	0			
1830–9	426	22	16	62
1840–9	860	34	6	60
1850–9	51	0	0	100
1860–9	1,480	70	0	30
1870–9	188	0	0	100
1880–9	1,464	14	0	86
1890–9	167	31	0	69
Spencer estate				
1800–9	169	0	100	0
1810–19	115	0	85	15
1820–9	1,809	0	93	7
1830–9	495	0	65	35
1840–9	722	2	5	93
1850–9	1,212	0	0	100
1860–9	1,005	18	0	82
1870–9	1,892	27	0	73
1880–9	2,206	100	0	0
1890–9	744	100	0	0

Sources: Buccleuch MSS: Barnwell and Boughton estate accounts, 1800–99; Spencer MSS: Estate acounts, 1800–26; Estate accounts and rentals, 1827–99; NRO, Fitzwilliam Higham Ferrers and Milton estate accounts, 1800–99

Of those Northamptonshire estates with runs of draining data, only those of the dukes of Bedford and Cleveland and of the Cartwright family displayed a more complete acceptance of landlord responsibility for the improvement. Before 1850 on the Bedford property, allowances had been made to tenants for draining they had undertaken with no estate supervision, resulting in, as Tycho Wing, the agent, described, 'a considerable want of method and economy in their manner of doing the work'.[9] From the middle of the century a draining superintendent was employed to carry out the improvement at the cost of the estate.[10] Full landlord control of draining on the Cartwright and Cleveland estates dated respectively from 1853 and 1848. Before, allowances and the supply of draining materials had been used to undertake the improvement, with no management of the systems adopted by the tenantry. On both estates the development of landlord responsibility was accompanied by a loan under the Public Money Draining Acts.[11]

Lordlord involvement in draining was more generally accepted on the sample estates in Devon. On the Fortescue estates, the provision of capital dated from 1830 and was maintained throughout the century. No mention can be found in the accounts of tenant financial contributions, while in the tenancy agreements that have survived no clause was included for tenants to take part in the draining of their holdings.[12] Such complete landlord control characterized draining on the duchy of Cornwall estates in the second half of the century and the estates of the dukes of Bedford from 1843, a mere £189 being distributed to tenants on the latter between 1843 and 1859 as draining allowances.[13] Although allowances were provided on the Sidmouth estate before 1850 to tenants who had drained their land to the satisfaction of the bailiff, from 1851 both the layout and the supply of capital of the improvement became landlord charges, again associated with the use of a Public Money Draining Acts loan.[14] Only on the estate of the earls of Devon was the landlord role in the improvement less clearly defined after 1850. Although the estate was prepared to meet the full cost of draining, in many cases only part of the capital was furnished, leaving the tenant with a major contribution to the improvement (Table 5.2).

The development of landlord control of the capital provision and management of the improvement occurred much earlier on most of the sample estates in Northumberland, usually in the first half of the nineteenth century. On the Blackett estate the financing of draining was a landlord charge from the first recording of the improvement in 1802, while on the Portland estate it became such from 1829.[15] Draining expenditure on the Carlisle estate, first noted in the accounts in 1815, initially took the form of allowances to tenants, a practice maintained until 1836. However, the provision of the full cost of the improvement had been introduced in 1833, and from 1837 the financing of draining became solely a landlord concern.[16] Although outlay had been made on land in hand on the Grey

Table 5.2 *Forms of draining capital on the Devon estate, Devon 1848–1899*

Decade	Landlord outlay (£)	Percentage on		
		full cost	allowances	materials
1848–9	500	71	27	2
1850–9	1,989	61	36	3
1860–9	754	0	34	66
1870–9	2,240	61	2	37
1880–9	1,147	53	16	31
1890–9	809	92	0	8

Source: DRO, Courtenay MSS: 1508M/14/B/111/2–39

estate from 1806, capital to cover the cost of materials and labour of draining tenanted land was made available from 1840 onwards. Large-scale landlord capital supply for the improvement on the estate of the dukes of Northumberland dated from 1844.

Furnishing all capital, the estates exercised considerable control over the draining systems applied to the land, there being little tenant involvement. As G. A. Grey, agent to the Carlisle estate, noted, 'the tenants are not asked or wished to look after the drainage – or to have any voice in the matter – if they were, every farm would be drained on a different principle according to the whim or ignorance of the tenant instead of the method most approved by persons of large and varied experience'.[17] The role of the landlords and agents in establishing new draining systems on the Blackett, Carlisle, Grey and Portland estates has already been discussed. As each estate appointed draining superintendents and workmen, there was assurance that these systems were implemented.

Such landlord management of the improvement may be seen at its most thorough on the estate of the dukes of Northumberland. Before 1844, stone draining had dominated on the estate for the purpose of controlling springs. A new system was introduced in 1844 by the third duke based on the use of tiles and thorough draining. A circular letter was sent to tenants in July of that year explaining the method of organization of the improvement. The layout, cutting and filling of drains were to be done by the estate drainer, John Loraine, and his workmen, with tiles being used throughout. Tenant participation was restricted: if a tenant wished to drain his land, it had to be carried out under the supervision of Loraine, with the tenant first having obtained permission from the duke's commissioner.[18] By 1844 estate control of the undertaking of the improvement was absolute.

Although there was a change to deep draining with the succession of the fourth duke in 1847, estate management was maintained. Gangs of drainers under a foreman supervised by an estate drainer were appointed to execute the work over the estate. In 1849 seven draining gangs employing 96 men were active on the bailiwicks of Alnwick, Longhoughton, Rothbury, Shilbottle and Warkworth, while in 1866 there were still 100 drainers on the estate.[19] Not only was the estate intent on ensuring that draining was properly carried out; it was also concerned with the efficiency of the improvement. Reports were made by bailiffs of bailiwicks of the state of draining: in June 1847, Mr Tate commented on the draining on Rothbury bailiwick, describing how effectively the land had been rendered dry.[20] Once the draining system had been established, the estate resolved to maintain it in functioning form. When draining failure began to be reported, it took responsibility for the repairing and cleaning of drains and, if needed, the redraining of land for the rest of the century.[21]

While draining has been regarded as part of the fixed equipment of farming provided by landlords, the sample estates indicate that there was no uniform acceptance of that responsibility. A range of approaches to financing the improvement was evident not only between counties but between estates in a single county. The extent of such variation suggests that draining fails to conform to the general relationship between landlord and tenant capital in the undertaking of permanent agricultural improvement in the nineteenth century described by D. Spring and F. M. L. Thompson.[22] Thus, on most of the estates in the Northamptonshire sample including many of the largest, draining remained an improvement jointly financed by landlords and tenants for the greater part of the century, full landlord control not emerging until after 1880. Draining did not evolve as a landlord-financed improvement on the remaining estates in Northamptonshire and on the majority of those in Devon until the 1840s and 1850s, in the wake of technical developments and often at the instigation of a loan. Only the Northumberland sample estates approximated the relationship defined by Spring and Thompson, the responsibility for draining outlay being clarified by the 1830s with the improvement becoming a landlord charge.

The existence of full landlord control, whilst allowing estate direction of the method of draining, facilitated expansion in investment in the improvement. On all the sample estates for which data runs are available, the periods of major outlay on draining occur when the improvement had become a landlord charge. The earlier that draining was accepted as part of the fixed capital of agriculture provided by landlords, the greater and more effective could be expenditure on the improvement. Thus, the relatively low per acre outlay on draining in Northamptonshire in the 1840s and 1850s was partly a product of four estates included in the sample, the Buccleuch, Ellesmere, Fitzwilliam Higham Ferrers and Spencer properties, furnishing but a proportion of the improvement's capital. The increase in investment in the county in the 1880s, whilst a reaction to falling rents and prices, was also a function of the adoption of draining as a landlord improvement by the Buccleuch and Spencer estates (Fig. 4.2). The precocious development of the improvement as a landlord responsibility on the sample estates in Northumberland must have contributed significantly to the intensity of draining recorded in that county.

The provision of landlord capital and estate size

Even with the recognition of draining as a landlord improvement, the intensity of adoption was prescribed by the funds that individual estates were prepared and able to allocate. Capital provision involved a degree of planning and, where evidence is available among the sample estates, the landlord determined with advice from his agent the sums to be allotted to draining over a given period. In Northumberland, with the introduction of a new draining programme in 1844, the third duke of Northumberland authorized £6,000 to be spent annually on the improvement, £500 per bailiwick.[23] The fourth duke enlarged the sum to £10,000 a year in 1848 to aid the adoption of his deep-draining system.[24] By 1857, when the 1855 survey of draining on the estate by the chief commissioner Hugh Taylor had shown that much of the land in need of the improvement had been treated, the amount was reduced.[25] With the succession of the fifth duke, annual allocation was raised once more to £7,500 for a period of six years to deal with the additional land requiring draining identified in 1866 by the new chief commissioner, J. Snowball.[26] From 1872, it was anticipated that draining expenditure should not exceed £3,000 a year.[27] A similarly regulated system was evident on the duke of Portland's neighbouring estate. After the adoption of thorough draining in 1834, the fourth duke agreed from 1836 to allow £500 a year for the improvement, instructing the agent, William Sample, to 'take care to disperse it generally over the estate'.[28] Sample found difficulty in keeping to that sum and in 1849 the duke increased the allocation to £1,000.[29] In spite of changes in ownership, this amount was made available until 1885, when it was reduced to £700.[30] On the Grey and Carlisle estates in Northumberland, assessments were made by their agents in the 1840s and 1860s respectively of land in need of draining so that landlord expenditure on the improvement could be ordered, a policy also applied by earl Fortescue in 1847 to his north Devon estate. Even on small estates such planning was evident and, to deal with the wet land on the Sidmouth property in Devon, F. Thynne suggested a programme for undertaking a certain amount each year.[31]

Decisions had also to be made on the manner of financing draining expenditure. Estates could obtain capital for the improvement either from their own resources or by means of a draining loan. The two methods were not mutually exclusive and both were employed on a number of the sample estates in varying degree (Tables 3.10, 3.13 and 3.16). On some of the larger estates, such as those of the dukes of Bedford and Northumberland and the earls Spencer, the improvement was financed from current income, Christopher Haedy, auditor to the duke of Bedford, noting in 1858 that when such revenue surpassed expenditure 'the preferable mode of dealing with the excess of income so long as a necessity exists for outlay [is] the permanent improvement of the estate'.[32] However, even on larger properties, loan capital formed part of financing draining programmes. Thus, on the Portland estate in Northumberland, the fifth duke on his succession in 1854 suggested the abandonment of the existing system of financing the improvement, replacing it by a loan to achieve the complete and immediate draining of the land. The agent, William Sample, argued against such a policy, reporting that he lacked the resources to undertake such a large-scale operation and preferring that the work should proceed gradually over a 20-year period.[33] Although Sample's view prevailed, a loan of £4,168 was eventually made by the Land Loan Company in 1883 to finance draining but, having borrowed that amount, the estate returned to providing its own funds.[34] Greater reliance on loan capital was evident on the Blackett estate. Between 1802 and 1849 estate revenues provided £10,141 for draining. Sir Edward Blackett considered that such outlay was becoming excessive and in 1854 he wrote to his agent, William Sample, 'that unless the expenses on the estate are reduced and also the expenses of draining per acre it will be quite impossible for me to drain any more land at present'.[35] To continue draining but to limit use of estate income, Blackett and Sample turned to the sources of loan capital. In 1856, Sample was inquiring about the rates of repayment of Lands Improvement Company loans. As their rates were higher than the uniform $6\frac{1}{2}$ per cent for 22 years under the Public Money Draining Acts, £5,000 was borrowed from the government fund.[36] The sum did not complete the draining necessary on the estate. As landowners were restricted to £5,000 under the Public Money Draining Acts, Sample approached the improvement companies, borrowing £6,374 in 1862 from the Lands Improvement Company, the body recommended by G. A. Grey, agent to the nearby Carlisle estate, who had acted as inspector for the Inclosure Commissioners for Blackett's first loan.[37] A further loan of £6,253 was obtained in 1867, this time from the Land Loan Company, from whom £2,009 was also borrowed in 1885.[38] From the introduction of thorough draining on the estate in 1840, loans financed 82 per cent of expenditure and paid for all new draining carried out after 1856.

The amounts that could be set aside for draining from current income were much less on smaller estates. For these properties, both independent units and outliers of larger estates, draining loans may have appeared an attractive source of capital with which to undertake the improvement. Thus, the estate of the Baker-Baker family in Northumberland covered 2,067 acres in 1847, most of which needed draining. It contained eight farms, which yielded a rent of £1,574 in 1872. Between 1851 and 1854, the sum of £4,079 was borrowed under the Public Money Draining Acts, accomplishing the draining of 43 per cent of the estate.[39] The capital for such a scale of improvement could not have been produced from estate revenue. Similarly a loan of £4,903, also under the Public Money Draining Acts, allowed Sir Henry Dryden to drain about 44 per cent of his Northamptonshire estate of 2,530 acres between 1858 and 1866, a level of capital investment and draining activity that would have been difficult to provide from rents, which were no more than £4,427 in 1844.[40] Nevertheless, a number of small estates in the sample were prepared to finance the improvement from current income, as on the Bedford and Ellesmere estates in Northamptonshire, although they tended to be outliers of larger properties.

Loans, by providing immediate capital, were of value to estates of all sizes in remedying long neglect or indifference to the improvement. On the small, badly managed Sidmouth estate in Devon, Thynne in his report in 1850 recommended heavy outlay on agricultural improvement to prevent 'loss of rent and neglect of land'. Yet he recognized that the task would cause 'the absorption of large sums of money totally unreasonable to expect that any owner would or could engage in' on an estate yielding a rent of £3,179.[41] To rectify the need for draining on the property, £3,762 was borrowed in 1851 under the Public Money Draining Acts and £8,660 in 1863 from the General Land Drainage Company.[42] Between 1851 and 1866, this body of capital allowed the draining of most of the land in need of the improvement on the estate, representing, if estate revenue had been used, 22 per cent of the gross rent due over the period. Again, the lack of interest shown by Sir Charles Monck from 1846 to his death in 1867 in draining his Northumberland estate forced his successor, Sir Arthur Middleton, to invest heavily in the improvement. From 1869 to 1884, the sum of £39,993 was spent on draining at £4.41 per acre, all financed by loans under the Public Money Draining Acts and from the Lands Improvement Company.[43] Similarly, the rapid increase in draining outlay on the Buccleuch estates in Northamptonshire from 1888 to 1894, in a period of rent fall, was made possible by a loan of £13,450 from the General Land Drainage Company[44] (Fig. 4.9b and c). However, the provision of such rescue capital was found on estates where loans were not undertaken, characterized, for example, by the draining activity on the duke of Bedford's north and mid-Devon properties, where the existence of life leases had allowed the farms to fall into serious neglect.[45] Overall, although the decision to employ loan capital was taken by individuals in the light of the financial position of different estates, the manner in which it was applied to draining seems to have varied little from the expenditure of private funds.

Table 5.3 *Land needing draining on 99 estates inspected by Andrew Thompson, 1857–1868*

Estate size (acres)	No. of estates	Total acreage of estates	Acreage needing draining	Percentage of estates in need of draining
under 1,000	54	13,187	8,581	65
1,000–2,999	23	41,940	21,440	51
3,000–9,999	15	86,330	39,700	46
10,000 and over	7	112,324	24,550	22
Total	99	253,781	94,271	37

Sources: KU, Sneyd MSS: Reports of Andrew Thompson, vols. 4–8, 1857–68

While regulation of capital was important in controlling the spread of draining on individual properties, the amounts invested by landlords on the improvement were ultimately dependent on the financial resources of their estates. Contemporaries accepted estate size as a reasonable measure of landlord capacity to finance agricultural improvement but were uncertain of the degree, if any, of variation that existed between different sized estates in the provision of capital for such purposes.[46] However, the reports of Andrew Thompson to the Inclosure Commissioners on applications for draining loans suggest that estate size had a direct effect on draining activity. For 99 of the 133 estates for which he provided draining reports between 1857 and 1868 (Fig. 3.13), Thompson noted the area in need of the improvement. He regarded this as wet land, leaving aside the question of whether or not draining would produce a return on the outlay. Acreages described as requiring draining were estimates given in rounded form. Although lacking precision, the figures indicate that, with decreasing estate size, the proportion of land to be drained increased, pointing to a higher degree of neglect on smaller properties, where there was less capital available to spend on agricultural improvement in general (Table 5.3).

The sample properties in the three counties are both too few in number and too heavily orientated towards large holdings to permit a satisfactory discussion of the relationship between estate size and draining outlay. However, the draining-loan data provide an opportunity to analyse the relationship more thoroughly throughout the country. The registers of certificates of loans record individual landowners and their estates and therefore allow examination of draining-loan capital by estate size. As the use of loan capital seems to have varied little from private funds, the results should broadly reflect general trends. Estates that borrowed for draining have been allocated on a county basis to the size categories developed by F. M. L. Thompson from John Bateman's *The Great Landowners of Great Britain and Ireland*.[47] Individual estate size has been obtained by reference to the 1873 'Return of owners of land in England and Wales' and to Bateman, a process aided by the frequent recording of estate area on the loan certificate. Although the expression of draining-loan expenditure over the period 1847–99 in terms of estate acreages identified around 1875 involves a degree of artificiality, no account being taken of changes in estate area, the data in the 1873 return and in Bateman's study represent the only comprehensive source on land ownership in the second half of the nineteenth century.

Between 1847 and 1899, draining loans under the land-improvement legislation were made on 2,703 estates. Of this total, some 118 estates cannot be traced, involving 3.8 per cent of all draining-loan capital. The remaining 2,585 estates covered 19.8 per cent of the area of England held by owners with one acre or more of land. The smallest estates, those sized between 1 and 999 acres, formed the largest group, 61.7 per cent of the total falling into this category. Increasing estate size was accompanied by a decline in numbers so that the largest, estates over 10,000 acres, represented only 4.6 per cent of the total (Table 5.4). Although not evident in all counties, the absolute amount borrowed by landowners for draining over the whole period grew with estate size, the average loan capital for estates over 10,000 acres in the country being 15 times greater than that for estates under 1,000 acres. Such a tendency is not surprising, being no more than a function of estate area. However, the intensity of draining expenditure with loan capital was consistently higher on the smaller properties. For those estates under 1,000 acres that borrowed, the loan capital produced an average draining outlay of £2.23 per acre, a rate that fell progressively to £0.53 per acre for estates in excess of 10,000 acres (Table 5.5.). Such a trend would suggest that reliance on loan capital for draining mounted as estate size and therefore financial resources declined.

Yet the value of draining loans was little appreciated by smaller estates, for only a slight proportion adopted them. Leaving aside properties under 100 acres, which would have included many units that would have been difficult to class as agricultural, only 4.4 per cent of owners of estates between 100 and 999 acres made use of draining loans. With successive estate size, the proportion of landowners employing loan capital increased, rising to over 44 per cent of those estates over 10,000 acres (Table 5.6). In addition to their general neglect of such capital, those small estates that made use of loans were slower to do so than larger properties. The rate of adoption of draining loans in the country as a whole increased with estate size, being most readily accepted on estates over 10,000 acres (Fig. 5.1). Of the 124 estates of such size

Table 5.4 Number of landowners using draining-loan capital and average amount of loan by estate size, 1847–1899

County	Under 1,000 acres		1,000–2,999 acres		3,000–9,999 acres		10,000 acres and over		Untraceable	
	no. of owners	average loan (£)	no. of owners	average loan (£)	no. of owners	average loan (£)	no. of owners	average loan (£)	no. of owners	average loan (£)
Bedfordshire	14	1,929	8	2,829	4	5,847	0		1	295
Berkshire	11	458	6	850	4	2,994	1	3,024	3	1,233
Buckinghamshire	30	460	8	2,436	3	12,106	0		2	1,390
Cambridgeshire	18	601	6	1,419	4	4,359	0		1	487
Cheshire	21	758	14	2,893	10	12,682	4	11,774	2	446
Cornwall	6	193	1	117	2	1,364	0		0	
Cumberland	167	361	12	1,131	7	5,314	4	5,578	6	529
Derbyshire	16	816	5	2,988	4	6,915	0		3	622
Devon	33	278	14	599	8	3,587	7	7,161	2	1,812
Dorset	29	739	16	2,009	6	3,765	4	13,193	2	1,603
Durham	77	907	16	2,758	10	5,802	4	20,600	5	1,141
Essex	21	707	7	2,926	6	2,417	1	771	5	663
Gloucestershire	74	574	20	2,275	9	3,760	2	9,441	8	1,169
Hampshire	10	480	11	2,890	6	6,124	3	6,293	1	131
Herefordshire	34	901	14	4,190	11	6,047	1	15,065	4	1,535
Hertfordshire	16	767	4	2,186	5	2,937	1	1,482	0	
Huntingdonshire	34	835	6	2,731	6	6,192	1	11,961	2	3,056
Kent	19	775	12	2,288	10	4,211	1	5,811	1	1,219

County										
Lancashire	31	574	7	3,898	15	5,194	4	13,185	3	5,213
Leicestershire	89	696	15	3,333	0		1	8,290	2	3,773
Lincolnshire	111	611	16	1,022	12	3,065	5	4,426	7	661
Middlesex	16	855	4	3,829	0		0		0	
Norfolk	5	190	4	1,068	3	1,008	3	1,887	0	
Northamptonshire	67	584	9	2,174	11	4,218	2	9,078	1	817
Northumberland	124	818	42	3,645	36	6,909	19	12,958	11	2,010
Nottinghamshire	40	444	8	2,369	6	2,437	5	6,498	1	242
Oxfordshire	30	547	10	2,387	4	6,781	1	14,921	3	2,958
Rutland	5	376	0		2	5,089	0		0	
Shropshire	31	846	27	2,761	25	6,356	7	6,465	5	567
Somerset	49	432	8	2,765	10	3,906	3	9,889	1	6,563
Staffordshire	34	639	14	2,273	8	8,450	4	8,818	0	
Suffolk	4	921	2	2,482	2	8,050	0		0	
Surrey	20	1,089	9	1,460	5	4,591	0		2	2,130
Sussex	19	1,237	11	1,746	9	4,218	3	5,165	3	3,530
Warwickshire	53	525	12	2,300	8	6,159	3	11,493	2	2,726
Westmorland	17	378	3	189	2	551	2	14,416	3	1,186
Wiltshire	24	618	11	2,929	13	3,608	5	18,391	0	
Worcestershire	66	524	12	2,829	10	6,715	1	20,481	5	2,172
Yorkshire	203	531	47	2,786	46	6,239	22	8,184	21	2,496
Total	1,668	626	451	2,522	342	5,547	124	9,780	118	1,766

Sources: PRO, 1R3/6–38; MAF 66/1–6, 8, 9, 11, 13–22, 25–39, 43–7; BPP, 1874, LXXII, parts 1 and 2, 'Return of owners of land in England and Wales, 1873'; J. Bateman, The Great Landowners, 1883; F. M. L. Thompson, English Landed Society, 1963, 32, 113–17.

Table 5.5 *Per acre intensity of draining-loan capital by estate size, 1847–1899*

County	Under 1,000 acres		1,000–2,999 acres		3,000–9,999 acres		10,000 acres and over	
	estate acreage	loan outlay per acre (£)	estate acreage	loan outlay per acre (£)	estate acreage	loan outlay per acre (£)	estate acreage	loan outlay per acre (£)
Bedfordshire	9,068	2.98	13,112	1.73	24,461	0.96	0	
Berkshire	2,360	2.14	10,861	0.47	17,447	0.69	19,225	0.16
Buckinghamshire	7,514	1.83	15,164	1.29	19,560	1.86	0	
Cambridgeshire	3,675	2.95	11,472	0.74	19,729	0.88	0	
Cheshire	6,219	2.56	26,780	1.51	58,675	2.16	67,059	0.70
Cornwall	800	1.45	2,029	0.06	8,299	0.33	0	
Cumberland	38,702	1.56	18,840	0.72	33,965	1.10	102,330	0.22
Derbyshire	5,412	2.41	8,478	1.76	21,087	1.31	0	
Devon	7,382	1.24	23,151	0.36	41,935	0.68	155,253	0.32
Dorset	10,199	2.10	27,648	1.16	28,777	0.79	65,493	0.81
Durham	25,143	2.78	28,427	1.55	56,819	1.02	101,347	0.81
Essex	5,808	2.56	11,022	1.86	33,699	0.43	19,086	0.04
Gloucestershire	18,023	2.36	33,033	1.38	37,208	0.91	34,874	0.54
Hampshire	3,316	1.45	21,281	1.49	37,777	0.97	32,610	0.58
Herefordshire	12,353	2.48	25,491	2.30	58,219	1.14	10,559	1.43
Hertfordshire	4,548	2.70	8,155	1.07	25,056	0.59	10,100	0.15
Huntingdonshire	8,367	3.39	12,557	1.30	28,571	1.30	13,835	0.86
Kent	7,854	1.88	23,566	1.16	48,131	0.96	16,209	0.36

Lancashire	7,297	2.44	12,895	2.12	73,396	1.06	60,538	0.87
Leicestershire	24,841	2.49	24,769	2.02	0		30,188	0.27
Lincolnshire	32,947	2.06	27,697	0.59	64,391	0.57	69,467	0.32
Middlesex	5,883	2.33	6,095	2.51	0		0	
Norfolk	1,442	0.66	9,057	0.47	15,720	0.19	46,102	0.12
Northamptonshire	18,026	2.17	14,995	1.30	56,816	0.82	33,590	0.54
Northumberland	39,974	2.54	85,663	1.78	204,568	1.22	337,180	0.73
Nottinghamshire	10,472	1.69	15,221	1.24	30,954	0.47	123,585	0.26
Oxfordshire	6,613	2.48	16,342	1.46	24,432	1.11	21,944	0.68
Rutland	797	2.36	0		18,179	0.56	0	
Shropshire	12,761	2.05	53,267	1.40	128,141	1.24	132,967	0.34
Somerset	12,036	1.76	15,952	1.34	45,366	0.86	53,760	0.55
Staffordshire	11,544	1.88	23,586	1.35	47,066	1.44	63,972	0.55
Suffolk	1,617	2.28	3,629	1.37	14,401	1.12	0	
Surrey	6,306	3.45	12,997	1.01	30,632	0.75	0	
Sussex	6,932	3.39	19,779	0.97	51,788	0.73	43,558	0.36
Warwickshire	14,429	1.93	22,535	1.22	41,180	1.20	39,965	0.86
Westmorland	5,391	1.19	5,307	0.11	13,439	0.08	52,080	0.55
Wiltshire	7,249	2.05	20,588	1.57	67,329	0.70	118,276	0.78
Worcestershire	17,068	2.03	22,345	1.52	50,771	1.32	14,698	1.39
Yorkshire	46,998	2.29	88,170	1.48	258,744	1.11	406,737	0.44
Total	467,336	2.23	821,956	1.38	1,836,728	1.03	2,296,587	0.53

Source: As for Table 5.4

Table 5.6 *Proportion of landowners by estate size using draining-loan capital, 1847–1899*

County	1–99 acres		100–999 acres		1,000–2,999 acres		3,000–9,999 acres		10,000 acres and over	
	total no. of owners	percentage using loans	total no. of owners	percentage using loans	total no. of owners	percentage using loans	total no. of owners	percentage using loans	total no. of owners	percentage using loans
Bedfordshire	1,825	0.3	269	3.3	27	29.6	10	40.0	2	0
Berkshire	2,315	0.1	397	2.0	40	15.0	17	23.5	3	33.3
Buckinghamshire	2,672	0.3	489	4.5	29	27.6	23	13.0	2	0
Cambridgeshire	5,373	0.1	721	1.7	39	15.4	9	44.4	2	0
Cheshire	5,296	0.1	431	3.7	39	35.9	32	31.3	7	57.1
Cornwall	4,028	0	923	0.7	48	2.1	36	5.6	8	0
Cumberland	4,497	1.1	1,185	9.9	51	23.6	13	53.8	6	66.7
Derbyshire	6,017	0.1	528	2.3	38	13.2	25	4.0	3	0
Devon	7,509	0.1	2,053	1.3	108	13.0	47	17.0	13	53.8
Dorset	2,794	0.1	360	7.2	59	27.1	22	27.3	10	40.0
Durham	2,376	0.2	492	14.8	33	48.5	15	66.7	5	80.0
Essex	5,476	0.1	1,340	1.2	87	8.0	38	15.8	1	100.0
Gloucestershire	7,107	0.3	848	6.5	55	36.4	37	24.3	4	50.0
Hampshire	5,102	0.1	733	1.2	78	14.1	51	11.8	9	33.3
Herefordshire	3,781	0.1	625	5.1	49	28.6	26	42.3	2	50.0
Hertfordshire	2,184	0.1	375	4.0	39	10.3	12	41.7	3	33.3
Huntingdonshire	1,612	0.3	243	11.9	8	75.0	11	54.5	2	50.0
Kent	6,062	0.1	1,144	1.3	75	16.0	43	23.3	3	33.3

County										
Lancashire	10,845	0.1	949	2.1	79	8.9	38	39.4	8	50.0
Leicestershire	3,823	0.3	651	11.7	38	39.5	20	0	2	50.0
Lincolnshire	14,118	0.1	1,702	5.3	91	17.6	50	24.0	18	27.8
Middlesex	2,433	0.1	209	7.2	5	80.0	1	0	0	0
Norfolk	7,936	0.1	1,165	0.2	113	3.5	5	60.0	11	27.3
Northamptonshire	3,287	0.4	600	9.2	31	29.0	26	42.3	6	33.3
Northumberland	1,531	1.4	470	21.9	84	50.0	44	81.8	24	79.1
Nottinghamshire	3,838	0.1	391	8.9	25	32.0	18	33.3	7	71.4
Oxfordshire	2,493	0.3	468	4.7	40	25.0	20	20.0	1	100.0
Rutland	458	0.7	42	4.8	5	0	2	100.0	2	0
Shropshire	3,841	0.1	669	4.0	65	41.5	50	50.0	8	87.5
Somerset	10,831	0.1	1,152	3.3	67	11.9	36	27.8	10	30.0
Staffordshire	8,617	0.1	551	4.4	37	37.8	24	33.3	8	50.0
Suffolk	4,965	0	1,095	0.4	65	3.1	29	6.9	11	0
Surrey	3,813	0.1	492	3.3	41	21.9	15	33.3	2	0
Sussex	3,915	0.1	817	1.8	86	12.8	37	24.3	11	27.3
Warwickshire	3,519	0.4	650	6.2	32	37.5	22	36.4	4	75.0
Westmorland	2,055	0.1	461	3.5	20	15.0	8	25.0	3	66.7
Wiltshire	3,485	0.3	544	2.8	61	18.0	38	34.2	11	45.4
Worcestershire	4,803	0.4	605	8.1	25	48.0	18	55.6	3	33.3
Yorkshire	23,328	0.2	3,002	5.2	225	20.9	125	36.8	44	50.0
Total	199,960	0.2	29,841	4.4	2,136	21.1	1,148	29.8	279	44.4

Source: As for Table 5.4

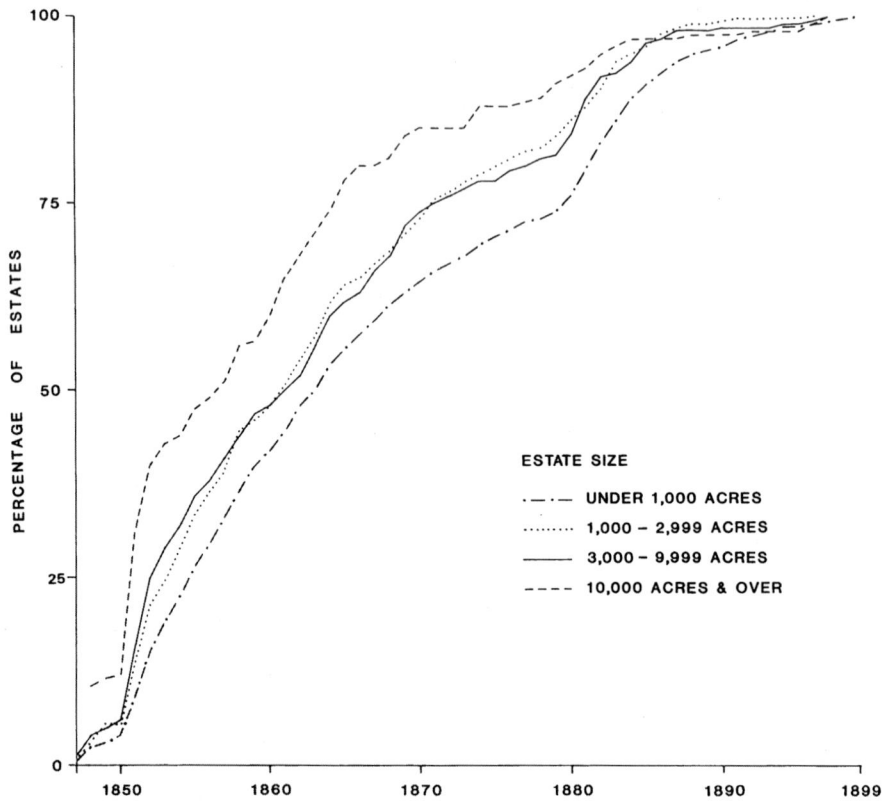

Figure 5.1 Adoption curves of draining loans in England by estate size, 1847–1899
(*Source:* As for Table 5.4)

involved, draining loans had been contracted on 50 per cent by 1857 and on
80 per cent by 1866. Only a small proportion of estates over 10,000 acres
adopted loan capital after 1875. For those estates under 1,000 acres, the 50
and 80 per cent levels were not attained until 1863 and 1881 respectively.
Unlike the largest estate category, the adoption curves reveal that a
significant proportion of estates under 1,000 acres were first employing loan
capital in the late 1870s and 1880s, when agricultural prices and rents were
beginning to fall. Late adoption in a period of agricultural difficulty would
indicate earlier indifference to the improvement on such estates.

   Although not uniform in each county, the draining-loan data demonstrate
a clear relationship between estate size and draining activity. The
improvement was adopted most speedily and most extensively on estates
over 10,000 acres. Little difference existed on estates in the categories 1,000
to 2,999 acres and 3,000 to 9,999 acres in the rate and intensity of draining

adoption, both being less advanced than the group of largest properties. Estates in the smallest size category, under 1,000 acres, formed the laggard group in the adoption process, in general failing to implement the improvement. However, each estate group occupied a considerable proportion of the total area of the country. Around 1875, estates over 10,000 acres covered 24 per cent of the total area of England excluding waste, those between 1,000 and 9,999 acres 29.4 per cent and those from 100 to 999 acres 26.5 per cent.[48] The varying adoption of the improvement on different sized estates may therefore be seen as having a major impact on the intensity of draining in the country in the second half of the nineteenth century. Given the physical need for draining, the loan data would suggest that estate structure played a significant part in determining the pattern of draining activity, the improvement being more likely to be adopted in areas dominated by large estates than in those where small properties proliferated.

Landlord return on investment in draining

As draining absorbed estate capital, the financial return that the improvement yielded was of concern to all landowners. Contemporary views of the return on draining in the form of increased rent varied over the second half of the nineteenth century. The security of an ample return from draining investment was widely accepted in the 1840s and 1850s, often reckoned at about 15 per cent on outlay.[49] Such returns were still being claimed for the improvement in 1863 by George Darby, one of the Inclosure Commissioners.[50] However, by 1873 the evidence given to the Select Committee of the House of Lords on the improvement of land revealed that opinions had begun to diverge on the value of draining as an investment. A small number of witnesses maintained that landlords could obtain an adequate return for the improvement, but the greater body of evidence did not endorse this view. As the agent, H. W. Keary, noted, 'we have learned that land is not altogether so expansive as some people thought it was'. In the report, draining was not regarded as a secure investment because it could break down. To compensate for this risk, landowners had to be assured of sufficient interest. In much of the evidence, draining was recorded as yielding a return of between 5 and 7 per cent on outlay expressed as increased rent. As such, the opinion was voiced in the report that by 1873 the improvement as an investment was not sufficiently lucrative to attract much capital, therefore resulting in a decline of the amount of draining undertaken.[51] Given the diversity of contemporary estimates, the amount and security of return on draining outlay warrant closer inspection, especially as they were regarded as factors of importance in the adoption of the improvement.

An indication of the return that landlords anticipated and were prepared to accept on capital outlay on draining may be derived from the sample estates. An initial distinction has to be made between partial and complete funding of the improvement by landlords. In the case of the former, exemplified by the Buccleuch, Ellesmere and Spencer estates in Northampton-shire, no interest was charged to tenants on draining allowances or the provision of draining materials. As such expenditure required a tenant capital input that was either equal to or greater than that of the landlord, the lack of direct return on such investment is not surprising.[52]

With the development of full estate responsibility for the provision of capital, landlords expected some direct return in the form of an interest payment on the outlay from the tenantry, who, as F. Thynne noted in drawing up tenancy agreements on the Sidmouth estate in Devon, were to receive the benefit of draining.[53] The improvement was undertaken only when a tenant agreed to the payment of interest.[54] Thus, on the Blackett and Portland estates in Northumberland, tenants paid interest on draining outlay from the introduction of the improvement in 1802 and 1829 respectively, while the promise of a tenant return formed an essential part of the draining programme of the third duke of Northumberland in 1844.[55] Again, the reports of Andrew Thompson indicate that on most of the estates he examined the willingness of tenants to pay interest was the main variable in determining the amount of draining carried out by landlords: he recorded that on Lord Bagot's Staffordshire estate 'preference will be given to the occupiers who are most eager to have their land drained and are willing to pay a fair percentage on the outlay'.[56]

Where difficulty existed in obtaining such interest payment, landlord capital was not made available, as may be demonstrated in an extreme form on the Bradninch estate of the duchy of Cornwall in Devon. Landlord draining began on the estate in 1854. Before that date, the estate was held on life leases, the holders of which paid a nominal rent and sublet the property. Under such circumstances, the estate put little effort into improvement as no return was possible. From 1854 the life leases started to expire. As these were replaced by annual tenancies or leases for fixed terms, landlord control of the estate increased and capital was provided for draining if new tenants agreed to meet the interest on the outlay (Table 5.7). Again, the later occurrence of the main period of capital supply for draining on the mid-Devon property of the dukes of Bedford in comparison with their Tavistock and north Devon estates was a product of a similar tenure pattern, outlay expanding only as the life leases began to fall in on a significant scale from 1857 (Fig. 4.7b, c and d).[57]

However, the rates of interest charged to tenants on landlord investment in draining were well below those advocated in the 1840s and 1850s. Although varying between estates, all fell within the range 4 to 7 per cent, rates between 5 and $6\frac{1}{2}$ per cent being most common. Most estates maintained a uniform charge with the acceptance of full financial responsibility for the

Table 5.7 *The incidence of landlord draining and tenure on the duchy of Cornwall's Bradninch estate, Devon, 1854–1899*

Farm	Date of draining	Acreage drained	Tenure		Interest charged
Park	1854	9	Life lease expired 1852:	10-year lease begun	5
Caseberry	1856	39	Life lease expired 1855:	31-year lease begun	5
Billingsmoor	1856	28	–		6
Wishay	1863	14	Life lease expired 1860:	annual tenancy, 1861	5
Netherstonhaies	1864	20	Life lease expired 1864:	21-year lease begun	6
Waterstave	1865	4	Life lease expired 1863:	annual tenancy, 1864	6
Chapelhaies	1869	32	Life lease surrendered, 1868		6
Park	1872	5	Lease expired 1864:	annual tenancy, 1864	6
Burnhaies	1881	18	Life lease expired 1880:	annual tenancy, 1880	0
Fordishaies	1882	10	Life lease expired 1882:	annual tenancy, 1882	0
Park	1884	32	Annual tenancy replaced by 26-year lease, 1878		0
Bowhill	1887	17	Life lease expired 1887:	annual tenancy, 1887	0
Coombe	1897	23	New tenant, 1897		0
Stokehouse	1898	3	New tenant, 1898		0

– : No data available

Sources: Cornwall MSS: Farm bundles for the manor of Bradninch; Inrolment book of patents and warrants, 1862–1902

improvement. The interest rate for draining on the Portland estate was established at 4 per cent in 1829 and was kept at that level; on the estates of the dukes of Northumberland and of the earls of Carlisle, the charge being recorded from 1844 and 1847 respectively was 5 per cent; and 6 per cent was levied from tenants on the Grey estate from the introduction of thorough draining in 1840.[58] The Blackett estate in Northumberland was one of the few in the sample that applied a varying interest charge for the improvement. However, the rate never exceeded 6 per cent, being 4 per cent on draining carried out between 1802 and 1839, increasing to 5 per cent with the spread of thorough draining after 1840, rising to 6 per cent in 1856 with the use of loan capital, and returning to 5 per cent from 1875 onwards.[59] Many estates, especially those small in size, that borrowed initially under the Public Money Draining Acts charged tenants the same rate as the rentcharge, $6\frac{1}{2}$ per cent, as on the Baker-Baker estate in Northumberland and on the Cartwright, Cleveland and Dryden properties in Northamptonshire. With subsequent loan capital obtained from the improvement companies, usually at slightly higher interest rates, the $6\frac{1}{2}$ per cent charge was normally retained, as on the Cartwright estate and on the Middleton/Monck estate in Northumberland.[60] Only the Sidmouth estate in Devon increased the interest charge to tenants on taking a General Land Drainage Company loan in 1863, levying 7 per cent, a level above the rentcharge.[61]

Where draining interest was charged, the amount was usually recorded as a separate item in estate accounts and rentals. With a change in tenant the charge was absorbed in the rent of the new letting and a fresh draining interest account opened.[62] Estates differed in maintaining interest payments on the improvement when agricultural prices and rents began to fall in the late 1870s. In Northumberland, they were applied to new draining, as distinct from repairing or redraining, on the Carlisle and Grey estates well into the 1880s, while on the Blackett, Northumberland and Portland properties they were in force until the end of the century.[63] The insistence on charging interest on new draining outlay tended to disappear on the sample estates in Devon and Northamptonshire after 1880. Thus, the expansion of landlord expenditure on the improvement on the Buccleuch and Spencer estates after 1880 was done free of interest to the tenant, while on earl Fortescue's north Devon property draining interest ceased to be charged after 1880, a practice adopted on the duchy of Cornwall's land from 1881.[64]

The sample estates reveal that the interest rates on landlord draining outlay were from the beginning of the widespread adoption of the improvement pitched at modest levels, averaging 5 or 6 per cent. The failure to achieve high returns on the improvement cannot therefore be regarded as either generally restricting or causing the reduction in the late 1860s and 1870s of landlord capital provision for draining. No estate demanded the rates recommended in the 1840s, and the pessimistic view expressed in the

1873 report of the Select Committee on the improvement of land that draining yielded insufficient return to attract capital should be seen as no more than a public reaction to the enthusiastic but untried claims made for the improvement in the wake of technical progress. Indeed, high returns on investment in the improvement were not anticipated as a matter of course and estates recognized that capital outlay was necessary to produce some form of rental return. Thus Christopher Haedy pointed to the north Devon estate of the dukes of Bedford, which had languished under life leases until the 1840s and which experienced high draining outlay, as 'a strong instance to show how small must be the value of land viewed as rawed [*sic*] material [which has] to be manufactured at such great cost into farms, a landlord estate…being made in great degree by the outlay of the landlord's capital'.[65] On many estates draining expenditure may have been undertaken for less spectacular but nevertheless important objectives than high returns. The draining of land rendered farms more productive units and made them attractive to tenants. The securing of tenants capable of paying rent was an advantage on all landed properties. Thus, Maxworthy, Wheatley and Trosswell farms covering 398 acres on the mid-Devon property of the dukes of Bedford were taken in hand between 1865 and 1873. They were described as excessively wet and £3,846 was spent largely on draining to improve the state of cultivation to command a good tenant.[66] Advertisements of vacant farms on the earl of Carlisle's estate in Northumberland throughout the 1850s were keen to inform prospective tenants that 'every encouragement will be given by the landlord in draining', while in the 1860s and 1870s they announced that 'most of the land has been recently drained'.[67] As late as 1888, T. Sample, agent to the Blackett estate, recommended draining as an aid to the future letting of any farm,[68] a hope no doubt that pervaded the large expenditure on the improvement on the Buccleuch estates in Northamptonshire after 1888.

The existence of interest charges on draining presented estates with the opportunity of recouping their capital outlay. Of all landlord improvements on the sample estates, draining alone was consistently subject to the formal levying of interest. As long as the draining system remained functional, estates that financed the improvement from current income and charged 5 per cent or more interest could expect to recover their investment in 20 years or less. Those estates that borrowed funds, charging tenants $6\frac{1}{2}$ per cent on the outlay, would repay the loan, depending on its source, in around 25 years. Thus, on the duke of Cleveland's Northamptonshire estate, tenants were paying £291 a year interest of an annual rentcharge of £325 for a £5,000 loan under the Public Money Draining Acts. In the same county on the Cartwright estate, draining loans amounting to £7,847 were made in the 1850s and 1860s under the Public Money Draining Acts and from the Lands Improvement Company, involving an annual rentcharge of £517, of which tenant interest provided £510.[69] On both estates the discrepancy between rentcharge and tenant interest was largely accounted for by the commission charged by the Inclosure Commissioners and the improvement company.

After securing the capital over these periods, effective draining provided the estate with clear profit. As interest charges were regularly applied until 1880, outlay on draining undertaken up to 1860 was likely not only to have been recovered by estates but also to have produced net return. After 1860, progressively less chance existed to recoup draining expenditure. With many estates abandoning interest charges on the improvement after 1880, the scope for recovery of outlay disappeared. Even on those properties that retained interest charges in this period, the effect would have been offset by rental falls.

In the light of the course of agricultural prosperity in the second half of the nineteenth century, early adoption gave estates the possibility of profiting from draining investment, late adoption little financial reward. The draining-loan data indicate that this benefit would have been enjoyed more on large than on small estates. Although unknown at the time, the delayed use of draining on many small estates must have prevented the gaining of return on outlay and placed severe strain on their resources. Thus, on the 1,049-acre estate of G. A. Ashby in Northamptonshire, £4,935 was borrowed between 1883 and 1887 from the Lands Improvement Company for draining and farm buildings. Draining accounted for £2,750 and was charged at 5 per cent to tenants, yielding £137 a year, with no interest being levied on the farm-building outlay. As the rentcharge for the whole loan amounted to £327, nearly £200 a year had to be found on an estate that produced a rent of £2,191 in 1887, falling in 1903 to £1,421. By the latter date, the estate had been put up for sale, Ashby's failure being attributed 'to losses in farming and especially to the fact that the various incumbrances upon the property involve outgoings in excess of receipts'.[70] On those estates that only partially funded the improvement, draining investment had the potential of being more costly than on those taking complete financial responsibility. Although expenditure was lower, there was no opportunity of directly recovering outlay as interest was not charged, while estate control over the draining system was less effective as tenants carried out the improvement.

The policy employed on many estates of charging interest on draining outlay had the effect of placing much of the financial burden of the improvement on the tenants. Landlords acted in varying degrees as suppliers of capital for draining, either from current income or from a loan, but tenants paid for the improvement through interest charges. As such a policy was maintained on the majority of estates until around 1880, the widespread adoption of draining between 1840 and 1870 may be seen as being largely funded from tenant resources. Even on those estates meeting only part of the improvement's costs, the tenant contribution was vital, matching in the case

Table 5.8 *Rental return on draining outlay on the Bedford mid- and north Devon estates, the Fortescue north Devon estate and the Sidmouth Devon estate, 1850–1895*

| Period | Changes in gross rent | | Draining outlay (£) | Percentage return |
	absolute (£)	percentage		
1850–69	+9,164	+59	42,201	+22
1860–79	+9,023	+47	38,838	+23
1870–89	+1,182	+5	13,688	+9
1880–95	−2,318	−8	2,345	−99

Sources: Bedford MSS: Annual reports, 1850–95; DRO, 1262M/E20/16–144; 152M/Accounts, 1850–99

of allowances or exceeding in the case of the provision of draining materials that of the landlord. This balance altered only after 1880, when with declining prices and rents such draining that was undertaken became a burden predominantly on landlord resources.

The precise return that estates achieved on draining outlay in the form of increased rent cannot be determined. Landlord return on overall investment in agricultural improvement is most effectively measured by expressing rental increase as a percentage of that investment over a given period. As draining was but one of the several improvements financed by landlords, being exceeded by farm buildings in the absorption of estate capital,[71] its contribution to the growth of rent cannot be isolated. While the percentage that rent increase formed of draining investment may be calculated, as has been undertaken for a number of the sample estates in Devon, Northamptonshire and Northumberland for 20-year periods from 1840[72] (Tables 5.8–5.10), the results provide an unsatisfactory indication of rental return, as the role of draining is exaggerated and the level of outlay on other improvements is ignored. The levels of rental return recorded for draining alone, particularly on the Devon and Northumberland estates, would suggest that between 1840 and 1879 landlords would have experienced low returns on their overall agricultural outlay. Nevertheless, the care with which estates identified and collected interest payments on draining would indicate that up to 1880 the return on investment in the improvement could not be significantly less than the rate of interest charged.

While producing a return in a period of agricultural growth, the effect of landlord draining expenditure on rent levels after 1880 is less clear. R. Perren has suggested a positive correlation between the amount spent on agricultural improvement in general on estates and the maintenance of rents from 1870

Table 5.9 *Rental return on draining outlay on the Buccleuch, Ellesmere and Spencer estates, Northamptonshire, 1840–1899*

Period	Changes in gross rent		Draining outlay (£)	Percentage return
	absolute (£)	percentage		
1840–59	+6,385	+15	12,472	+51
1850–69	+10,874	+25	12,329	+88
1860–79*	+6,937	+15	10,145	+68
1870–89*	−11,918	−23	17,127	−70
1880–99*	−15,416	−30	18,678	−83

* Buccleuch and Spencer estates only

Sources: Buccleuch MSS: Barnwell and Boughton estate accounts, 1840–99; Spencer MSS: Estate accounts and rentals, 1840–99; NRO, Ellesmere MSS: Box X.461, Rentals, 1837–74

Table 5.10 *Rental return on draining outlay on the Blackett, Carlisle, Grey, Northumberland and Portland estates, Northumberland, 1840–1899*

Period	Changes in gross rent		Draining outlay (£)	Percentage return
	absolute (£)	percentage		
1840–59	+14,728	+10	256,376	+6
1850–69	+29,802	+21	276,557	+11
1860–79	+23,316	+14	143,456	+16
1870–89*	−13,221	−8	75,689	−18
1880–99*	−14,366	−10	38,684	−37

* Blackett, Northumberland and Portland estates only

Sources: Northumberland MSS: General accounts, 1808–48; Ledger balances, 1847–65; Abstract of accounts, 1865–1907; Variation in rental vols., 1827–1912; Draining vols., 1844–1903; DPD, Grey MSS: Rentals, 1803–42, Cash and ledger books, 1806–92; Draining vols., 1841–86; Howard MSS: N. 73/2; N. 80/14–19; N. 101–6; Estate accounts, 1875–1900; Rental vols., 1879–1912; Draining vols., 1856–1901; NdRO Blackett MSS: ZBL/54/2 and 3; ZBL/274/6–10; ZBL/275/5 and 6, ZBL/282/2; Sample MSS: ZSA/8/1–5

onwards.[73] Such a consistent relationship between draining outlay and rent movements is not evident on the sample estates (Tables 5.8–5.10). Thus, the heavy draining expenditure on the group of Northumberland estates between 1840 and 1879 was associated with rental increases that proportionally were of the same order as on the Northamptonshire properties and less than on the estates in Devon, both with considerably smaller sums devoted to the improvement. Again, the growth in draining outlay on the Buccleuch and Spencer estates in Northamptonshire after 1880 did little to stem the fall in rents. And, in spite of the large investment in the improvement on the Northumberland estates up to 1880, the rate of rental decline thereafter exceeded that on the Devon estates, where much less had been spent. While there was an increase in draining outlay after 1880 on the sample estates in general (Fig. 4.2), it not only failed to yield any positive return but the amount in itself would seem to have had little direct influence on the pattern of rents, the changes that occurred reflecting broader trends in agricultural prosperity.

The tenant contribution to draining

Whilst contemporary agricultural literature reported draining as a tenant undertaking in parts of East Anglia before 1840, little reference can be found of tenant activity on the sample estates. Although some tenant draining had occurred, the examples were few and on a small scale. On the Northumberland estate of the dukes of Northumberland, the bailiff of Warkworth bailiwick, J. Reed, noted in 1845 that one tenant had drained thirty acres over the previous decade, but, that comment apart, the annual returns of the state of farms made by the bailiff of each bailiwick to the chief commissioner contained little other record of tenant activity in the period 1827–44.[74] William Sample, agent to the duke of Portland's estate, was unaware of any tenant work before the introduction of thorough draining in 1834.[75] Again, at the expiry of life leases on the Bedford and Cornwall estates in Devon, little tenant investment in draining was identified.[76] After the mid-century technical improvements, references to draining carried out solely by tenants are rare. Despite the passing of the Agricultural Holdings Acts of 1875 and 1883, which provided tenants with initially the possibility and subsequently the right to compensation for the unexhausted value of improvements they had undertaken,[77] none of the sample estates provides evidence of the use of the legislation in respect of draining.

If the quantity of tenant draining was slight, tenant involvement was to a large extent the active variable in the adoption of the improvement on estates after 1840. As tenants paid interest on the sums expended on draining their farms or provided major portions of the total cost, estates normally not supplying capital without these contributions, they defined what land was to be drained. Thus, from 1834 tenants on the Portland estate applied each year to the agent to have land drained, who allocated the funds made available for the improvement. William Sample 'regulated the quantity of draining to be done upon each farm in proportion to the amount of money' set aside, tenants not applying having no draining done.[78] On the duke of Northumberland's estate from 1844, tenants made applications for draining to the bailiff of each bailiwick. As on the Portland estate, a system of apportionment was developed to deal with the excess of claims. Initially, Hugh Taylor, the chief commissioner, reduced the applications proportionally to the sum available, but, as he considered some tenants overbidded in an attempt to secure the required amount for draining purposes, a new method of allocation was introduced in 1851. Applications under £50 were granted in full; a percentage, varying from 40 to 50 per cent, of applications between £50 and £200 was allowed; and those over £200 were ignored.[79] By the late 1860s when applications had fallen to lower levels, they were passed in full, a practice that was still in operation in 1883.[80] Clearly the pattern of demand for draining on the Northumberland estate was tenant lead (Fig. 5.2). Even where only draining materials were provided, tenants applied annually for the pipes and tiles they required, as on the Ellesmere, Overstone and Spencer estates in Northamptonshire.[81]

Tenants had to be convinced of the value of draining to their farms. Most relied on having seen the improvement at work. As William Sample wrote to the duke of Portland in 1836, 'since [thorough] draining has been introduced ... and its effects seen on the few specimens that have been done, many of the tenants are becoming anxious to try the system'.[82] Few tenants were given formal instruction in the improvement, unlike those of the third duke of Northumberland, who in 1843 on the eve of his draining programme arranged for Professor Johnston of Durham University to address them on the advantages of draining and manuring.[83] Tenants were reported to be aware that immediate benefit in the form of increased produce could be derived from draining and many were intent on applying the improvement to their land.[84] By 1845, Sample found difficulty in keeping 'within the sum ... to be expended' on draining on the Portland estate, while Taylor informed the duke of Northumberland in 1849 that the excess of tenant applications over the allocated draining sum was satisfactory evidence of 'the tenants' disposition to adopt this indispensable mode of improvement under the system now practised'.[85]

Tenant enthusiasm for the improvement was general and not restricted to particular groups as represented by farm size. For eight of the sample estates, mainly in Northumberland, draining activity can be identified on a farm basis (Tables 5.11–5.13). They reveal that a strong correspondence existed between the proportion a farm-size category occupied of an estate and the proportion of total draining activity on that farm-size category, especially on

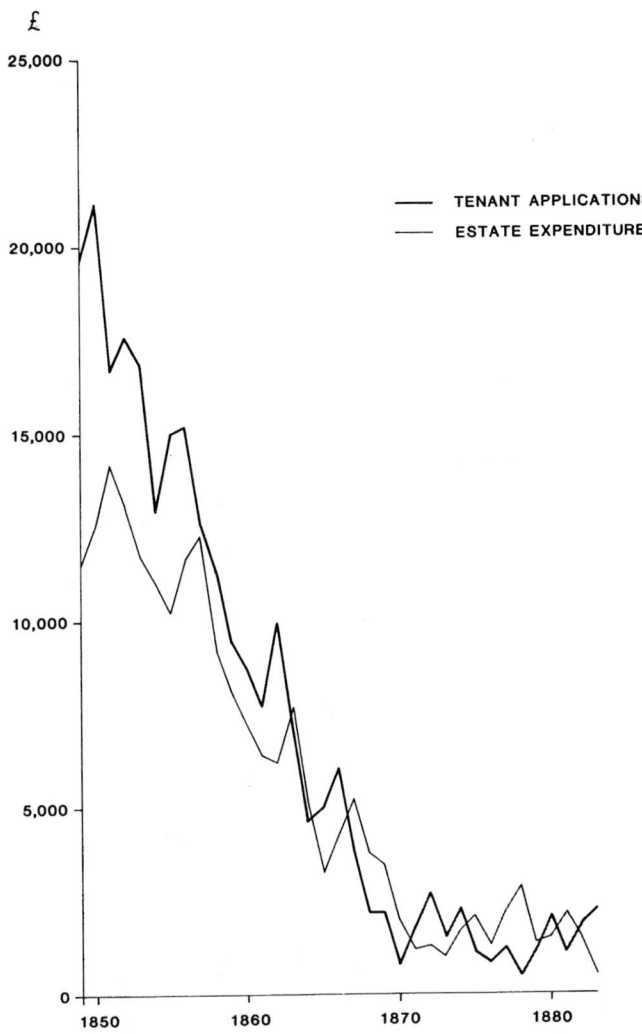

Figure 5.2 Tenant applications and landlord expenditure on new draining on the duke of Northumberland's estate, 1849–1883 (*Sources:* Northumberland MSS: Draining vols., 1–3; Business minutes, vols. 4–75, 1849–1884)

the Carlisle, Grey, Middleton/Monck and Ridley properties in lowland Northumberland and on the duke of Cleveland's Northamptonshire estate. Draining was not concentrated on the larger units, whose tenants in theory possessed more capital. Indeed, on the duke of Northumberland's and Sir Edward Blackett's estates in Northumberland and on the duke of Bedford's mid-Devon land, draining activity was less intense on farms over 500 acres.

Table 5.11 *Intensity of draining by farm size on estates in Northumberland,
1840–1899*

Farm size (acres)	No. of farms	Acreage of farms	Percentage of total farm area	Acreage drained	Percentage of total drained area
Blackett estate, 1840–99					
20–99	11	442	3	76	2
100–299	28	5,860	40	2,465	58
300–499	10	3,625	25	1,018	24
500 and over	6	4,680	32	654	16
Total	55	14,607		4,213	
Northumberland estate, 1844–65					
20–99	74	3,904	3	1,812	5
100–299	141	26,925	18	13,139	37
300–499	55	21,207	15	10,383	29
500 and over	53	93,520	64	10,305	29
Total	323	145,556		35,639	

Farm size (acres)	No. of farms	Acreage of farms	Percentage of total farm area	Draining outlay (£)	Percentage of total draining outlay
Carlisle estate, 1850–99					
20–99	1	80	1	69	0
100–299	32	6,538	55	18,410	56
300–499	8	2,932	25	8,360	25
500 and over	4	2,199	19	6,150	19
Total	45	11,749		32,989	
Ancroft, Burton, Carham and Chevington parts of the Grey estate, 1840–92					
20–99	0				
100–299	3	808	6	3,891	6
300–499	5	2,028	15	9,950	14
500 and over	14	10,697	79	54,171	80
Total	22	13,533		68,012	
Middleton/Monck estate, 1869–84					
20–99	4	272	3	1,191	3
100–299	24	5,255	58	19,997	50
300–499	6	2,899	32	13,985	35
500 and over	1	635	7	4,820	12
Total	35	9,061		39,993	

Table 5.11 (*cont.*)

Farm size (acres)	No. of farms	Acreage of farms	Percentage of total farm area	Draining outlay (£)	Percentage of total draining outlay
Ridley estate, 1847–85					
20–99	2	108	1	440	1
100–299	19	4,186	48	12,351	40
300–499	4	1,581	18	7,248	24
500 and over	3	2,866	33	10,732	35
Total	28	8,741		34,771	

Sources: Northumberland MSS: Business minutes, vol. 37, 16 February 1866; DPD, Grey MSS: Estate cropping book, 1845–78; Ledger books, 1840–92; Draining vols., 1841–86; Howard MSS: N. 99/2, Survey of the Northumberland estate, 1886; N. 101–6; Draining vols., 1856–1901; NdRO, Blackett MSS: ZBL/4/10; ZBL/54/2 and 3, ZBL/282/2; Belsay (Middleton) MSS: Box 12/1X/ Drainage accounts, 1869–84; B42/2, General cultivation book, 1868–72; Ridley MSS: ZR1/44/4; ZR1/49/11 and 12

Table 5.12 *Intensity of draining by farm size on the Cleveland estate, Northamptonshire, 1848–1871*

Farm size (acres)	No. of farms	Acreage of farms	Percentage of total farm area	Acreage drained	Percentage of total drained area
under 50	47	988	29	397	25
50–99	13	933	27	498	32
100–299	9	1,531	44	682	43
Total	69	3,452		1,577	

Sources: Raby MSS: Particulars of Brigstock and Sudborough rents, 1849; Sudborough draining volume, 1848–53; Draining abstracts, 1848–53; Draining volumes, 1861–72

As many of these large farms, particularly on the Northumberland properties, were located in areas of open moorland, the amount of land that would benefit from draining was limited. In broad terms, farm size cannot be regarded as a factor influencing the spread of the improvement on individual estates.

Nevertheless, the improvement was not adopted uniformly by tenants even on the same estate. Given similar soil conditions, farms frequently displayed

Table 5.13 *Intensity of draining by farm size on the Bedford mid-Devon estate, 1860–1889*

Farm size (acres)	No. of farms	Acreage of farms	Percentage of total farm area	Draining outlay (£)	Percentage of total draining outlay
20–99	21	1,226	30	3,201	13
100–299	10	1,935	48	14,066	60
300–499	1	334	8	4,780	20
500 and over	1	587	14	2,244	9
Total	33	4,102		24,291	

Sources: Bedford MSS: Annual reports, 1860–89; Annual report, 1867: report on Devon estates by G. Martin

differing intensities of draining, as can be seen for example on the Middleton/Monck and Ridley estates in Northumberland (Figs. 5.3 and 5.4). The decision to apply for money or materials to have land drained was individual to each tenant, and herein lies the basis of variation. Not all tenants were prepared to adopt the improvement or to the same level. On the Newburn bailiwick of the duke of Northumberland's estate, W. Glover, the bailiff, reported in 1850 that Hill Head farm, with 208 acres, 'might be brought into a higher state of cultivation if the tenant would drain' while in 1853 he regretted that the tenant of Walbottle Fill House farm, 210 acres in size, declined 'draining any part of the land'.[86] Sir Matthew Ridley informed the Newcastle upon Tyne Farmers' Club in 1859 that 'it was...disheartening to a landlord if a tenant would not allow his land to be drained', no doubt a reflection of the varying tenant interest in the improvement on his estate.[87] Again, G. Herriot, agent of the duchy of Cornwall's Bradford estate in Devon, noted in 1867 of Marsh farm covering 68 acres that 'thorough drainage of the land would doubtless improve it in many respects, but yet the results of such improvements in the neighbourhood do not appear to have been hitherto so beneficial as to induce tenants to offer to pay interest on outlay for drainage'.[88]

For most tenants, the adoption of the improvement depended on an assessment of its financial return. Draining would pay only if it increased yields beyond the interest charged or allowed the tenant to recoup his capital outlay within a short period when materials or allowances were provided. An indication of a tenant's possible return may be obtained by costing the average wheat-yield increases reported by J. Bailey Denton, Philip Pusey and James Smith in the 1840s.[89] Using their lowest value of yield increase of 15

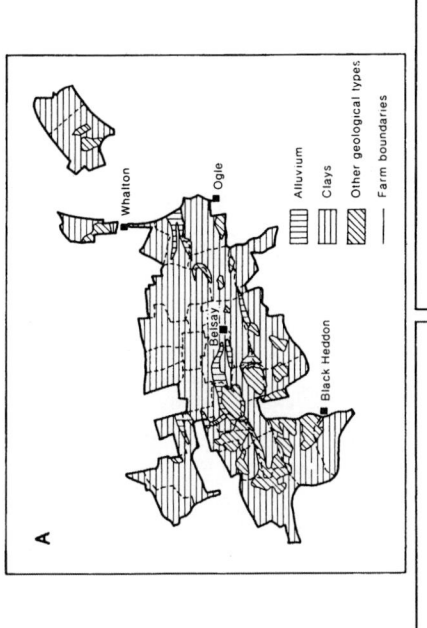

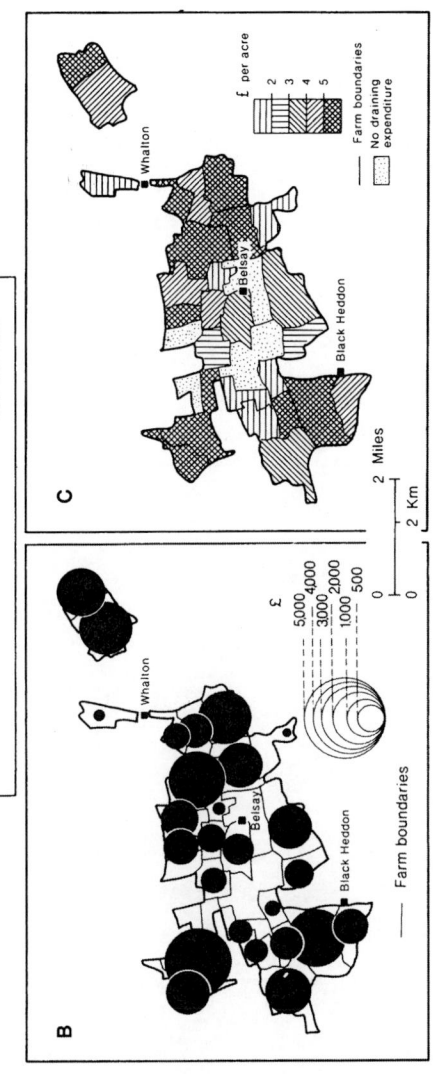

Figure 5.3 A. Drift geology and 1868 farm boundaries on the Middleton/Monck estate, Northumberland; B. Draining outlay, 1869–1884, per farm; C. Per acre draining expenditure, 1869–1884, by farm (*Sources:* NdRO, Belsay (Middleton) MSS: Part III (Supplemental), S. 18, Plans of estate, 1847–1862; Box 12/IX/Draining accounts, 1869–1884; B. 42/2, General cultivation book, 1868–1872; Geological Survey, 1 in drift sheet, 14)

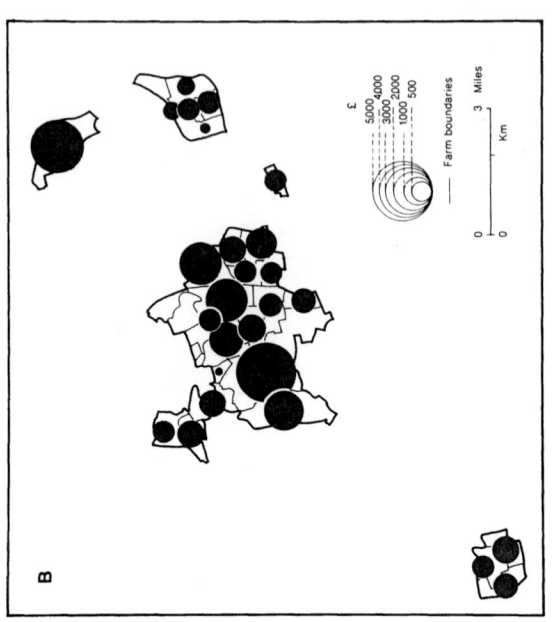

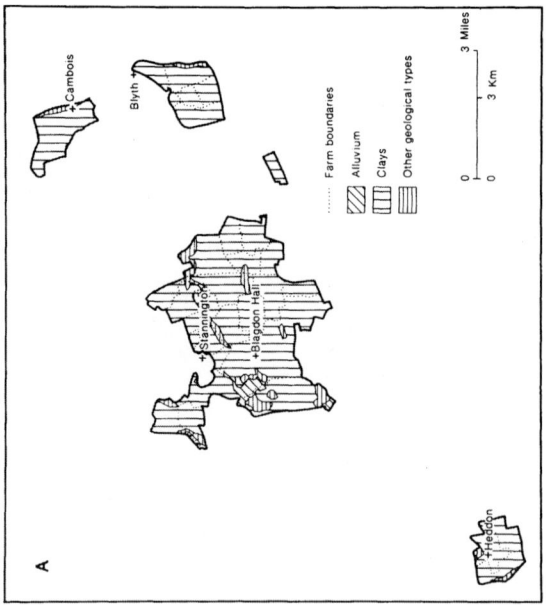

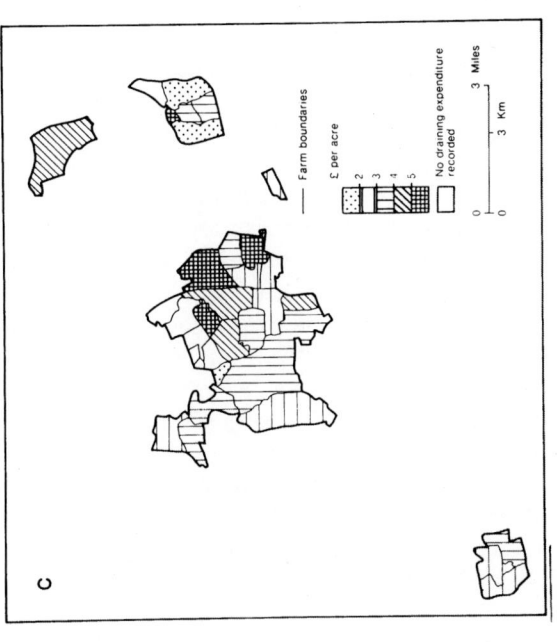

C

Farm boundaries

£ per acre

2
3
4
5

No draining expenditure
recorded

0   3 Km
0   3 Miles

Figure 5.4 A. Drift geology and 1889 farm boundaries on the Ridley estate, Northumberland; B. Draining outlay, 1847–1885, per farm; C. Per acre draining expenditure, 1847–1885, by farm (*Sources*: NdRO, Ridley MSS: ZR1/49/11 and 12; ZR1/44/4; Geological Survey, 1 in drift sheets, 14 and 15)

per cent, taking James Caird's 1850 estimate of average wheat yield of $26\frac{3}{4}$ bushels per acre and the average annual price of wheat over the decade 1850–9 of £2.67 per imperial quarter,[90] and assuming an average cost of draining of £6 per acre, a tenant charged $6\frac{1}{2}$ per cent interest on draining would pay £0.39 per acre on the outlay and receive through increased output £1.34. Excluding any additional costs involved in cultivation and processing and in maintaining the draining system, the improvement would have provided a direct return of around £1 if interest was charged or would have redeemed the labour costs of putting in materials supplied by the landlord within three to four years. Such a calculation, although artificial, offers a useful guide to the considerations that would have borne upon a tenant in deciding to employ the improvement.

The return on other cereals would not have matched that of wheat, for if draining produced the same proportional growth in yields the lower prices of barley and oats offset the greater output.[91] The fact that the draining of wheat offered the possibility of the highest return among the cereals from the 1850s to the 1870s provides clarification of the correlation between the supply of draining-loan capital and wheat prices. However, wheat was not grown continuously and tenants had to balance the level of return between it and other less profitable crops in the rotation when applying for draining to be done. In addition, cereal yields were not uniform throughout the country at the middle of the century. In Devon they were much lower than those in Northamptonshire and Northumberland, which approached the national average,[92] and made the achievement of tenant profit on draining more difficult in the southwest than in other districts.

At the same time, the cost of draining was not static, Caird and Denton suggesting a rise from £5 per acre in the 1850s to nearer £8 in the 1870s.[93] Not all estates experienced such a rate of growth in draining costs. Average acreage costs on the duke of Northumberland's estate rose from £4.93 in the 1850s to £6.25 in the 1870s and to £6.68 in the 1890s.[94] On the Blackett and Carlisle estates, the increase was less rapid, being respectively £5.42 and £5.10 in the 1850s and £5.69 and £5.16 in the 1870s.[95] Yet draining costs varied regionally, and on the sample estates in Devon they were higher. On the Sidmouth estate in the 1850s they had averaged just over £6, while on the north and mid-Devon properties of the dukes of Bedford they reached £7.19 per acre in the 1880s.[96] With increasing cost and a standard rate of interest, the tenant's margin of return on all crops would have been correspondingly reduced. This situation would have been aggravated by the easing in the trend of cereal prices over the third quarter of the century, the average annual price of wheat per imperial quarter in the decade 1870–9 at £2.52 being some 6 per cent lower than in the 1850s.[97] The fall in cereal prices after 1880, wheat being on average 44 per cent and barley and oats 33 per cent per imperial quarter lower in the 1890s than in the 1870s, would have rendered the level of return on the improvement more questionable, although the increased productivity associated with draining would still have offered some financial benefit to the tenant, particularly if free of interest charge.

Not all wet land, when drained, was capable of providing a sufficient increase in yield to cover outlay and tenants expressed little interest in improving such land. The bailiff of Tindale bailiwick on the duke of Northumberland's estate wrote in 1850 that 'the tenants fully appreciate the benefits to be derived from draining, though the stock farmers whilst very anxious to have their hay land drained ... are of opinion that the improvement of the sheep walks by thorough drainage would not counterbalance the percentage to be added'.[98] On the Bradford estate of the duchy of Cornwall in Devon, where wet land was inherently unproductive, whose farms were described by the agent as never being better than third-class, where cereal yields were below average and arable was subservient to cattle breeding and rearing, and where draining costs were higher than normal, the lack of widespread adoption of the improvement by tenants becomes understandable.[99] Tenants concentrated on applying for the draining of land which was likely to yield most return. Thus, Taylor ascribed the falling off in draining applications on the duke of Northumberland's estate in the late 1850s (Fig. 5.2) 'to the superior lands (which pay best) having been first drained'.[100] On similar lines, the agent of the duke of Bedford's Devon estates reported in 1864 that expenditure on the north Devon property had 'not been heavy, as the greater proportion of the land which required draining had been drained in previous years', while he was happy to record in 1867 that there was 'but very little more draining required upon the Tavistock estate'.[101]

The pattern of draining, both spatially and temporally, clearly reflected tenant commitment to the improvement. Tenant demand for draining was initially high in an effort to secure the upgrading of land that would provide the largest yield increments. As such land was increasingly drained, tenant request for the improvement waned. The gradual rise in the draining costs and the slight fall in cereal prices from the 1850s, by reducing overall return, would have further limited the amount of land on which the improvement would have been thought profitable by tenants. The decline in draining outlay after the late 1860s from the peaks of the 1840s and 1850s that characterized most of the sample estates (Fig. 4.2) may be seen largely as a product of most land considered to be worth draining by tenants having been treated.

Although the improvement was a joint enterprise between landlords and tenants, the sample estates establish that no standard formula was followed in undertaking it, considerable variation being evident in both the capital provision and management of draining. Some estates regarded the improvement very much as a combined activity, employing equal, if not greater, amounts of tenant capital and initiative, and retaining such a system until late in the century. A number abandoned direct tenant involvement with the advances in draining techniques at the middle of the century or with the availability of loan capital. Others had pursued a policy of taking sole responsibility for the supply of capital for the improvement from much earlier in the century. As landlords possessed a larger financial resource base than tenants, even though it may have been obtained by means of loans, the decision of estates to control capital provision was associated with a greater level of draining activity. While William Bearn reported that the practice of providing draining materials was popular with the tenantry on the Buccleuch, Overstone and Spencer estates in Northamptonshire,[102] the intensity of draining on those properties, even allowing for tenant contributions, never matched that where landlords were responsible for funding the improvement.

Estates not only differed in the method of financing draining but also over the decision to provide capital for the improvement at all. The draining loan data would suggest that landlord investment in the improvement had much to do with estate size. The improvement was generally and quickly appreciated on large properties, but with decreasing estate size draining was increasingly neglected so that on small estates, under 1,000 acres, it failed to be widely adopted. Such differential adoption by estates must be regarded as a major factor in the spatial pattern of draining in the second half of the nineteenth century based on the loan-capital data.

By affording a means of increasing yields, draining was regarded as a productive improvement. Where landlords decided to cover the full cost, they were intent to share in this return by charging interest on the outlay. The interest rates applied to draining were established by estates in the light of the amounts tenants were willing to pay.[103] These rates were low in comparison to contemporary returns from industrial investment. However, the argument put forward by E. J. T. Collins and E. L. Jones and by G. Hueckel that low returns inhibited landlord supply of capital for draining is to a great extent irrelevant,[104] because the rates represented the levels at which landlords were prepared to provide funds for the improvement. By charging interest, landlords were guaranteed some direct return on their outlay, irrespective of the important, indirect benefits that draining brought to their estates. Indeed, they could recover their investment and must have largely done so on effective draining carried out by 1870. Of all agricultural improvements financed by landlords before 1880, draining was the only one, unlike farm buildings to which much larger sums were devoted, for which tenants agreed formally to the regular payment of interest. Such a situation must have rendered landlords, who possessed the resources, less hostile to the provision of draining capital. Landowners on the sample estates failed to obtain financial return on draining only after 1880, such outlay that was made, as C. O'Grada has suggested,[105] attempting to aid hard-hit tenants.

Although desirous of having his agricultural land drained, the landowner in effect was little more than a supplier of capital.[106] On most estates, the tenant largely determined the extent of adoption of the improvement. By paying interest on outlay, the tenant identified land to be drained, choosing that which through increased yields would be financially rewarding. At the middle of the century, using contemporary estimates, the draining of wheat was likely to have bestowed a considerable return on a tenant's $6\frac{1}{2}$ per cent annual interest payment. However, the level of return varied in response to the quality of land in need of draining, the cost of the improvement and agricultural prices. As lands capable of providing the greatest increase in yield would have attracted immediate attention, leaving less productive land, and as draining costs gradually rose while cereal prices remained stable or fell slightly from the 1850s to the 1870s, increasing the marginality of the lower-quality land in terms of capital investment, tenant incentive for the improvement would have declined. Overall, the sample estates and the loan-capital data reveal that, within the area physically in need of draining, the detailed pattern of adoption in the second half of the nineteenth century clearly reflected the attitudes of individual estates and their tenants towards the improvement.

6

The success of underdraining as an agricultural improvement

With the revaluation in techniques around 1840, draining was promoted as a productive agricultural improvement. If excess water were effectively removed, draining was credited not only with the physical ability of making land more easily cultivable, valuable in its own right, but also with the potential to enhance crop yields and to transform farming systems, bestowing considerable economic advantages. As landlord and tenant investment in draining was based on the expectation of increasing agricultural output from land injured by waterlogging, the extent to which these benefits were realized had a significant influence on the spread of the improvement. At the same time, an examination of the changes that occurred in farming practice on the adoption of draining is essential for an assessment of the value of the improvement to nineteenth-century English agriculture.

The effectiveness of draining systems

To be a viable and productive agricultural improvement for both landlord and tenant, draining had not only to be effective but durable, with a life that extended beyond the period of interest payment. Contemporary opinion diverged over the improvement's claims to these qualities. While James Caird, George Darby and George Ridley, all Inclosure Commissioners, attested the general reliability of draining carried out from the 1840s to function beyond twenty-five years, the report of the Select Committee of the House of Lords on the improvement of land in 1873, an expression of landowning concern in capital investment in agriculture, carefully identified both the limited life and the likelihood of failure of draining systems.[1] These differing views reflected respective interests in the improvement, but in general the effectiveness of draining depended on the systems and materials employed and the level of subsequent maintenance.

For draining financed by loans under the land-improvement legislation, the Inclosure Commissioners from the beginning were intent on establishing

Table 6.1 *Drain intervals on the Blackett Northumberland, Dryden Northamptonshire and Fortescue north Devon estates, 1848–1872*

		Acreage drained with layout recorded	Drain interval percentage of area at			
Estate	Period		less than 30 ft	30–42 ft	43–54 ft	55 ft and over
Blackett	1857–72	3,134	3	95	2	0
Dryden	1858–66	1,112	6	94	0	0
Fortescue	1848–52	1,114	0	83	0	17

Sources: DRO, Fortescue MSS: 1262M/E1/103, Draining sheets and schedules, 1847–64; NRO, Dryden MSS: D (CA) 450, Draining schedules, 1858–66; NdRO, Blackett MSS: ZBL/54/2 and 3, Draining schedules, 1857–72

a standard system that would be applicable throughout the country and that would last as long as the rentcharge so as in theory not to become a burden on the estate. They adopted the system of deep draining developed by Josiah Parkes, insisting on a depth of at least $3\frac{1}{2}$ ft and preferring 4 ft where fall permitted.[2] Their inspectors applied the system rigorously, Andrew Thompson for example noting in 1862 that 'if the plan proposed were in the least degree a departure from deep and permanent draining, [he] should be the very last to consent to it'.[3] Although less control was exercised over drain intervals, much being left to the decisions of inspectors in the field in the light of soil conditions, they still revealed a high degree of uniformity. Thus, on the estates Thompson inspected between 1857 and 1868 all the draining he recommended for clays and marls fell between 24 and 36 ft.[4] The Inclosure Commissioners also demanded the use of pipes as the most effective draining material, although there was no insistence on the 1 in size popularized by Parkes. An indication of the standardization achieved by the Inclosure Commissioners may be obtained by comparing the draining systems carried out by means of government- and improvement-company loans on three estates in widely different locations over the period 1848–72, the Dryden estate in Northamptonshire, the Fortescue north Devon estate and the Blackett estate in Northumberland (Table 6.1). On the first two no drain was less than 4 ft deep, while on the Blackett property 1 per cent of the drained area was at less than 4 ft, the remainder being at 4 ft or more; and the drain interval on all three was predominantly between 30 and 42 ft.

Many estates not employing loan capital also adopted the deep-draining system of the Inclosure Commissioners. The third duke of Northumberland introduced shallow thorough drains, 3 ft deep and from 18 to 24 ft apart, on

Table 6.2 *Drain depths and intervals on the Northumberland estate, North-umberland, 1844–1899*

Period	Acreage drained with layout recorded	Drain depth percentage of area at				Drain interval percentage of area at		
		under 3 ft	3 and 3½ ft	4 ft	4½ and over	under 30 ft	30–42 ft	43 ft and over
1844–9	7,342	1	65	30	4	56	40	1
1850–9	22,558	0	5	83	12	41	58	1
1860–9	9,888	0	5	86	9	25	69	6
1870–9	2,771	1	17	81	1	29	53	18
1880–9	1,905	3	59	35	3	32	55	13
1890–9	452	22	78	0	0	63	36	1

Source: Northumberland MSS: Draining vols., 1–3

Table 6.3 *Drain depths on the Carham and Chevington parts of the Grey estate, Northumberland, 1840–1854*

Period	Acreage drained with layout recorded	Drain depth percentage of area at			
		under 3 ft	3 ft and 3½ ft	4 ft	4½ ft and over
1840–7	1,982	35	54	11	0
1850–4	1,071	8	35	57	0

Source: DPD Grey MSS: Boxes 550 and 551, Draining papers, 1841–69

his Northumberland estate in 1844. This system was abandoned on the succession of the fourth duke and his employment of Parkes in 1848 to report on the draining of the estate. Parkes advocated 4 ft deep drains, using 1 in pipes, 30 to 40 ft apart.[5] Not all the recommendations on drain layout were applied and a significant proportion of land was drained at less than 30 ft intervals. Nevertheless, 4 ft deep drains dominated on the estate from 1848 to 1879 (Table 6.2). A similar move to deep draining was evident on the neighbouring Grey and Portland estates. For example, on the Carham and Chevington portions of the Grey estate, drains had been laid mainly at shallow depths from 1840 when thorough draining was introduced, but from

Table 6.4 *Drain depths and intervals on the Cleveland estate, Northampton-shire, 1848–1871*

Period	Acreage drained with layout recorded	Drain depth percentage of area at				Drain interval percentage of area at			
		under 3 ft	3 and 3½ ft	4 ft	4½ ft and over	under 30 ft	30–42 ft	43–54 ft	55 ft and over
1848–52	1,160	0	9	79	12	1	55	44	0
1861–71	415	1	10	89	0	1	11	88	0

Sources: Rabys MSS: Sudborough draining vol., 1848–52; Draining abstracts, 1848–53; Draining vols., 1861–71

1850 depths of 4 ft became more important (Table 6.3). Some estates, having accepted the Inclosure Commissioners' design in connection with a draining loan, retained the layout when financing the improvement from their own resources. Thus, the draining system on the duke of Cleveland's Northamptonshire estate under a Public Money Draining Acts loan from 1848 to 1852 was largely 4 ft deep and between 30 and 54 ft apart. When further draining was financed by the estate itself between 1861 and 1871, this system was broadly maintained, the increase in interval reflecting a change in the nature of the land (Table 6.4). On the sample estates in all three counties, where the full cost of the improvement was provided and for which drain layout is detailed, deep-draining systems were predominant between 1850 and the late 1870s.

Fewer data are available on the draining systems on those estates where the policy of providing materials, leaving tenants to find labour, was pursued. Agents recommended drain depths and layouts for tenants to follow. John Beasley in 1840 on the Overstone estate in Northamptonshire favoured shallow depths for drains, suggesting that on clays 'no drain should be less than 2 ft deep'.[6] On the Ellesmere estate, James Loch had by 1846 moved towards deep draining, advising drains at least 3 ft deep and not more than 60 ft apart.[7] Yet there is little evidence that tenants adopted these recommendations. Increasing depth and decreasing interval both succeeded in raising labour costs in draining. On the Cartwright and Dryden estates in Northamptonshire, the labour for a rod of drains at 3 ft deep averaged 2.5p, and at 4 ft deep 3.1p under Public Money Draining Acts loans between 1856 and 1866.[8] Using these costs, the labour involved in draining an acre 3 ft deep at 21 ft intervals totalled £3.15 as against £3.94 at 4 ft deep. At 36 ft apart, the cost of labour for 3 and 4 ft deep drains would have been respectively £1.84 and £2.30. With materials being provided, as was the case at various times on the Buccleuch, Ellesmere, Fitzwilliam, Overstone and Spencer estates in Northamptonshire, there must have been real temptation on the part of tenants to employ shallower and wider draining systems. The results would have possessed less efficiency and durability, and greater variability than those sanctioned by the Inclosure Commissioners, as Caird reported from various parts of the country in 1850–1.[9] In Northamptonshire, Loch noted ruefully in 1846 that tenants on the Ellesmere estate displayed 'too much anxiety to drain cheaply rather than to drain well', while William Bearn commenting specifically on the Buccleuch, Overstone and Spencer estates pointed to the need for care, 'some persons being unwilling to incur the expense of putting [the drains] deep into the ground'.[10]

A greater degree of uniformity can be identified in the materials used for draining on the sample estates from 1840. During the 1840s, most estates which had not previously done so substituted stones with tiles, which in their turn were replaced by pipes around the middle of the century. Stones and tiles had been provided as draining material to tenants up to 1845 on the Ellesmere estate. From 1846 to 1852 tiles dominated, with pipes becoming the sole draining material from 1853, a date at which the change from tiles to pipes had also largely taken place on the Fitzwilliam and Spencer estates.[11] The adoption of pipes on the duke of Cleveland's estate dated from 1848 with the purchase of two drain-pipe machines for the Sudborough tile works and the taking up of a Public Money Draining Acts loan.[12] In Northumberland, on the duke of Northumberland's estate, tiles had become the chief drain material with the employment of F. W. Etheredge in 1843 to establish a number of tileries, only Tindale bailiwick without a tilery continuing to use stones as fill.[13] From 1847, with the development of the large-scale deep-draining programme on the estate, there was a determined effort to ensure a constant supply of drain materials. Hugh Taylor, the chief commissioner, required the bailiffs to report the number of tileries in each bailiwick, their ownership and clay-supply, and the type, number and price of tiles made. At the same time, he inquired of the bailiffs' preference for pipes or tiles. The results showed that estate tileries could produce tiles for eight of the bailiwicks and that all but two of the bailiffs favoured pipes. After 1848, pipes became the main drain material on the estate, and for those bailiwicks without tileries long-term contracts were negotiated with independent manufacturers to supply the necessary quantities.[14]

While pipes had become the dominant drain material by the middle of the century, their efficiency was largely dependent on their size. As part of his deep-draining system, Parkes had considered that pipes of 1 in diameter were capable of effectively removing soil water. However, pipes of that size were not widely adopted on the sample estates.[15] A number employed larger pipes from the beginning. Thus, the duke of Portland informed William Sample, agent to his Northumberland estate, in 1847 that nothing smaller than 2 in pipes should be used in draining, and from 1856 to 1899, when a record of pipe production from the estate's tileries is available, no pipe less than $2\frac{1}{2}$ in was made, that size forming 84 per cent of the 11.5 million manufactured.[16] Again, although only supplying pipes to tenants, Loch advised the use of those of 2 in diameter on the Ellesmere estate in Northamptonshire.[17] Where initially small-sized pipes had been employed on estates, they were abandoned in favour of those with larger diameters. On the duke of Northumberland's estate, despite Parkes' patronage of 1 in pipes, the fourth duke insisted that there should be no pipes smaller than $1\frac{1}{2}$ in in use. On those bailiwicks where Parkes did not directly control the undertaking of the improvement, the draining superintendent, John Loraine, on Taylor's advice applied 2 in pipes, which Taylor noted in 1850 should be increased to $2\frac{1}{2}$ in.[18] Pipes with diameters of $1\frac{1}{2}$ in and less were employed in the first year of draining on the duke of Cleveland's Northamptonshire estate under the Public Money Draining Acts loan. By 1849, the size most commonly used had grown to $1\frac{3}{4}$ in, which in the following year rose to 2 in. Of the 1.295 million pipes used on the estate between 1848 and 1852 in connection with the loan, 6 per cent possessed a diameter of $1\frac{1}{2}$ in or less, 37 per cent $1\frac{3}{4}$ in and 57 per cent 2 in and over.[19] A similar trend was recorded on the Blackett estate in Northumberland. From the adoption of thorough draining on the estate in 1840 to 1846, tiles were the main draining material, being replaced in 1847 by pipes after the purchase of a drain-pipe machine in 1846. Up to 1847 many tiles were produced whose size was equivalent to a 1 in diameter pipe, but thereafter pipes of 2 in and greater size were dominant.[20] By the early 1850s, virtually all the sample estates, for which data exist, were draining with pipes no smaller than 2 in in diameter, reducing the possibilities of blockage associated with the use of the 1 in pipe. The larger size of pipe was retained for the rest of the century, as can be seen, for example, on the Blackett and Carlisle estates in Northumberland (Table 6.5). Although alternative systems of draining were discussed in the contemporary agricultural literature, few estates attempted methods of draining other than with pipes in this period. Despite John Fowler's development of a steam-powered draining plough from 1850, there is little evidence of the use of the equipment on the sample estates.[21] Steam draining was reported on earl Spencer's estate at Strixton in Northamptonshire in 1885–6 but was not carried further.[22] Mole draining also received little landlord support, none of the sample estates exceeding the £41 spent on the technique on the Barnwell property of the duke of Buccleuch in 1884 and 1893.[23]

Although there was little change in draining material in the second half of the nineteenth century, some relaxation in draining methods occurred from the late 1870s. New draining in this period was undertaken at shallower

Table 6.5 *Pipe size used for draining on the Blackett and Carlisle estates, Northumberland, 1840–1899*

Period	No. of pipes in millions	Percentage at			
		under 2 in	2 and $2\frac{1}{2}$ in	3 in	4 in
Blackett estate					
1840–9	2.551*	30	63	4	3
1850–9	1.914	0	83	5	12
1860–9	3.023	0	87	6	7
1870–9	0.409	0	71	11	18
1880–9	0.889	0	79	8	13
1890–9	0.205	0	78	8	14
Carlisle estate					
1856–9	3.178	0	91	2	7
1860–9	4.102	0	87	3	10
1870–9	0.461	0	90	2	8
1880–9	0.498	0	86	5	9
1890–9	–				

–: No data available

* up to 1847, this figure includes equivalent-sized tiles

Sources: DPD Howard MSS: Draining vols., 1–3; NdRO, Blackett MSS: ZBL/78, Valuation of Matfen tilery, 1847; ZBL/268/25, East Matfen tilery, 1848; ZBL/281/1, 4 and 5, Matfen and Melkridge tilery accounts, 1840–1908

depths and closer intervals. On the Blackett, Carlisle and Northumberland estates in Northumberland in the last two decades of the century, drains at depths of 3 and $3\frac{1}{2}$ ft and under 30 ft apart became the prevailing system, displacing those at 4 ft deep and with intervals of between 24 and 36 ft (Tables 6.2, 6.6 and 6.7). This change in practice was also recorded on the sample Northamptonshire estates. W. T. Scarth, agent to the duke of Cleveland, informed tenants in 1881 that he was prepared to allow drains 3 ft deep if laid no further apart than 27 ft.[24] The draining carried out on the Overstone estate between 1880 and 1885 was at $2\frac{1}{2}$ and 3 ft depths and 24 and 36 ft intervals.[25] Even loan-financed draining sanctioned by the Inclosure Commissioners and their successors moved to shallower systems. Thus, on the Ashby estate in Northamptonshire, drains were approved at 3 ft deep and 18 ft apart in 1883, while on the Buccleuch estates the land drained under the General Land Drainage Company's loan from 1888 was at the depths of 3

Table 6.6 *Drain depths and intervals on the Blackett estate, Northumberland, 1857–1889*

Period	Acreage drained with layout recorded	Drain depth percentage of area at				Drain interval percentage of area at		
		under 3 ft	3 and 3½ ft	4 ft	4½ and over	under 30 ft	30–42 ft	43 ft and over
1857–9	605	0	0	100	0	3	97	0
1860–9	2,286	0	0	100	0	4	94	2
1870–9	284	0	9	81	0	0	98	2
1880–9	359	0	100	0	0	82	12	6

Source: NdRO, Blackett MSS: ZBL/54/2 and 3, Draining schedules, 1857–86

Table 6.7 *Drain depths on the Carlisle estate, Northumberland, 1856–1899*

Period	Total no. of chains of drains with depth recorded	Percentage of total at			
		under 3 ft	3 and 3½ ft	4 ft	4½ ft and over
1856–9	47,587	0	24	71	5
1860–9	78,520	0	22	73	5
1870–9	7,167	0	21	74	5
1880–9	9,247	2	71	26	1
1890–9	458	4	96	0	0

Source: DPD, Howard MSS: Draining vols., 1–3

and 3½ ft and intervals of 21 to 36 ft.[26] The reduction in depth cannot be seen solely as an attempt to reduce labour costs in draining, for the savings that accrued would have been offset largely by the increased frequency of drains and the greater number of pipes needed. Largely, it reflected changing attitudes to the purpose of draining. From the late 1870s, a number of draining treatises emphasized the importance of concentrating on the removal of surface water alone, which could be more effectively achieved by shallower, closer drains.[27] And a series of wet seasons in the late 1870s and early 1880s experienced on all the sample estates must have convinced not only landlords and tenants but also the Inclosure Commissioners of the need to adopt this policy.

On estates that accepted full responsibility for the improvement, few examples can be found of complete failure of the draining systems employed after 1840. Where such occurred, they acquired a degree of notoriety and were widely publicized. Thus, the breakdown of the drains on earl Fortescue's north Devon estate around 1850 was still recounted in 1882 by W. C. Little to the Royal Commission on the depressed condition of the agricultural interest.[28] Nevertheless, such reports could involve an element of exaggeration, as can be seen from the Fortescue case. In 1848, there was an expansion in draining on the estate in connection with a Public Money Draining Acts loan, with Parkes acting as inspector for the Inclosure Commissioners. He advised the use of 4 ft deep drains at $41\frac{1}{4}$ ft intervals with $1\frac{1}{4}$ to $1\frac{1}{3}$ in pipes. By 1850 Lord Fortescue notified the Inclosure Commissioners that the system was not functioning effectively on all the land drained and that he wished to add an intermediate drain 2 ft deep between those laid by Parkes, a system recommended to him by the earl of Wharncliffe from his Yorkshire estate. The Inclosure Commissioners were not prepared to accept such modification to their deep-draining system and sent another inspector, Hewitt Davis, to report on the improvement. Davis suggested that the distance apart of the drains should be reduced to 33 ft, but the real problem stemmed from the use of too small a pipe. As the Inclosure Commissioners would not yield to intermediate drains, Lord Fortescue forfeited the loan and from 1853 resorted to draining by his own methods. Yet, examination of Davis' report in detail reveals that failure was limited to no more than 50 acres of some 1,000 acres drained by Parkes, and the system applied on the estate after 1853 was little different, with drains 4 ft deep and pipes less than $1\frac{3}{4}$ in in diameter being largely employed.[29]

Draining systems on the remaining sample estates, where capital provision for the improvement rested with the landlord, appear to have created fewer problems. The draining volumes of the duke of Northumberland's and earl of Carlisle's estates in Northumberland allow the area of land redrained to be calculated respectively for the periods 1844–99 and 1856–99. On both the acreage redrained amounted to no more than 2 per cent of the total (Table 6.8). As the estate interest was best served by having a working draining system, attempts were made by landlords to remedy blocked or defective drains. On the Dryden estate in Northamptonshire, landlord responsibility for the repair of the improvement was written into tenancy agreements from the 1880s.[30] Drains that had been laid with pipes of too small a bore on parts of the north Devon estate of the duke of Bedford in the 1840s and had become choked were opened up and refilled with larger pipes between 1869 and 1873.[31] Fields drained at 45 ft intervals on the Bradford estate of the duchy of Cornwall from 1861 to 1865 were becoming increasingly wet by

Table 6.8 *Acreage redrained on the Carlisle and Northumberland estates, Northumberland, 1844–1899*

Period	Acreage drained	Acreage redrained	Redrained area as percentage of total
	Carlisle estate		
1856–9	1,912	0	0
1860–9	3,149	26	1
1870–9	320	0	0
1880–9	528	85	16
1890–9	20	0	0
Total	5,929	111	2
	Northumberland estate		
1844–9	7,643	0	0
1850–9	23,117	4	0
1860–9	10,105	90	1
1870–9	2,882	123	4
1880–9	1,943	504	25
1890–9	536	43	8
Total	46,226	764	2

Sources: Northumberland MSS: Draining vols., 1–3; DPD, Howard MSS: Draining vols., 1–3

1879 and intermediate drains were added over the following six years, a practice found also on the Sidmouth estate.[32] Some estates instituted regular inspections of drain outlets, a procedure that dated from the 1860s on the duke of Cleveland's Northamptonshire estate and from 1885 on the duke of Bedford's Devon properties.[33] Outlay in repairing drains and redraining increased as a percentage of total draining expenditure in the last two decades of the century, as can be seen on the Carlisle, Grey, Northumberland and Portland estates in Northumberland (Table 6.9), no doubt partly as a result of the wet seasons around 1880. Repair expenditure after 1880 on the duke of Northumberland's estate was greatest on the Alnwick, Longhoughton, Shilbottle and Warkworth bailiwicks. These had been the areas under Parkes' direct supervision from 1848 to 1855, being largely drained with $1\frac{1}{2}$ in pipes. By 1880, many of these small-sized pipes had become blocked, requiring replacement. Nevertheless, overall repair expenditure formed a small proportion of total draining outlay, not exceeding the figure of 8 per cent recorded on the duke of Northumberland's estate, and it is

Table 6.9 *Expenditure on redraining and repairing drains as a percentage of total draining outlay on the Carlisle, Grey, Northumberland and Portland estates, Northumberland, 1840–1899*

Period	Total outlay (£)	Repairing and redraining outlay (£)	Repairing as percentage of total outlay	Total outlay (£)	Repairing and redraining outlay (£)	Repairing as percentage of total outlay
	Carlisle estate			Grey estate		
1840–9	–	–	–	29,805	0	0
1850–9	9,744[a]	0	0	28,501	29	0
1860–9	16,090	145	1	10,986	139	1
1870–9	1,650	38	2	2,531	154	6
1880–9	2,383	328	14	10,400	552	5
1890–9	104	0	0	234[b]	40	17
Total	29,971	511	2	82,457	914	1
	Northumberland estate			Portland estate		
1840–9	38,634[c]	0	0	–	–	–
1850–9	114,027	87	0	5,166[a]	96	2
1860–9	52,394	1,091	2	13,324	52	0
1870–9	19,809	2,558	13	10,418	56	1
1880–9	19,303	10,648	55	5,540	249	4
1890–9	8,881	5,588	63	2,994	592	20
Total	253,048	19,972	8	37,442	1,045	3

–: No data available
a: 1856–9 only; b: 1890–2 only; c: 1844–9 only
Sources: Northumberland MSS: Draining vols., 1–3; DPD, Grey MSS: Ledger books, 1840–92; Draining vols., 1841–86; Howard MSS: Draining vols., 1–3; NdRO, Sample MSS: ZSA/8/2–5, Estate accounts 1856–99

difficult not to conclude that effective and functioning draining systems were maintained on these estates throughout the second half of the nineteenth century.

However, on estates where tenants provided labour for the improvement, draining systems possessed a lower level of durability. R. Hunter Pringle reported that in Northamptonshire much of such draining failed to withstand the series of wet seasons around 1880.[34] On the estate of the earl Spencer in 1879 alone there were demands from at least nine farms covering 2,775 acres for land to be redrained.[35] Again, the level of outlay on the Buccleuch and Overstone estates after 1880 would suggest that, in addition to treating undrained land, an extensive need existed for redraining land previously improved by tenants. In general, the policy of limited responsibility for draining that was adopted on these Northamptonshire estates until the end of the 1870s proved detrimental to the landlord interest. While restricting capital outlay in a period of agricultural prosperity, devolution of responsibility prevented full control of the *methods* of draining. The deficiencies in methods were revealed around 1880, necessitating considerable remedial expenditure to be made on the improvement in a period of growing depression.

The use and quality of land drained

With the development of technical efficiency around 1840, draining was recognized by contemporary agriculturalists as an improvement primarily for arable land, especially cereal production. J. Bailey Denton and J. F. W. Johnston, for example, suggested that, with the aid of draining, wheat output could be so increased as to convert the country into an exporter of cereals.[36] Many saw the increased production arising from the draining of arable as a means to offset price falls that might result from the repeal of the Corn Laws, a connection made by Sir Robert Peel himself in 1846 in introducing the government draining loans in his repeal speeches.[37] Far less attention was paid to the draining of pasture in the agricultural literature in the middle of the century, a neglect explained by Hewitt Davis in 1848 by the fact that the draining of arable produced a more direct and immediate return but that was still reported in 1865 by J. C. Morton.[38]

Although this dichotomy in the draining of agricultural land was not as absolute on the sample estates as contemporary agriculturalists suggested, the advantages of improving arable were widely appreciated. In Northumberland on the Portland estate, William Sample, the agent, noted that land was drained whilst in fallow so that tenants could reap the benefit from the increased yield of subsequent arable crops.[39] Tenants on the Blackett estate expressed a preference for having arable drained: in 1858 J. R. Spraggon, tenant of Willimontswyke farm at West Water wrote that although the agent, T. Sample, proposed to drain grassland he 'would much rather have some of the tillage land drained first', while the condition that J. Smith made in 1853 in taking Fenwick North farm at Matfen was for land 'each fallow quarter to be thorough drained when required as it comes in course'.[40] John Beasley in outlining the procedure for undertaking draining on the Overstone estate in Northamptonshire in 1840 regarded the provision of tiles essentially for arable purposes, leaving tenants largely to their own devices on pasture.[41] And in his report on the proposed draining of the Fortescue north Devon estate in 1848, Parkes noted that the main intention was to improve arable land.[42]

This emphasis on draining arable may be demonstrated more fully on

Table 6.10 *Land use of fields drained on the Grey estate, Northumberland, 1851–1868*

Locale	Field acreage drained	At time of draining percentage in tillage	grassland	Cultivated acreage 1851	Percentage of cultivated area in tillage	grassland	Percentage drained of 1851 area in tillage	grassland
Ancroft	1,204	92	8	2,029	60	40	91	12
Burton	1,164	80	20	2,192	58	42	73	25
Carham	1,042	91	9	4,153	60	40	38	6
Chevington	2,891	79	11	4,160	62	38	89	38
Howick	264	85	15	2,441	55	45	17	4

Sources: DPD, Grey MSS: Estate cropping book, 1845–78; Boxes 550 and 551, Draining papers, 1841–69; Draining vols., 1841–86

those estates which possess not only records of individual fields drained but also surveys and cropping books detailing specific land uses. On the Grey estate in Northumberland, draining undertaken between 1851 and 1868 was focussed on arable, the proportion of land drained in arable being 79 per cent and over on all parts of the property (Table 6.10). The same predominance in draining arable was evident on the Lucker, Newburn, Shilbottle and Warkworth bailiwicks of the duke of Northumberland's estate, where of the land drained between 1844 and 1869 the proportion was in excess of 72 per cent (Table 6.11). Such arable concentration may have been no more than a reflection of the overall land-use structure on these estates. However, on those properties where grassland was dominant, the preference for draining arable can still be detected. Thus, grassland covered 63 per cent of the cultivated area of the Blackett Matfen estate in 1862, but 57 per cent of the land drained from 1857 to 1872 was in arable (Table 6.12). Again, on the Alnwick bailiwick of the duke of Northumberland's estate, where in 1850 pasture formed 55 per cent of the cultivated area, arable represented 60 per cent of the drained acreage in the period 1844–69 (Table 6.11). This arable preponderance can be clearly seen by expressing the area of arable and pasture drained as a percentage of their respective total acreages on each estate. On the three properties in Northumberland so far mentioned, a consistently higher proportion of the total arable area was drained up to 1870 than of total grassland (Tables 6.10–6.12). This situation was found even where the total arable acreage was small, as on the Barrasford bailiwick of the duke of Northumberland's estate, an upland area with pasture occupying

Table 6.11 *Land use of fields drained on six bailiwicks of the Northumberland estate, 1844–1869*

Bailiwick	Field acreage drained	Drained area by 1850 use percentage in		Cultivated acreage 1850	Percentage of cultivated area in		Percentage drained of 1850 area in	
		arable	pasture		arable	pasture	arable	pasture
Alnwick	5,366	60	40	9,284	45	55	77	42
Barrasford	454	45	55	8,022	19	81	13	4
Lucker	5,662	72	28	9,038	54	46	84	38
Newburn	1,956	87	13	3,148	81	19	67	42
Shilbottle	2,921	77	23	5,132	68	32	64	41
Warkworth	2,617	80	20	4,350	68	32	71	38

Sources: Northumberland MSS: T. Bell and sons, Survey and terrier... 1850; Draining vols., 1–3

Table 6.12 *Land use of fields drained on the Matfen section of the Blackett estate, Northumberland, 1857–1872*

Field acreage drained	1,997
At time of draining percentage in:	
tillage	57
grassland	43
Cultivated acreage 1862	6,978
Percentage of cultivated area in:	
tillage	37
grassland	63
Percentage drained of 1862 area in:	
tillage	44
grassland	20

Sources: NdRO, Blackett MSS: ZBL/54/2 and 3, Draining schedules, 1857–86; ZBL/287/8, Matfen cropping book, 1862–88

81 per cent of the cultivated land. In Northamptonshire, although arable formed only 15 per cent of the cultivated area of the duke of Cleveland's estate in 1849, the drained proportion was much greater than that of pasture (Table 6.13).

In addition to intensity, priority was given to the draining of arable land. Draining carried out on the duke of Cleveland's estate occurred in two

Table 6.13 *Land use of fields drained on the Cleveland estate, Northampton-shire, 1848–1871*

Field acreage drained	2,357
At time of draining percentage in:	
arable	21
pasture	79
Cultivated acreage 1849	3,413
Percentage of cultivated area in:	
arable	15
pasture	85
Percentage drained of 1849 area in:	
arable	96
pasture	64

Sources: Raby MSS: Sudborough draining vol., 1848–52; Draining abstracts, 1848–53; Draining vols., 1861–71; Field books, 1849–78

periods, 1848–52 and 1861–71 (Fig. 3.21). A Public Money Draining Acts loan financed the improvement in the first period and the duke wished to use these funds to drain the arable of the estate, attaching little importance to its value on pasture. The agent, T. F. Scarth, differed from this view, believing that pasture could benefit from the improvement. However, only when all the arable that required draining had been treated was Scarth allowed to apply the remnants of the loan to pasture areas.[43] When the improvement was re-introduced a decade later, the draining of pasture was dominant, 97 per cent of the field area drained, as there was little undrained arable remaining on the estate. Such precedence in the treatment of arable characterized the selected bailiwicks on the duke of Northumberland's estate and the Matfen part of the Blackett estate, even where, as in the case of the Barrasford bailiwick and Matfen, grassland was more extensive (Tables 6.14 and 6.15).

The draining of grassland, although not ignored on these estates, was less intense before 1870. The return obtainable from grassland could never be so high or as quickly realized as that from arable. On most farms, given the need to pay interest on the improvement, tenant insistence on draining arable, producing the more immediate return, was understandable. Only after the benefit from arable land had been largely reaped was attention turned to wet grassland. As a result the draining of grassland was a secondary and subsequent activity and in the period 1840–70 draining must be seen primarily, although not exclusively, as an improvement for arable.

This arable priority would seem to have disappeared towards the end of the century. With the decline in draining activity from the 1870s, less information is available on the agricultural use of land drained. However, for

Table 6.14 *Agricultural use of land drained on the Matfen section of the Blackett estate, Northumberland, 1857–1889*

Period	Acreage drained with land-use data	At time of draining percentage in	
		tillage	grassland
1857–9	568	67	33
1860–9	1,268	52	48
1870–9	161	51	49
1880–9	124	12	88

Source: As for Table 6.12

Table 6.15 *Area drained on six bailiwicks of the Northumberland estate, Northumberland, 1844–1869, in terms of 1850 land use*

Period	Acreage drained	Percentage in arable	pasture	Acreage drained	Percentage in arable	pasture	Acreage drained	Percentage in arable	pasture
	Alnwick			Barrasford			Lucker		
1844–9	916	90	10	–	–	–	1,362	94	6
1850–9	3,627	60	40	320	55	45	2,938	82	18
1860–9	823	21	79	134	19	81	1,362	29	71
	Newburn			Shilbottle			Warkworth		
1844–9	340	96	4	415	86	14	340	90	10
1850–9	967	90	10	2,018	79	21	1,831	85	15
1860–9	649	79	21	488	61	39	446	50	50

–: No data available
Source: As for Table 6.11

the greater part of the Overstone (Fig. 3.20) and for the Buccleuch Boughton estates in Northamptonshire, both of which displayed an expansion in landlord investment in the improvement after 1880, the use of land drained in the periods 1880–5 and 1888–96 respectively may be gauged (Table 6.16). Despite their different land-use structures, both offer little evidence that draining was concentrated on arable land. Indeed, arable and pasture were drained with equal intensity, suggesting that after 1880 wet land was treated to improve general cultivation, irrespective of land use. In 1894, Lord Wantage, the then owner of the Overstone estate, explained the aim as to put property 'in the best condition and best order ... perhaps not entirely with the

Table 6.16 *Area drained on the Overstone estate, 1880–1885, and on the Buccleuch Boughton estate, 1888–1896, Northamptonshire, in terms of repective land uses in 1880 and 1896*

Estate	Acreage drained	Percentage in		Cultivated acreage	Percentage in		Percentage drained of cultivated area in	
		arable	pasture		arable	pasture	arable	pasture
Overstone	1,825	68	32	10,455	67	33	19	18
Buccleuch	1,416	42	58	10,151	38	62	15	13

Sources: Buccleuch MSS: Numerical reference of the Boughton estate, 1896; 25 in Ordnance Survey maps (1885 edition) of the Boughton estate; NRO, Overstone MSS: Ov. 196–8, Estate cropping books, 1850–1919; Ov. Maps, 184–92, 194–5, 327, 329–31, 342, 354

view of making simply a profit'.[44] As tenants paid no interest on the improvement on either estate in this period, the draining of all waterlogged land was to their advantage.

Besides agricultural usage, the occurrence of the improvement was related to land quality. Up to 1880 tenants would have been prepared to drain only land that would be capable of covering either the payment of interest or their contribution to the cost of the improvement. In effect, such economic considerations would have rendered draining most profitable on land of high agricultural value whose potential had been limited by excess soil water. That this relationship was widely recognized by tenants may be demonstrated by plotting parishes subject to draining rentcharge arising from loans under the land-improvement legislation over the period 1847–99 against the classification of land quality devised in the first Land Utilization Survey[45] (Fig. 6.1). Loan-financed draining was located predominantly in areas of good-quality agricultural land and to a lesser extent on medium-quality land, there being little evidence of draining loans being applied to land of poor quality. The use of draining to realize the productivity of good-quality land damaged by water may be illustrated in detail on the sample estates. Where value can be determined, draining was employed not on the lowest valued land but on that slightly below the estate average. Thus, of the 32 acres drained in 1868 on Chapelhaies farm on the duchy of Cornwall's Bradninch estate, the arable was assessed in 1855 at £0.90 and the pasture at £1.025 per acre, rates that were just under the average for arable and pasture on the farm, being respectively £0.975 and £1.15 per acre.[46] Again, on the duke of Cleveland's Northamptonshire estate, the 1846 value of the fields covering 1,444 acres to

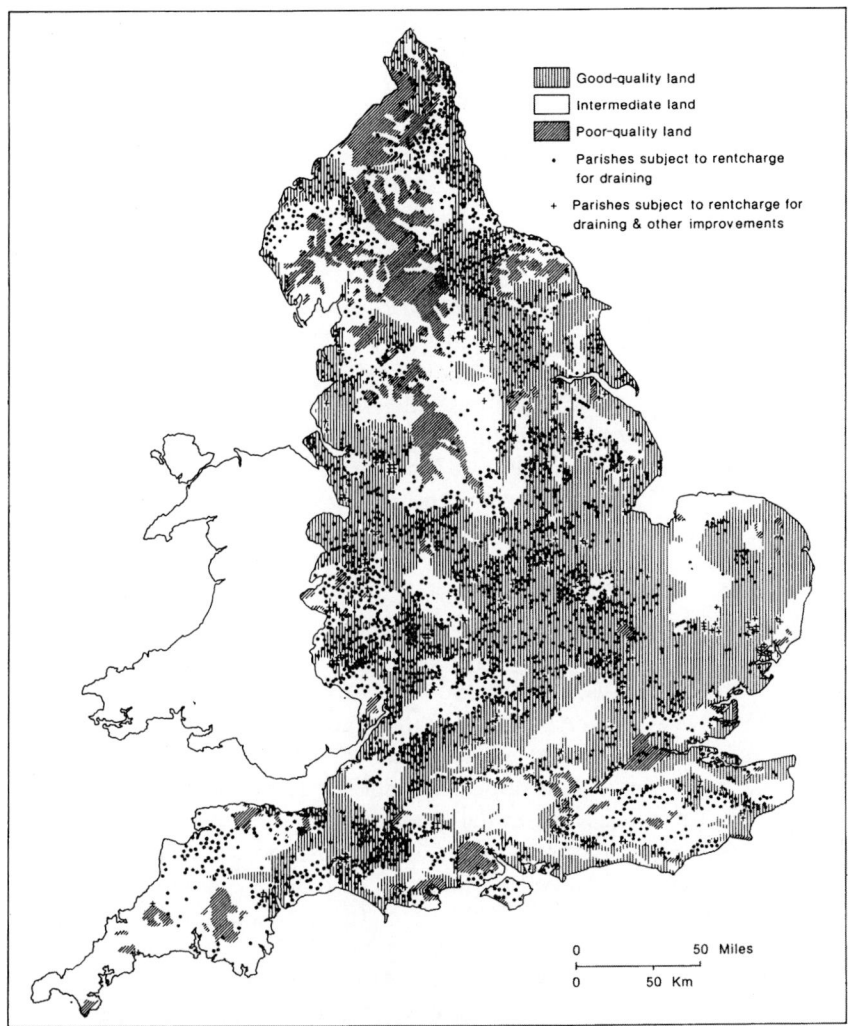

Figure 6.1 Parishes subject to draining rentcharge, 1847–1899, and land quality after the first Land Utilization Survey (*Sources:* As for Fig. 3.3; Land Utilization Survey, *Land Classification, 1:625,000*, 1945, sheets 1 and 2)

be drained between 1848 and 1852 was £1.11 per acre in comparison to an estate average of £1.27. By 1855, the value of the drained land had risen to £1.35 per acre, close to the new estate average of £1.375, indicating that on this property draining had enhanced land quality and therefore value.[47] There was little attempt to drain wet land that was also poor and low valued, even on estates with expansive draining programmes. Thus, although

J. Snowball, chief commissioner to the duke of Northumberland, identified 1,356 acres in need of draining on Prudhoe bailiwick in 1866, at least 500 were valued at no more than £0.35 per acre and such land, described by the bailiff as being unable to pay for the improvement, remained undrained.[48] Concentration on better-quality land restricted the financial risk that may have attended the adoption of the improvement for both landlord and tenant.

Draining and agricultural productivity

In that the draining systems employed after 1840 were generally successful in drying land, the adoption of the improvement was likely to have resulted in a growth in agricultural output and hence in overall productivity. Contemporary agriculturalists identified a variety of ways by which draining achieved improvement in output. At the most basic level, the provision of draining was considered to increase the cultivable acreage, particularly in areas with ridge and furrow. Philip Pusey estimated that on undrained ridged land, 10 per cent of the cultivable area was lost through furrows acting as open drains.[49] Whether or not ridges were lowered after draining, as Parkes advocated as part of his deep-draining scheme,[50] the laying of drains in furrows made more land available for cultivation. The practice was reported on many of the sample estates. On the duke of Northumberland's property, Parkes in his 1848 report recommended the gradual reduction of ridges after draining, not less than 3 in at each fallowing. Although supported by the fourth duke, there was tenant opposition and by 1850 Parkes was prepared to lay drains in furrows as long as they were straight and of sufficient distance apart.[51] The placing of drains along the lines of furrows was also common on the Matfen part of the Blackett estate from 1840 and became the dominant draining system in the 1880s.[52] Such use of furrows similarly characterized the draining systems that had developed by the 1880s on the Overstone and Spencer estates in Northamptonshire.[53] Although not capable of precise calculation, an element of the increased agricultural output arising from the use of draining must have been the product of rendering all land in a field suitable for cultivation.

At the same time, many agriculturalists ascribed to this process of drying land by effective draining other improvements in cultivation. J. H. Charnock and Hewitt Davis, for example, suggested that draining could be expected to make cultivation in general more secure and regular, to reduce cultivation costs and to improve crop quality.[54] The realization of these benefits was reported from many of the sample estates. R. Mein informed Christopher Haedy in 1852 of 'the economy of seed and labour' that invariably accompanied the draining of arable on the Northamptonshire estate of the duke of Bedford, while the agent to their mid-Devon property was able to record in 1864 an upgrading in the composition of grassland on the drained farms.[55] Despite their disagreement over method, both Lord Fortescue and Parkes attested to the improvement in the condition of arable crops and pasture on the greater part of the drained area of the north Devon estate.[56] And the marked improvement that draining had brought to the general state of cultivation on Waterloo farm on the Bedford mid-Devon estate particularly pleased the agent, for as the farm was the first on entering the property from Cornwall it was 'very desirable that it should present a good appearance'.[57] Such changes, beyond quantification, must have contributed significantly to the overall productivity of individual farms.

The impact of draining on agricultural output, however, was most frequently assessed by crop yield. The contemporary agricultural literature of the 1840s and 1850s agreed that draining would enhance yields but varied considerably in predicting the scale of that increase, which was expressed in terms of arable crops, especially cereals, there being little discussion of the value of the improvement to grassland. Assessment of yield increment tended to take two forms, statements of general expectation, which for cereals ranged from 8 to 100 per cent improvement,[58] and reports of outstanding returns, usually at the level of individual fields, which were widely publicized in the literature to encourage adoption.[59] There was little attempt to establish the differences in yield that had been experienced by the generality of farmers, a more useful guide to the productive nature of the improvement.[60]

The ability of draining to augment yields was widely appreciated. William Sample informed tenants on the Portland estate in Northumberland thinking of applying for the improvement that they could count on a considerable increase in yield.[61] However, the overall yield increments would appear much smaller than those suggested in the literature. Sample, for example, illustrated to the duke of Portland that, while the level of increase was subject to much variation, reflecting the nature of individual fields, high returns were exceptional, there being few instances on the estate where draining had doubled the produce.[62]

More detail on crop yields and draining is available for the duke of Northumberland's estate. Arising from a wish 'to possess some statistical information relative to the annual produce upon the drained and undrained lands', the bailiff of each bailiwick was required in 1850 to report on the yield of arable crops on the two states of land by farm.[63] Although relating only to a single year and neglecting grassland, the extent of the survey provides a basis for establishing aggregate data on the effect of draining on crop output. By 1850, only 9,838 acres of the estate had been drained, 21 per cent of the total area improved over the period 1844–99, and not every farm in the reports possessed drained land. To make an effective comparison, only those farms which grew crops on both drained and undrained land have been

Table 6.17 *Average per acre yield by farm of arable crops on drained and undrained land on the Northumberland estate, Northumberland, 1850*

Bailiwick	Total no. of farms	Wheat			Barley			Oats			Turnips		
		farms with data	bushels/acre on land drained	undrained	farms with data	bushels/acre on land drained	undrained	farms with data	bushels/acre on land drained	undrained	farms with data	tons/acre on land drained	undrained
Alnwick	28	22	24.9	22.4	8	28.8	31.6	11	34.1	31.8	14	18.7	15.4
Barrasford	26	3	16.0	15.7	1	30.0	28.0	3	32.0	23.3	7	10.1	11.9
Chatton	21	10	25.0	23.4	10	32.1	30.9	10	29.0	33.6	12	13.8	12.3
Longhoughton	27	16	26.0	24.6	2	30.0	30.0	5	36.0	36.0	12	14.3	14.6
Lucker	23	13	27.3	23.5	7	32.9	30.0	14	32.5	29.2	11	12.0	12.6
Newburn	17	7	22.3	21.7	1	36.0	30.0	3	43.3	43.3	2	15.5	14.5
Prudhoe	29	10	24.0	17.2	0			6	34.0	30.7	1	14.0	12.0
Rothbury	26	9	21.1	18.8	2	38.0	31.0	5	21.8	20.1	7	14.7	12.9
Shilbottle	23	12	21.8	17.9	1	34.0	32.0	8	31.3	26.8	8	17.8	16.9
Tindale	36	3	24.0	19.3	4	33.5	26.5	6	37.5	28.8	6	17.3	11.9
Tynemouth	36	12	26.0	20.6	1	40.0	40.0	5	44.0	32.4	4	17.3	16.3
Warkworth	21	7	22.3	20.0	4	29.0	24.0	4	35.5	33.5	4	14.3	11.0
Estate average	313	124	24.2	21.2	41	31.6	29.9	80	33.5	30.5	88	15.0	13.7

Sources: Northumberland MSS: Statistical reports for Barrasford, Chatton, Longhoughton, Lucker, Newburn, Prudhoe, Tynemouth and Warkworth; NdRO, Bell Collection: ZAN 65/18, Statistical report for Tindale; ZHE 1/2, Statistical reports for Alnwick, Rothbury and Shilbottle

included in the analysis. The average yield per acre of wheat, barley, oats and turnips on these farms has been calculated for each bailiwick (Table 6.17). The results indicate that, while output varied greatly over the estate, the yield for all four crops on drained land was consistently higher than on undrained. Indeed, for the estate as a whole the difference in yield of all crops on drained and undrained land on these farms was statistically significant. The Wilcoxon matched-pairs signed-test, which measures the direction and magnitude of differences between pairs,[64] produced z values of -6.18, -2.52, -2.83 and -2.99 respectively for wheat, barley, oats and turnip yields on drained and undrained land by farm. These values allow the null hypothesis, that there was no statistical difference in yield on drained and undrained land, to be rejected at the 0.01 level for barley, oats and turnips and at the 0.001 level for wheat.

While yields were heavier on drained land, the difference in the level of increase fluctuated widely among the bailiwicks for all the crops. However, output from drained land did not exceed 40 per cent for wheat, 26 per cent for barley, 37 per cent for oats and 45 per cent for turnips. For the estate as a whole, the yield differential between drained and undrained land was more modest. The largest yield increment occurred between drained and undrained land growing wheat. Output on average was 14 per cent more on drained land, emphasizing the value of the improvement to wheat production. The differences in the yield of barley, oats and turnips were smaller, drained land possessing respectively 6, 10 and 9 per cent higher output per acre. To some extent the data may disguise the amount of yield increase of certain crops after draining. Not all the undrained land would have required the improvement and such may have been the case on much of the land that produced barley and turnips, crops more suited to drier, lighter soils. While the absolute difference in yield of these two crops on drained and undrained land was less than for wheat and oats, the adoption of draining had raised output on formerly wet land to levels comparable to and slightly above those obtained on naturally dry land. As W. Laws, bailiff of Barrasford bailiwick, noted on Chishillways farm:

The land upon this farm which has been drained is all of very poor quality, the naturally dry land is ... what may be called a thin turnip soil. It will be seen that the produce from the drained and undrained land does not differ much in quantity ... The former however is much improved by draining and but for that process the difference of produce would have been much greater.[65]

Even with these allowances, the yield data for the estate reveal that the average level of increase approximated the lower ranges of contemporary expectations. Nevertheless, although consistently high returns may have been rare, they clearly establish that draining improved output by a significant amount.

These yield values may be used to determine a more realistic measure of the financial return to tenants on draining. The average levels of yield increase recorded on the estate in 1850 demonstrate that for cereals tenant participation in the improvement would have paid. Wheat afforded the largest return on outlay. If constant throughout the 1850s, the average difference in wheat yield on drained land on the estate would have produced an additional return of £1 per acre on the average price of the cereal of £2.67 per imperial quarter in the decade 1850–9. The cost of draining averaged £4.93 per acre on the estate in the same period, resulting for the tenant in an annual interest payment of £0.25. The improvement, therefore, provided a balance of £0.75 per acre when wheat was grown, a sum representing a threefold return on the tenant outlay, the annual interest payment. As the average proportional yield increases and the current prices were lower for barley and oats, the margin of profit from these grains was smaller but none the less significant. Thus, over the same decade using average prices, drained land in oats would have produced additional receipts of £0.43 per acre on the estate. Draining interest absorbed £0.25 of this amount, the residue forming a 72 per cent return on the tenant's charge for the improvement. While the scale of yield increase after draining on the duke of Northumberland's lands would suggest that the financial return fell below the level that many tenants and landowners may have been led to expect from the contemporary literature, it was sufficient to confirm William Sample's assurance to the duke of Portland in 1848 that on his Northumberland estate 'in every instance [he had] reason to believe the produce has been increased beyond the ratio of the draining rent'.[66]

Changes in farming systems after draining

The technical efficiency of the new methods of draining of the 1840s persuaded many agriculturalists that changes in farming systems must accompany the adoption of the improvement. By drying land, soil texture was modified, converting, according to James Smith, the most sterile soils into deep rich loams.[67] On this new land, farmers were presented with the opportunity of increasing their prosperity by abandoning traditional practices and by introducing more productive systems. Of these, the adoption of mixed farming was most frequently predicted in the literature, draining allowing the practices pursued on the light soils to be transferred to the heavy lands. Thus, draining would lead to the removal of fallow, permit the growth and extension of green crops, especially turnips, and facilitate an increase in livestock numbers which could then be folded on the green crops.[68] The value of draining achieving such a transformation in cropping was widely recognized, John Grey for example noting in 1841 'what an immense increase of wealth to individuals and of produce to the public might be obtained'.[69]

Table 6.18 *Cropping on the Portland estate, Northumberland, 1841–3 to 1864–6*

Period	Cultivated acreage	Average percentage for three years in		
		cereals	fallow	pulses
1841–3	11,480	33	18	3
1849–51	11,480	32	20	3
1864–6	11,480	30	15	3

Source: NdRO, Sample MSS: ZSA/18/5, Tillage accounts, 1841–52, 1864–6

Table 6.19 *Cropping on the Cleveland estate, Northamptonshire, 1850–1899, by five-year averages*

Period	Cultivated acreage	Percentage in		Arable crops as a percentage of the cultivated area						
		pasture	arable	wheat	barley	oats	pulses	roots and green crops	fallow	rotation grasses
1850–4	3,405	84	16	6	2	0	3	0	4	1
1855–9	3,320	84	16	6	3	0	3	0	3	1
1860–4	3,283	84	16	6	2	0	4	0	4	0
1865–9	3,382	81	19	6	3	0	4	0	5	1
1870–4	3,361	81	19	6	3	0	4	3	2	1
1875–9	3,380	81	19	5	4	0	3	3	2	2
1880–4	3,313	83	17	5	3	1	2	1	3	2
1885–9	3,315	88	12	3	2	1	2	2	0	2
1890–4	3,355	88	12	3	2	1	1	2	1	2
1895–9	3,283	89	11	3	1	2	1	2	1	1

Source: Raby MSS: Field books, 1849–1909

However, not all commentators saw draining accomplishing such fundamental changes in farming systems, and several suggested that the improvement would enhance prosperity by the more straightforward means of allowing an extension of cereal cultivation. Thus, Parkes could report in 1848 that large parts of England, previously unable to carry wheat because of excessive wetness, were now capable of producing superior crops.[70] Overall, the potential changes in farming systems deriving from draining were perceived in terms of arable production and, although Caird argued in 1850–1 that the future of heavy-land agriculture lay with grass and green

Table 6.20 *Cropping on six bailiwicks on the Northumberland estate, Northumberland, 1840–1870*

			Percentage of cultivated area in			
Year	Cultivated acreage	Percentage drained after 1844	grass 2 yrs old and over	tillage including clover	fallow	roots
Alnwick						
1840	–	–	–	–	–	–
1845	–	–	–	–	–	–
1850	–	–	–	–	–	–
1855	8,845	27	55	45	13	12
1860	8,845	39	56	44	8	17
1865	9,505	46	59	41	9	17
1870	9,505	51	63	37	5	24
Barrasford						
1840	8,836	–	76	24	17	10
1845	8,881	0	73	27	17	10
1850	8,022	1	81	19	15	14
1855	–	–	–	–	–	–
1860	8,024	12	78	22	6	11
1865	7,992	16	77	23	3	13
1870	7,301	17	83	17	–	–
Lucker						
1840	9,117	–	50	50	15	12
1845	9,117	0	48	52	14	12
1850	9,038	15	46	54	10	14
1855	9,023	37	46	54	10	16
1860	8,978	46	49	51	10	17
1865	8,985	57	50	50	7	20
1870	9,044	60	51	49	8	21
Newburn						
1840	3,110	–	23	77	20	5
1845	3,110	2	19	81	19	5
1850	3,148	9	19	81	18	7
1855	3,133	34	18	82	15	8
1860	2,919	38	20	80	13	9
1865	3,221	50	23	77	11	9
1870	3,221	59	29	71	10	12

Table 6.20 (*cont.*)

Year	Cultivated acreage	Percentage drained after 1844	Percentage of cultivated area in			
			grass 2 yrs old and over	tillage including clover	fallow	roots
Shilbottle						
1840	5,805	–	30	70	20	5
1845	5,809	1	34	66	21	5
1850	5,120	12	34	66	18	9
1855	5,235	42	41	59	15	8
1860	5,237	50	47	53	18	10
1865	5,175	61	51	49	14	6
1870	5,217	64	53	47	12	14
Warkworth						
1840	4,201	–	35	65	18	8
1845	4,201	1	34	66	17	9
1850	4,184	7	31	69	16	10
1855	4,235	25	33	67	12	12
1860	4,247	44	38	62	14	14
1865	4,228	52	41	59	12	15
1870	4,228	57	48	52	10	17

–: No data available

Data not available for Barrasford in 1845, those for 1846 used; data not available for Alnwick, Newburn, Shilbottle and Warkworth for 1860, those for 1859 used; data not available for Warkworth for 1870, those for 1869 used

Source: Northumberland MSS: Annual returns of state of farms, 1840–70

crops,[71] no contemporary case was made for drained land to be converted to intensive grassland husbandry.

Anticipation of a revaluation in farming systems on the adoption of the improvement was evident on many of the sample estates. On the north Devon property of the earls Fortescue, the aim of the draining programme begun in 1848 was to aid the spread of turnip husbandry, while F. Thynne in 1850 hoped that the improvement on the Sidmouth estate would allow the conversion of its dairy system into one of good corn farming.[72] However, for those estates with cropping books, the changes that draining brought to agricultural practices may be established in detail.[73]

The results in general indicate that the divergence between prediction and reality was considerable. The ability of the improvement to promote the introduction of mixed farming was exaggerated, draining failing to provide a uniform response in cropping, and the degree of adoption of roots and green crops on drained land largely reflected soil quality. By 1864–6 on the duke of Portland's Northumberland estate, about 35 per cent of the cultivated area had been drained. However, the overall fallow area had altered little from 1841–3, occupying between 15 and 20 per cent of the cultivated acreage (Table 6.18). The greater part of the estate lay on clay soil with an average rental value of £0.73 per acre in 1859–61.[74] The agent, William Sample, acknowledged that the heaviest soils on the property had been excessively cropped and that a bare fallow was essential to their cultivation. Yet, with draining, the land was still too heavy for turnips and the associated husbandry, and the need for fallowing remained. On Cockle Park farm, Sample could note in 1852 that the land drained was 'of too strong a nature to produce green crops [and] it cannot be got into so good a tilth as it will be by a naked fallow'.[75] Again, on the duke of Cleveland's Northamptonshire estate, nearly all the arable had been drained between 1848 and 1852. However, there was no reduction in fallow nor corresponding increase in the area of roots and green crops until after 1870, and therefore draining could not have been the direct cause (Table 6.19).

An extension of the root area may be observed on some estates as the drained acreage expanded, although falling short of the full introduction of mixed farming. Thus, on six of the bailiwicks on the duke of Northumberland's estate, the proportion of the cultivated area devoted to roots, which included potatoes, increased between 1840 and 1870 as more land was drained (Table 6.20). Although there was a fall in the fallow acreage, the adoption of draining failed to eradicate this feature, which still formed around 10 per cent of the cultivated area on most of the bailiwicks by 1870. That turnip husbandry could follow on draining may be illustrated from the Grey estate in Northumberland. The Carham part of the property, located in a district of light soils where there was little need for draining, had developed that system of cultivation by 1845 (Table 6.21). While wet, the estate at Ancroft and Burton lay mainly on loam soils and as these were drained from 1840 the removal of fallow was attended by a growth in the acreage of roots and green crops and of barley. The Chevington section of the estate possessed heavier clay soils, similar to those of the neighbouring Shilbottle and Warkworth bailiwicks on the duke of Northumberland's property. Like those bailiwicks, despite intensive draining, although the area growing roots and green crops increased, fallow still remained an important element of the tillage by 1870.

Draining did not produce the thorough revaluation of farming on heavy soils that had been prophesized by many contemporary agriculturalists, whose enthusiasm for the technical efficiency of the improvement no doubt coloured their assessment of its agricultural potential. The full development

Table 6.21 *Cropping on constituent parts of the Grey estate, Northumberland,*
1845–1870

Year	Cumulative draining outlay per acre from 1840 (£)	Cultivated acreage	grass	tillage	wheat	barley	oats	pulses	root and green crops	fallow
Ancroft										
1845		1,663	33	67	15	7	16	8	11	10
1850	2.37	2,029	40	60	15	5	16	5	11	8
1855		1,981	38	62	11	7	17	4	18	5
1860	3.99	2,493	44	56	8	10	14	3	18	4
1865		2,493	44	56	6	13	14	3	13	7
1870	5.18	2,493	41	59	4	16	18	2	19	0
Burton										
1845		2,193	40	60	18	6	9	4	17	6
1850	1.95	2,192	43	57	13	8	16	3	12	5
1855		2,192	39	61	12	10	15	4	15	5
1860	3.52	2,189	43	57	14	8	15	2	16	2
1865		2,189	48	52	12	10	11	3	15	1
1870	4.07	2,189	48	52	10	10	11	2	18	1
Carham										
1845		4,199	45	55	9	11	15	1	18	1
1850	0.64	4,153	43	57	14	9	12	1	21	0
1855		4,218	37	63	11	12	14	5	21	0
1860	1.48	4,218	39	61	6	16	14	5	20	0
1865		4,218	39	61	6	16	15	5	19	0
1870	2.07	2,631	44	56	4	16	16	2	18	0
Chevington										
1845		4,175	32	68	21	2	16	3	5	21
1850	2.46	4,160	35	65	21	2	13	3	7	19
1855		4,148	39	61	17	3	12	5	7	17
1860	6.22	4,148	44	56	15	3	13	6	7	13
1865		4,144	48	52	12	4	11	6	9	10
1870	6.86	3,213	47	53	14	4	12	5	7	11

Sources: DPD, Grey MSS: Ledger books, 1840–73; Draining vols., 1841–86;
Cropping book, 1845–78

of mixed-farming systems after draining was limited in extent, being
restricted to what T. Sample described as 'the better class of land'.[76] Thus,
the 132 acres of tillage on Pegswood Middle Moor farm on the Portland
estate were reported in 1864 capable of growing turnips after draining, but
only 40 of these, being light gravelly soil, could allow the crop to be eaten off

the ground by sheep.[77] Nevertheless, even on land too heavy for the light-land systems, the adoption of draining was associated with improvements in cropping. The reduction in fallow and the expansion of roots that characterized the duke of Northumberland's estate would have made a major contribution to an overall increase in farm productivity. On estates with poorer soils, draining provided less dramatic but none the less important benefits to farming practices. Although no significant fall in the fallow acreage occurred on the Portland estate between 1841–3 and 1864–6, William Sample was pleased to admit that the removal of excess water from strong land had allowed sheep to be fed on it with safety when in grass, avoiding poaching, and permitted some replacement of clover with tares to augment the supply of winter fodder.[78]

If the realization of mixed farming on drained land was imperfect, little evidence is available to support the contention that the improvement encouraged an extension of cereal production in the 1840s and 1850s. Wheat may have provided the largest financial return on draining, but no increase in its acreage occurred on those sample estates with cropping data in this period. Indeed, most of the estates displayed some decline in the proportion that arable or tillage formed of the total cultivated area between 1840 and 1870. Such a trend would seem to endorse the view proposed by R. W. Sturgess that on the claylands of the north and west of the country in the 1850s and 1860s there was a positive policy to convert newly drained land from corn to grass to enhance livestock production.[79] However, on closer inspection the cropping data available for Northumberland estates would suggest little direct connection between draining and the extension of grassland.

Although the grassland area increased in the 1850s and 1860s, the extent of the expansion was slight on the sample estates. The largest proportional growth was recorded on the Shilbottle bailiwick on the duke of Northumberland's estate, where grassland rose from 34 per cent of the cultivated area in 1850 to 53 per cent in 1870 (Table 6.20). Correspondingly, while the cereal and arable areas fell over the same period, the decline was small. On the Portland estate, cereals as a percentage of the cultivated area decreased by only two percentage points from 1849–51 to 1864–6 (Table 6.18). No fall in cereal acreage occurred between 1850 and 1870 on the Grey estate at Ancroft, while at Burton and Chevington the decline was limited, cereals forming about 36 per cent of the cultivated area in 1850 and about 30 per cent in 1870 (Table 6.21). Clearly, there was no major movement from cereals to grassland on these estates.

While the improvement was concentrated on arable, large-scale conversion of such land to grass on draining cannot be detected. On the Shilbottle bailiwick, with the largest proportional swing to grassland, tillage fell and grassland rose by 720 acres between 1850 and 1860. Over that period, 2,018 acres were drained of which 1,594 were in tillage. Assuming that all tillage converted to grassland had been drained, over 800 acres of drained tillage would have remained unchanged in its land use. There was less correspondence between acreage drained and the growth of grassland on the Newburn bailiwick on the same estate. From 1850 to 1870, grassland increased by just over 330 acres at the expense of tillage. At the same time, 1,616 acres were drained of which 1,374 were in tillage, leaving over 1,000 acres unconverted.[80] On all the estates the rise in grassland acreages was well below the arable areas drained.

However, the assumption that the growth of grassland occurred on drained arable land must be questioned. Many tenants would have been reluctant to put newly drained land to grass because of the costs involved. The draining of arable, especially when sown with cereals, produced an immediate increase in output and financial benefit. The conversion of drained land to grass required time for the crop to be effectively established, with a slower resulting return. Nevertheless, over that growing period the tenant would have been committed to paying the landlord interest on the improvement. Indeed, rather than encouraging the conversion to grassland, the nature of the return on draining and the need to pay interest, while not enlarging the arable area, served to retain land in arable production. Thomas Sample writing in 1885 of land use changes on the Portland estate in the 1850s noted that arable became focussed on the drained land, leaving inferior land to be laid to grass to reduce cultivation costs.[81] An indication of the maintenance of arable on drained land may be seen on the Matfen section of the Blackett estate. Between 1862 and 1885 tillage on the whole estate declined from 37 to 18 per cent of the cultivated area (Table 6.22). However, on land drained between 1857 and 1869 the proportion in both tillage and cereals remained throughout above the estate average. In general, on these Northumberland estates, the adoption of draining did not result in an extensive conversion of corn to grass in the 1850s and 1860s. The increase in grassland that occurred would seem to derive from limited adjustments on poorer-quality land to trends in agricultural prices towards livestock products rather than from a planned programme of arable conversion based on draining.

After 1870, discussion of the effect of draining on farming systems disappeared from the agricultural literature. The cropping and draining data that are extant for the sample estates in this period suggest that changes in land use tended to occur independently of the amount drained. Thus, although the adoption of the improvement on the duke of Bedford's mid-Devon estate continued well after 1870, unlike the north Devon and Tavistock properties where draining activity had been largely completed by that date (Fig. 4.7b, c and d), the trends in land use on the mid-Devon estate of a continuous decline in tillage and a corresponding growth in grassland after 1870 mirrored closely those on the north Devon and Tavistock

Table 6.22 Cropping on land drained, 1857–69, on the Matfen section of the Blackett estate, 1862–1885

Year	Acreage drained 1857–69	Percentage of drained area in					Total cultivated estate acreage	Percentage of total cultivated area in				
		grass	tillage	cereals	roots and green crops	fallow		grass	tillage	cereals	roots and green crops	fallow
1862	1,836	48	52	37	8	7	6,978	63	37	25	8	4
1865	1,836	54	46	26	15	5	6,978	68	32	21	9	2
1870	1,836	63	37	24	10	3	6,978	73	27	18	8	1
1875	1,836	71	29	19	9	1	6,978	77	23	15	7	1
1880	1,836	72	28	21	6	1	6,978	77	23	16	6	1
1885	1,836	79	21	16	5	0	6,978	82	18	12	6	0

Source: As for Table 6.12

Table 6.23 *Cropping on the Bedford Tavistock and mid- and north Devon estates, 1867–1895*

Year	Cultivated acreage	Cumulative acreage drained from 1867	Percentage of cultivated area in					
			grass	tillage	wheat	oats and barley	fallow and green crops	orchard
Mid-Devon								
1867	3,427		58	42	8	15	16	3
1870	3,828	505	54	46	11	17	16	3
1875	4,254		59	41	8	17	14	2
1880	4,353	1,579	60	40	8	18	12	2
1885	4,616		67	33	4	18	9	2
1890	4,621	1,653	69	31	3	18	8	2
1895	4,618	1,668	70	30	1	19	8	2
North Devon								
1867	1,167		54	46	6	18	16	6
1870	1,760	75	53	47	9	16	18	5
1875	1,824		60	40	9	12	14	5
1880	1,907	285	64	36	9	12	12	4
1885	1,935		65	35	5	15	12	4
1890	1,933	321	66	34	7	13	11	4
1895	1,963	324	70	30	3	14	9	4
Tavistock								
1867	9,224		71	29	3	12	9	5
1870	10,072	35	74	26	4	11	9	2
1875	10,418		79	21	3	9	7	3
1880	10,644	151	82	18	2	9	6	2
1885	10,743		84	16	1	9	6	2
1890	10,891	230	84	16	1	8	6	2
1895	11,091	245	84	16	1	9	5	2

Source: Bedford MSS: Annual reports, 1867–95

properties (Table 6.23). Again, on parts of the Overstone estate in Northamptonshire, the movements in land use from 1870 to 1890 on land drained between 1880 and 1885 differed little from the overall changes in cropping (Table 6.24). While draining improved the productivity and cultivability of land, it would seem to have exerted little influence on farming systems practised towards the end of the century, cropping responding to broader economic factors.

The adverse assessment made by E. J. T. Collins and E. L. Jones of draining as an agricultural improvement needs considerable revision in the light of the evidence of the sample estates.[82] As a technique to remove excess water and to dry land, the systems of draining that evolved from the 1840s

Table 6.24 *Agricultural use of land drained, 1880–1885, on the Braybrook, Clipstone, Hackleton, Pytchley and Stanwick parts of the Overstone estate, Northamptonshire, 1870–1890*

Year	Acreage drained 1880–5	Percentage in		Total cultivated acreage	Percentage of cultivated area in	
		arable	pasture		arable	pasture
1870	1,241	63	37	5,036	54	46
1880	1,241	63	37	5,024	51	49
1890	1,241	35	65	4,983	34	66

Sources: NRO, Overstone MSS: Ov. 196–8, Estate cropping books, 1850–1919; Ov. Maps, 187, 190, 191, 327, 331

were generally successful. Few cases of failure were reported in the main period of adoption of the improvement between 1840 and 1870 on those estates that accepted draining as a landlord charge. Much of this success arose from the standardization of draining systems on these estates, initially on the principles of deep draining and subsequently after the late 1870s at slightly shallower depths, and from the widespread use of pipes of sensible diameter for fill. Most estates were careful to avoid the extreme recommendations of the proponents of new draining systems. The greater part of the draining established in the period 1840–70 remained functional to the end of the century, the amount of land redrained on these estates forming a small proportion of total activity. Nevertheless, with increasing age, draining systems required attention and landlords were prepared to provide this service as maintenance of existing drains reduced the need for more costly redraining. The achievement of coherent draining systems on estates that only provided materials was more difficult, tenants who put the pipes into the ground not necessarily following a uniform scheme. The technical limitations of much of this draining was revealed by the wet seasons around 1880 on the Buccleuch, Overstone and Spencer estates in Northamptonshire. Protracted water-logging compelled these landlords to undertake extensive redraining to prevent deterioration in land quality and capability.

Although draining systems were effective and estates attempted to ensure their working order, not all waterlogged, cultivated land was treated. In the period of most active adoption, draining was regarded as an improvement for arable land. It was applied with less intensity and later to grassland. For a tenant with interest to pay or labour costs to recover and a concern for profit, the amount and immediacy of the return on draining were important. Thus, land of good rather than of poor agricultural quality tended to be selected for the improvement, for when dry the former would be naturally capable of providing higher yields than the latter. And arable was favoured in preference to grassland in that its return on draining was greater and more direct.

Despite contemporary assurances, adoption of the improvement brought little radical change to farming systems on heavy land between 1840 and 1870. Sturgess' view that draining accomplished a conversion of cornland to grassland in the north and west of the country to take advantage of secular trends in agricultural prices towards livestock cannot be substantiated from the cropping data on the Northumberland sample estates.[83] The increase in grassland on these properties was slight, much drained arable remaining unaltered in land use. And much of the conversion from arable to pasture that was recorded was likely to have taken place on undrained rather than drained land. At the other extreme, the introduction of light-land mixed-farming systems did not follow automatically in the wake of draining. Indeed, little drained land on the sample estates witnessed the full-scale development of turnip husbandry. Its occurrence was largely restricted to land that, although wet, possessed soils of a lighter character. Draining could not so improve the texture of most clay soils as to allow the complete integration of grain and livestock farming.

Although failing to effect a revolution in farming systems, the assumption of Collins and Jones that draining did little to improve clayland agriculture cannot, however, be sustained.[84] The changes that resulted may not have been spectacular but none the less they were real and contributed to increasing the productivity of heavy-land farming. Thus, on a number of the sample estates the adoption of the improvement was accompanied by some reduction in fallow and some extension of the root- and green-crop area, producing an overall growth in output. Where there was less alteration to cropping, the effective removal of excess water was of value in itself. Draining made the cultivation of all land easier and therefore less costly, and by modifying the effect of wet weather it reduced the risk to those crops already grown. More significantly, by drying land draining improved crops yields. On the duke of Northumberland's estate, the yield of the main cereals was on average between 6 and 14 per cent higher on drained than on undrained land. Although well short of the extravagant claims of many contemporary agriculturalists, these orders of increase are comparable to those reported from present-day draining.[85] If the yield increments on the Northumberland estate were representative of the general experience of draining in England after 1840, the improvement's adoption must have played a major part in the marked rise in average wheat yields that M. J. R. Healy and E. L. Jones identified as taking place throughout the 1840s and 1850s.[86] In addition, these average levels of yield increase, although moderate by contemporary standards, would have provided a financial return to tenants in excess of interest payments on the improvement on all cereals well into the 1870s. The availability of such a profit must have led many tenants to keep drained land in arable in spite of prices moving in favour of livestock products. Indeed, the increased productivity that efficient draining systems brought to arable farming would have contributed to the preservation of the importance of cereals, especially wheat, in much heavy-land agriculture.[87]

7

Findings about underdraining

This study has been primarily based on data arising from the draining loans under the mid-nineteenth-century land-improvement legislation, relating to the whole country, and from the adoption of draining on estates in Devon, Northamptonshire and Northumberland. The data represent a marked advance on the sources, secondary contemporary accounts and questionable surrogate measures of land drained, employed in much present-day literature to determine the extent and effect of the improvement. As with so many aspects of nineteenth-century agriculture, too much emphasis has been placed on anecdotal evidence on draining, with little attempt being made to accumulate relevant and precise data on the improvement.[1] For draining in particular, reliance on anecdotal material poses real problems, for the perception of the improvement underwent radical revaluation in nineteenth-century agricultural literature. The enthusiasm for the new draining systems and materials around 1840 stimulated extravagant and untried claims for the improvement in the fields of cost, permanence, rental return, yield output and farming practice. When these expectations failed to materialize, there was a reaction to the improvement, its efficiency and productivity being doubted most notably in the 1873 report of the Select Committee of the House of Lords on the improvement of land. Such biases inherent in much of the contemporary literature preclude the attainment of any satisfactory assessment of draining as a nineteenth-century agricultural improvement and merely serve to perpetuate the divergence of present-day opinion. However, the draining-loan and estate data, although not allowing a full reconstruction of draining activity, provide at both the national and local level a consistent and reliable basis initially to clarify the adoption of the improvement and subsequently to assess its value in the development of nineteenth-century English agriculture.

As an innovation, the adoption of draining was prescribed by technical, physical and economic forces. Draining was perfected as an agricultural technique for removing excess water from soils in the 1840s, with the introduction of coherent draining systems and layouts, which experienced

241

only limited modification in the second half of the nineteenth century, and the mass-production of tiles and more importantly pipes for fill, superseding in effectiveness previous methods of surface and underdraining. These technical developments were widely adopted on the sample estates, there being a high degree of standardization in the drain format used on most properties after 1840. The great majority of these draining schemes were successful in drying land and, as most estates made efforts to maintain them in working order through inspection, cleaning and repairing, and limited redraining, they remained functional to the end of the century.

The physical need for draining agricultural land in nineteenth-century England was extensive. About 55 per cent of the cultivated acreage of the country in the 1870s possessed soils with impeded drainage that led to seasonal waterlogging, so inhibiting cultivation practices and crop growth. Although generally characteristic of and coincident with clays, land with impeded drainage was not restricted to those formations and included significant areas of finer-textured soils that could be classed as loams. Before 1840, available evidence suggests that the underdraining of waterlogged land was limited not only in extent but also in effectiveness and permanence. As a result, the technical advances in draining that occurred in the 1840s were potentially of benefit to the greater part of the total agricultural area with difficulties of drainage.

However, despite physical need and technical capability, the new draining systems were not applied to all land with impeded drainage after 1840. From the draining-loan and estate data, the inference may be made that about 4.5 million acres were drained between 1845 and 1899, a sum that represented around 35 per cent of the total wet-land area. Adoption of the improvement was most rapid between 1840 and 1870 and, despite some revival of activity in the early 1880s, thereafter investment in draining gradually declined to the end of the century. There was a lack of uniformity in the pattern of adoption of the improvement on soils in need of drainage, with the result that the intensity of draining varied widely throughout the country. Capital investment in the improvement made least progress in eastern and to a lesser extent southeastern and extreme southwestern England, but was most developed in northeastern, west-midland and western counties. The draining-loan data indicate that less than 15 per cent of soils in need of drainage in Essex, Norfolk and Suffolk were treated between 1845 and 1899, compared with over 50 per cent in Durham and Northumberland and in Herefordshire, Shropshire and Worcestershire. In general, draining activity was most intense in those parts of the country, the north and west, that in the third quarter of the nineteenth century were marked by the highest per acre incomes for landowners and farmers, and experienced, according to James Caird, the greatest rise in land values.[2] The adoption of the improvement may be seen not only as a product but also as an element of the prosperity of these areas. The failure of underdraining to be widely adopted in East

Anglia demands comment, especially as soils with impeded drainage formed
a significant proportion of the region, occupying 40 per cent of the area of
Norfolk and Suffolk. The region was renowned for agricultural innovation[3]
and tenant systems of underdraining practised there had been frequently
described in the literature up to the 1840s. However, the neglect of draining
is not inconsistent with other evidence of a faltering in agricultural advance
in the region in the second half of the nineteenth century: E. H. Hunt
suggested that labour productivity in agriculture in the area was amongst the
lowest in the country, while B. A. Holderness identified a general slackening
in landlord expenditure on agricultural improvement in Norfolk and Suffolk
from the 1830s.[4] Clearly, the tarnishing of the agricultural reputation of East
Anglia after 1850 warrants closer examination.

The specific pattern of the adoption of underdraining after 1840 can best
be understood in the context of landlord and tenant attitudes towards the
improvement. Most nineteenth-century agricultural improvements represented joint enterprises between landlords and tenants, and the delineation
of the relationships behind those undertakings offers the most rewarding but
largely untried opportunities to appreciate the process of their diffusion. For
draining, the increase in the average cost to over £5 per acre and the need for
a degree of care and supervision for effective implementation[5] that were the
consequences of the technical advances of the 1840s made the improvement
ideally suited to landlord finance and control. As a result, underdraining was
most extensively and successfully developed on estates, the majority in the
sample, that recognized and accepted the financial and administrative
implications of the new draining systems. However, not all estates identified
the need or were willing to take responsibility for the improvement. A
number of those examined employed the alternative methods of carrying out
draining of providing half the cost through allowances or of distributing
materials, leaving tenants to find in the first case an equal capital sum or in
the second the labour for the improvement. While curtailing landlord outlay
in the short run, in the long term the amount and efficiency of draining on
these estates were lower than on those where the improvement was a landlord
charge.

Although most effective as a landlord improvement, the intensity and
occurrence of draining were prescribed by the capital sums estates were
prepared to provide. While not discounting the importance of the attitude of
individual landowners, the draining-loan data indicate that capital supply for
the improvement was strongly related to estate size. Underdraining was most
readily and widely adopted on large estates, which overall possessed greater
financial resources for the support of agricultural improvement. The
provision of draining capital would seem to be generally ignored on smaller
properties, especially those under 1,000 acres, where rental incomes offered
less scope for extensive agricultural investment. On these, which formed a
considerable part of the cultivated area of the country, waterlogged land

must have remained in a largely unimproved state. On the evidence of draining, estate size emerges as a crucial element in understanding both the spatial and temporal diffusion of landlord-financed agricultural improvements in the nineteenth century.

Draining was the only one of the landlord improvements on which tenants consented formally to the regular payment of interest on outlay. In the provision of draining capital, whether from current income or from a draining loan, estates had the opportunity of obtaining a direct return on the improvement. Despite the varying rates of interest levied, as the draining programmes were in general effective, the greater part of the capital expenditure made by landlords on the improvement between 1840 and 1870 was likely to have been recovered before rents and agricultural prices began to move downwards after 1880. Moreover, although rent trends before and after 1880 would seem to be little influenced by the level of draining outlay, the improvement by upgrading land quality brought to estates the discreet financial advantage of making farms more attractive, competitive and secure to existing and prospective tenants. Landlord expenditure on draining should not be judged a financial miscalculation:[6] estates were aware of the interest payments tenants were prepared to provide and were willing to supply capital for the improvement at those rates. Although not comparable with industrial investment, draining through the interest charged yielded up to 1880 a real and clearly defined return. The much greater outlay that estates made on farm buildings over the same period and for which the payment of interest was rare represents a more questionable and at present largely unexplored use of landlord funds.

Yet landlord involvement in draining was generally confined to the supervision and capital supply of the improvement. In the main period of draining activity, 1840–70, as capital was normally only provided for the improvement when there was an agreement to pay interest or to contribute to the cost, the tenants must be seen as the active regulators of the amount and type of land drained. Tenants were forced to assess the adoption of draining largely in commercial terms.[7] The improvement presented a means to increase crop output and income from waterlogged land, but at the same time necessitated the payment of interest or the outlay of tenant capital. Only land that when drained could produce through increased yields a return that not only would cover their financial commitments to the improvement but also and, more importantly, provide a profit would be attractive to tenants. As a result, the improvement was concentrated on good-quality agricultural land, which when cleared of excess water had a high natural fertility, and on arable, which on draining afforded a greater and more direct return than grassland. Of the arable crops, wheat offered up to the 1870s the largest return on interest payment and a strong correlation existed at the national level in the second half of the nineteenth century between the provision of draining-loan capital and wheat prices. As the supply of suitable land declined and as arable prices began to fall after 1880, draining became a less viable commercial proposition for tenants and demand for the improvement dwindled. Such underdraining carried out towards the end of the century tended to be landlord sponsored, being essentially remedial action to preserve land capability and quality and thereby, hopefully, tenants.

The pattern and process of adoption provide the necessary framework for an appreciation of the role of draining in the nineteenth-century agricultural economy, thereby permitting some resolution of the conflicting claims of agricultural historians and others for the improvement. As a productive improvement, draining failed to work a revolution in the agricultural systems of the wet and heavy lands of the country after 1840. Only a proportion of the area with impeded drainage benefited from the improvement. And where undertaken, draining was not accompanied either by the large-scale adoption of mixed-farming systems found on light, free-draining soils as contemporaries anticipated or by the widespread conversion of arable to intensive grassland husbandry as R. W. Sturgess has more recently predicted.[8]

Yet it would be wrong to underestimate the impact of draining on heavy-land farming systems and productivity. The improvement represented a major capital input into English agriculture in the second half of the nineteenth century, the draining-loan and estate data suggesting that it absorbed at least £27.5 million between 1845 and 1899. In that 'high farming' implied the increasing of inputs to farming in an attempt to offset falling or stable prices by increasing output,[9] draining symbolized the application of that concept to wet and heavy lands in the country. Clearly, many heavy-land areas were attractive of large-scale agricultural capital: their investment record could not be regarded as poor.[10]

As with many agricultural innovations, the value of draining must be seen on a regional basis.[11] In those areas of adoption, although draining could not transform soil texture, the drying of waterlogged land allowed improvements to be made in farming systems, the degree of change being largely dependent on the inherent nature of the land.[12] On lighter soils, draining could be accompanied by the development of mixed farming; on land too heavy for the introduction of this system, the improvement could encourage a reduction in fallow and an expansion in roots and green crops; and, on soils lacking the potential for cropping modification, draining made the growth of existing crops more certain. Although not matching the levels on free-draining land with no requirement for capital outlay on the improvement,[13] draining was important in bringing to those wet and heavy lands a relative advance in agricultural productivity. Indirectly, besides making cultivation easier and more economical, the removal of excess water raised the effectiveness of all fertilizers, the use, range and quantity of which grew rapidly from the 1830s.[14] Directly, the drying of land resulted in an expansion of output through increasing yields, the evidence from the duke of

Northumberland's estate suggesting on average a growth of around 10 per cent for cereals. At these rates of yield increase, draining would have provided a return on all cereals in excess of interest payment throughout the greater part of the period 1840–75, mitigating the relative stability of the prices of these products. Such a definite and reliable return would have fostered an understandable reluctance amongst tenants to abandon the arable usage of their drained land. The efficiency with which draining increased output from arable on heavy land should be regarded as among those factors which, in spite of relative price trends, impeded the movement to livestock production in the third quarter of the nineteenth century.[15] Indeed, given the emphasis on improving arable, in those areas of most extensive activity, draining was likely to have produced an increase in total cereal output in this period.[16] Overall, given the rise in output resulting from the improvement and the area of land drained, draining must be seen as a major component in increasing agricultural productivity in the middle decades of the nineteenth century.[17]

A longer-term effect on agricultural productivity may be seen as arising from the pattern of capital investment in draining, largely established between 1840 and 1870. Adoption of the improvement presented to wet and heavy land a degree of flexibility in farming practice lacking on undrained land. When cereal prices began to fall from the late 1870s, the concentration of drained land in northern and western counties provided a sound base from which to seek alternatives to arable farming. However, the neglect of the improvement in much of eastern and southeastern England limited the range of opportunities for change on wet, heavy-land arable, making effective conversion to productive grassland difficult, and added to the general farming problems of these areas at the end of the century.[18]

The findings of this study should not be seen in isolation and the approach used has relevance for future discussions of nineteenth-century agricultural improvement. Draining was but one of a range of productive agricultural innovations, encompassing amongst others artificial fertilizers, machines, seed selection and farm layout, employed by landlords and tenants in the nineteenth century to increase output. The relative importance of these innovations in agricultural change over that period can be fully appreciated only by the provision of reliable and systematic data both on the timing and pattern of their adoption and on their effect on farming systems and productivity. Without such pedigrees, simple assertions of the value of a technical improvement in agriculture can add little to our understanding of agricultural change, offering at best a basis for interesting speculation, at worst a source for confusion, as exemplified in the debate on draining. Such rigorous, data-based analyses of new agricultural methods present the opportunity for a more satisfactory and precise foundation from which to examine both the nature and rate of agricultural change in the nineteenth century and, as important, their regional and spatial manifestations.

Notes

1 Debates about underdraining

1 Lord Ernle, *English Farming Past and Present* (London, 6th edn, 1961), 365–7; W. Smith, *An Historical Introduction to the Economic Geography of Great Britain* (London, 1968 reprint), 58; H. C. Darby, 'The draining of the English claylands', *Geographische Zeitschrift*, 52 (1964), 190; E. L. Jones, *Agriculture and Economic Growth in England 1650–1815* (London, 1967), 8–10, and *The Development of English Agriculture 1815–1873* (London, 1968), 14–15; J. D. Chambers and G. E. Mingay, *The Agricultural Revolution 1750–1880* (London, 1966), 207; E. J. T. Collins and E. L. Jones, 'Sectoral advance in English agriculture, 1850–80', *AHR*, 15 (1967), 66–7; A. D. M. Phillips, 'A study of farming practices and soil types in Staffordshire around 1840', *North Staffordshire Journal of Field Studies*, 13 (1973), 27–52, and 'Agricultural land use, soils and the Nottinghamshire tithe surveys *circa* 1840', *East Midland Geographer*, 6 (1976), 284–301; J. R. Walton, 'Agriculture 1730–1900', in R. A. Dodgshon and R. A. Butlin (eds.), *An Historical Geography of England and Wales* (London, 1978), 251.

2 Jones, *Agriculture and Economic Growth*, 7–12, and *Development of English Agriculture*, 14–15; Collins and Jones, 'Sectoral advance in English agriculture, 1850–80', 66–7.

3 R. W. Sturgess, 'The agricultural revolution on the English clays', *AHR.*, 14 (1966), 105–7, 110, 114–16; Jones, *Development of English Agriculture*, 15.

4 For general accounts of the agricultural technique of underdraining, see C. H. J. Clayton, *Land Drainage from Field to Sea* (London, 1919); B. W. Adkin, *Land Drainage in Britain* (London, 1933); H. H. Nicholson, *The Principles of Field Drainage* (Cambridge, 1946); R. G. Kendall, *Land Drainage* (London, 1950); M. C. Livesley, *Field Drainage* (London, 1960). For non-English examples, see Q. C. Ayres and D. Scoates, *Land Drainage and Reclamation* (New York, 2nd edn, 1939); M. M. Weaver, *History of Tile Drainage in America prior to 1900* (Waterloo, New York, 1964); A. D. M. Phillips and H. D. Clout, 'Underdraining in France during the second half of the nineteenth century', *Transactions of the Institute of British Geographers*, 51 (1970), 71–94; W. C. Found, A. R. Hill and E. S. Spence, *Economic and Environmental Impacts of Land Drainage in Ontario* (Toronto, 1974).

5 A. J. Thomasson and B. D. Trafford, 'Introduction', and B. D. Trafford, 'Drainage design', in A. J. Thomasson (ed.), *Soils and Field Drainage* (Harpenden, 1975), 1–4 and 5–17.

6 B. D. Trafford, 'Field drainage', *JRASE*, 131 (1970), 146–50, and 'Farm drainage', *Journal of the Royal Society of Arts* (vol. for 1973), 139–45; Trafford, 'Drainage design', 9; B. D. Trafford and R. A. Walpole, 'Drainage design in relation to soil series', in Thomasson, *Soils and Field Drainage*, 61.

7 Trafford, 'Farm drainage', 137–8; Trafford and Walpole, 'Drainage design in relation to soil series', 60–1; B. D. Trafford, 'Recent progress in field drainage', *JRASE*, 138 (1977), 29–30. See also R. Daubney, 'The influence of good drainage in relation to certain parasitic diseases of stock', *Journal of the Ministry of Agriculture*, 31 (1924), 616–21.

8 H. G. Richardson and G. E. Fussell, 'The beginnings of field drainage', *Journal of the Ministry of Agriculture*, 29 (1922), 585–91; G. E. Fussell, 'The evolution of field drainage', *Journal of the Bath and West and Southern Counties Society*, sixth series, 4 (1929–30), 59–72; Darby, 'The draining of the English clay-lands', 192–3; Ministry of Agriculture, Fisheries and Food, *History of Agricultural Drainage* (Ministry of Agriculture, Land Drainage Service, 1977), 1–6.

9 See, for example, J. Parkes, *Work on Draining* (Worksop, 1847); J. Trimmer, *On the Improvement of Land as an Investment for Capital* (London, 1847); L. de Lavergne, *The Rural Economy of England, Scotland and Ireland* (Edinburgh and London, 1855), 183–4; J. Caird, *The Landed Interest and the Supply of Food* (London, 4th edn, 1880), 87–8; J. B. Denton, *Agricultural Drainage* (London, 1883).

10 J. T. Coppock, 'The changing face of England: 1850–*circa* 1900', in H. C. Darby (ed.), *A New Historical Geography of England* (Cambridge, 1973), 606; See also, amongst many, Ernle, *English Farming Past and Present*, 365–7; G. E. Fussell, 'The dawn of high farming in England', *AH*, 22 (1948), 83; L. Hoelscher, 'Improvements in fencing and drainage in mid-nineteenth-century England', *AH*, 37 (1963), 75; B. A. Holderness, 'Capital formation in agriculture', in J. P. P. Higgins and S. Pollard (eds.), *Aspects of Capital Investment in Great Britain 1750–1850* (London, 1971), 174; R. A. C. Parker, *Coke of Norfolk* (Oxford, 1975), 142.

11 Chambers and Mingay, *The Agricultural Revolution*, 14; G. E. Mingay, 'The agricultural revolution in English history: a reconsideration', *AH*, 37 (1963), 130; F. M. L. Thompson, 'The second agricultural revolution, 1815–1880', *EcHR*, second series, 21 (1968), 63–4; E. Kerridge, *The Agricultural Revolution* (London, 1967), 37.

12 Darby, 'The draining of the English clay-lands', 190; Sturgess, 'The agricultural revolution on the English clays', 105, 110; Collins and Jones, 'Sectoral advance in English agriculture, 1850–80', 66–9, 80–1.

13 O. R. McGregor, 'Introduction: after 1815', in Ernle, *English Farming Past and Present*, cxiii.

14 A. D. M. Phillips, 'Underdraining and the English claylands, 1850–80: a review', *AHR*, 17 (1969), 45; J. B. Harley, 'England *circa* 1850', in Darby, *A New Historical Geography of England*, 548; F. H. W. Green, 'Aspects of the changing environment: some factors affecting the aquatic environment in recent years', *Journal of Environmental Management*, 1 (1973), 378; 'Ridge and furrow,

mole and tile', *Geographical Journal*, 141 (1975), 89; and 'Field, forest and hill drainage in Scotland', *Scottish Geographical Magazine*, 95 (1979), 161; G. E. Mingay, *The Gentry* (London, 1976), 168–9.

15 J. H. Clapham, *An Economic History of Modern Britain: 2, Free Trade and Steel, 1850–1886* (Cambridge, 1932), 270; Fussell, 'The dawn of high farming in England', 85; Kerridge, *The Agricultural Revolution*, 37.

16 See, for example, G. E. Fussell, 'Home counties farming, 1840–80', *Economic Journal*, 57 (1947), 321–45; 'High farming in the west midland counties 1840–1880', *Economic Geography*, 25 (1949), 159–79; and 'High farming in the east midlands and East Anglia, 1840–1880', *Economic Geography*, 27 (1951), 72–89; C. S. Davies, *The Agricultural History of Cheshire 1750–1850* (Manchester, 1960), 109; D. Grigg, *The Agricultural Revolution in South Lincolnshire* (Cambridge, 1966), 143–4; Walton, 'Agriculture 1730–1900', 252–3.

17 G. E. Fussell, 'High farming in the north of England 1840–1880', *Economic Geography*, 24 (1948), 298; Hoelscher, 'Improvements in fencing and drainage in mid-nineteenth-century England', 79.

18 Collins and Jones, 'Sectoral advance in English agriculture, 1850–80', 69–70, 73–5, 80.

19 R. W. Sturgess, 'The agricultural revolution on the English clays: a rejoinder', *AHR*, 15 (1967), 84–5.

20 Trafford, 'Field drainage', 131–2.

21 F. H. W. Green, 'Field under-drainage before and after 1940', *AHR*, 28 (1980), 120–2; M. Robinson, 'The extent of farm underdrainage in England and Wales, prior to 1939', *AHR*, 34 (1986), 80–2.

22 BPP, 1870, LXVIII, 'Agricultural returns, 1870', 16.

23 BPP, 1873, XVI, 'Select Committee of the House of Lords on the improvement land', qq. 586–9.

24 'Select Committee... on the improvement of land', qq. 798–807.

25 BPP, 1881, XV and XVII, 'Royal Commission on the depressed condition of the agricultural interest', qq. 4707–22.

26 'Royal Commission on the depressed condition of the agricultural interest', qq. 6324, 6385–6.

27 Select Committee... on the improvement of land', q. 4126.

28 BPP, 1873, LXIX, 'Agricultural returns, 1873', 20; 'Agricultural depression', *Edinburgh Review*, 151 (1880), 29.

29 'Agricultural depression', 29.

30 'Select Committee... on the improvement of land', q. 856; 'Royal Commission on the depressed condition of the agricultural interest', qq. 4719–21; 'Agricultural depression', 30.

31 Holderness, 'Capital formation in agriculture', 176; D. Taylor, 'The English dairy industry, 1860–1930', *EcHR*, second series, 29 (1976), 598; G. Hueckel, 'Agriculture during industrialisation', in R. Floud and D. McCloskey (eds.), *The Economic History of Britain since 1700* (Cambridge, 1981), vol. 1, 194; F. M. L. Thompson, 'Free trade and the land', in G. E. Mingay (ed.), *The Victorian Countryside* (London, 1981), vol. 1, 105; D. Grigg, *The Dynamics of Agricultural Change* (London, 1982), 127.

32 Sturgess, 'The agricultural revolution of the English clays: a rejoinder', 84–5.

33 Phillips, 'Underdraining and the English claylands, 1850–80: a review', 49; *London Gazette*, 1846–78, *passim*.

34 T. W. Fletcher 'The agrarian revolution in arable Lancashire', *Transactions of the Lancashire and Cheshire Antiquarian Society*, 72 (1962), 116.

35 Notices of application in the *London Gazette* of 2, 3, 5, 11, 12, 13 and 15 February 1847.

36 PRO, 1R3/6–38, Certificates of draining advances under the Public Money Draining Acts.

37 J. B. Denton, 'The effect of underdrainage on our rivers and arterial channels', *JRASE*, 24 (1863), 573–89, and *Agricultural Drainage*, 39.

38 Trafford, 'Field drainage', 131.

39 *Ibid.*

40 *Post Office Directory...of Northamptonshire, Huntingdonshire, Bedfordshire, Buckinghamshire, Berkshire and Oxfordshire* (London, 1869); *Kelly's Directory...of Bedfordshire, Huntingdonshire and Northamptonshire* (London, 1885 and 1890); *Kelly's Directory...of Berkshire, Buckinghamshire and Oxfordshire* (London, 1887 and 1891); *Post Office Directory...of Cambridgeshire, Norfolk and Suffolk* (London, 1869); *Kelly's Directory...of Cambridgeshire, Norfolk and Suffolk* (London, 1888); *Kelly's Directory...of Birmingham, Staffordshire, Warwickshire and Worcestershire* (London, 1880, 1888 and 1896).

41 'Royal Commission on the depressed condition of the agricultural interest', q. 4722.

42 Green 'Field under-drainage before and after 1940', 120–2; Robinson, 'The extent of farm underdrainage in England and Wales, prior to 1939', 79.

43 *Ibid.*, 80–1, 83.

44 L. D. Stamp, *The Land of Britain: Its Use and Misuse* (London, 3rd edn, 1962), 377–80; H. M. E. Holt, 'Upland farming in northern England, *circa* 1840 to *circa* 1880: some evidence from Cumbria and Northumberland', unpublished University of Exeter Ph.D. thesis (1985), 214–17, 226, 377–88.

45 Ministry of Agriculture, Fisheries and Food, *Underdrainage Information Sheet for Draining Advisers' Reports* (Ref. CG4, rev. 1977).

46 The help of the Ministry of Agriculture's draining advisers at Alnwick and Stafford is gratefully acknowledged.

47 Darby, 'The draining of the English clay-lands', 191.

48 Sturgess, 'The agricultural revolution on the English clays: a rejoinder', 84–5.

49 Robinson, 'The extent of farm underdrainage in England and Wales, prior to 1939', 83–4. See also Walton, 'Agriculture 1730–1900', 252.

50 Thompson, 'Free trade and the land', 105.

51 Mingay, *The Gentry*, 170; H. C. Prince, 'Victorial rural landscapes', in Mingay, *The Victorian Countryside*, vol. 1, 21.

52 S. Macdonald, 'The development of agriculture and the diffusion of agricultural innovation in Northumberland, 1750–1850', unpublished University of Newcastle Ph.D. thesis (1974), 355–68; Davies, *Agricultural History of Cheshire*, 109–10; R. E. Porter, 'Agricultural change in Cheshire during the nineteenth century', unpublished University of Liverpool Ph.D. thesis (1974), 40–6; Fletcher, 'The agrarian revolution in arable Lancashire', 112–16; J. Thirsk,

English Peasant Farming (London, 1957), 301; Grigg, *The Agricultural Revolution in South Lincolnshire*, 142–4; T. W. Beastall, *The Agricultural Revolution in Lincolnshire* (Lincoln, 1978), 175, 180; J. Thirsk and J. Imray, *Suffolk Farming in the Nineteenth Century* (Suffolk Records Society, 1958), 26–7; A. G. Parton, 'Town and country in Surrey *c.* 1800–1870: a study in historical geography', unpublished University of Hull Ph.D. thesis (1973), 116, 193–4; B. M. Short, 'Agriculture in the High Weald of Kent and Sussex 1850 to 1953', unpublished University of London Ph.D. thesis (1973), 93–7; B. R. Dittmer, 'An agricultural geography of northwest Wiltshire, 1773–1840', unpublished University of London M.A. thesis (1963), 71–4; F. M. L. Thompson, 'Agriculture since 1870', in *Victoria County History of Wiltshire* (London, 1959), vol. 4, 92–114; J. R. Walton, 'Aspects of agrarian change in Oxfordshire, 1750–1880', unpublished University of Oxford D. Phil. thesis (1976), 84–99; R. C. Gaut, *A History of Worcestershire Agriculture and Rural Evolution* (Worcester, 1939), 316–18.

53 B. W. Avery, D. C. Findlay and D. Mackney, *Soil Map of England and Wales, 1:1,000,000* (Southampton, 1975); Soil Survey of England and Wales, *Soil Map of England and Wales, 1:250,000* (Southampton, 1983). See also Darby, 'The draining of the English clay-lands', 192; E. H. Whetham, 'Sectoral advance in English agriculture 1850–80: a summary', *AHR*, 16 (1968), 47.

54 F. H. W. Green, 'Quantification of areas of agricultural and forestry drainage, as they affect extrapolation of the results of representative and experimental basins', in *The Influence of Man on the Hydrological Regime: Proceedings of the Helsinki Symposium* (1980), 387; A. J. Thomasson, 'Other site factors; climate and land use', in Thomasson, *Soils and Field Drainage*, 30–1; Robinson, 'The extent of farm underdrainage in England and Wales, prior to 1939', 84–5.

55 E. L. Jones, *Seasons and Prices* (London, 1964), 155–78; C. S. Orwin and E. H. Whetham, *History of British Agriculture 1846–1914* (London, 1964), 95–6; R. Perren, 'The landlord and agricultural transformation, 1870–1900', *AHR*, 18 (1970), 41; P. J. Perry, *British Farming in the Great Depression 1870–1914* (Newton Abbot, 1974), 56.

56 Among the exceptions, see A. D. M. Phillips, 'The development of under-draining on a Yorkshire estate during the nineteenth century', *Yorkshire Archaeological Journal*, 44 (1972), 195–206; S. Wade Martins, *A Great Estate at Work* (Cambridge, 1980), 86.

57 Derived from Ernle, *English Farming Past and Present*, 365–7; J. H. Clapham, *An Economic History of Modern Britain: 1, the Early Railway Age, 1820–50* (Cambridge, 1926), 134; Hoelscher, 'Improvements in fencing and drainage in mid-nineteenth-century England', 76–9; Darby, 'The draining of the English clay-lands', 193–6; Orwin and Whetham, *History of British Agriculture*, 100; Chambers and Mingay, *The Agricultural Revolution*, 64–5, 175–6; Trafford, 'Field drainage', 129–30; Green, 'Aspects of the changing environment: some factors affecting the aquatic environment in recent years', 378; A. Harris, 'Changes in the early railway age: 1800–1850', in Darby, *A New Historical Geography of England*, 484; G. E. Mingay, *The Agricultural Revolution: Changes in Agriculture 1650–1880* (London, 1977), 48–9.

58 Ernle, *English Farming Past and Present*, 366; Clapham, *The Early Railway Age, 1820–50*, 134; Darby, 'The draining of the English clay-lands', 193; Chambers and Mingay, *The Agricultural Revolution*, 65; Holderness, 'Capital formation in agriculture', 176.

59 D. Spring, 'A great agricultural estate: Netherby under Sir James Graham, 1820–1845', *AH*, 29 (1955), 77–80; J. T. Ward, *Sir James Graham* (London, 1967), 60–1; Fletcher, 'The agrarian revolution in arable Lancashire', 110–12; Thirsk, *English Peasant Farming*, 287; Grigg, *The Agricultural Revolution in South Lincolnshire*, 141–2.

60 Clapham, *The Early Railway Age, 1820–50*, 134; Fletcher, 'The agrarian revolution in arable Lancashire', 111–15; Trafford, 'Field drainage', 131.

61 Ernle, *English Farming Past and Present*, 367; Orwin and Whetham, *History of British Agriculture*, 100; Chambers and Mingay, *The Agricultural Revolution*, 175; Harris, 'Changes in the early railway age: 1800–1850', 484.

62 Jones, *Seasons and Prices*, 126; Hoelscher, 'Improvements in fencing and drainage in mid-nineteenth-century England', 79; Green, 'Recent changes in land use and treatment', 15; Sturgess, 'The agricultural revolution on the English clays: a rejoinder', 87; Clapham, *Free Trade and Steel, 1850–1886*, 270; Adkin, *Land Drainage in Britain*, 226; Mingay, *The Agricultural Revolution*, 48–9; Trafford, 'Field drainage', 131; Robinson, 'The extent of farm underdrainage in England and Wales, prior to 1939', 79.

63 Thompson, 'The second agricultural revolution 1815–1880', 64; Holderness, 'Capital formation in agriculture', 174; D. Cannadine, 'Aristocratic indebtedness in the nineteenth century: the case reopened', *EcHR*, second series, 30 (1977), 641.

64 R. J. Thompson, 'An enquiry into the rent of agricultural land in England and Wales during the nineteenth century', *Journal of the Royal Statistical Society*, 70 (1907), reprinted in W. E. Minchinton (ed.), *Essays in Agrarian History* (Newton Abbot, 1968), vol. 2, 71–5; Smith, *An Historical Introduction to the Economic Geography of Great Britain*, 52–3.

65 F. M. L. Thompson, 'English great estates in the nineteenth century, 1790–1914', *Contributions to the First International Conference of Economic History* (Paris, 1960), 394, and *English Landed Society in the Nineteenth Century* (London, 1963), 231–6.

66 D. C. Moore, 'The Corn Laws and high farming', *EcHR*, second series, 18 (1965), 456; D. B. Grigg, 'A note on agricultural rent and expenditure in nineteenth-century England', *AH*, 39 (1965), 151; B. A. Holderness, 'Landlord's capital formation in East Anglia 1750–1870', *EcHR*, second series, 25 (1972), 439.

67 Thompson, *English Landed Society*, 240–2, 315; Perren, 'The landlord and agricultural transformation, 1870–1900', 41–2; C. O'Grada, 'Agricultural decline 1860–1914', in Floud and McCloskey, *The Economic History of Britain since 1700*, vol. 2, 188.

68 Thompson, *English Landed Society*, 226.

69 Wade Martins, *A Great Estate*, 97–8.

70 Holderness, 'Landlord's capital formation in East Anglia 1750–1870', 439.

71 Ernle, *English Farming Past and Present*, 323–4; Perry, *British Farming in the Great Depression 1870–1914*, 19; Chambers and Mingay, *The Agricultural*

Revolution, 163; Jones, *Development of English Agriculture*, 29; J. V. Beckett, *The Aristocracy in England 1660–1914* (Oxford, 1986), 179–80.

72 Thompson, 'An enquiry into the rent of agricultural land in England and Wales during the nineteenth century', 78; Cannadine, 'Aristocratic indebtedness in the nineteenth century: the case reopened', 641.

73 Jones, *Development of English Agriculture*, 15–16.

74 Thompson, 'English great estates in the nineteenth century, 1790–1914', 393; D. Spring, 'English landed society in the eighteenth and nineteenth centuries', *EcHR*, second series, 17 (1964), 149.

75 Mingay, *The Gentry*, 168; Beckett, *The Aristocracy in England*, 178–80.

76 Holderness, 'Landlord's capital formation in East Anglia 1750–1870', 436.

77 F. M. L. Thompson, 'The social distribution of landed property in England since the sixteenth century', *EcHR*, second series, 19 (1966), 506; Holderness, 'Landlord's capital formation in East Anglia 1750–1870', 436.

78 BPP, 1874, LXXII, parts 1 and 2, 'Return of owners of land in England and Wales, 1873'; J. Bateman, *The Great Landowners of Great Britain and Ireland* (London, 4th edn, 1883); Thompson, *English Landed Society*, 32, 113–17; H. A. Clemenson, *English Country Houses and Landed Estates* (London, 1982), 20–6.

79 Thompson, *English Landed Society*, 252.

80 Thompson, *English Landed Society*, 250–1; D. Spring, *The English Landed Estate in the Nineteenth Century: Its Administration* (Baltimore, 1963), 49–50; P. A. David, *Technical Choice, Innovation and Economic Growth* (Cambridge, 1975), 257, 264; Wade Martins, *A Great Estate*, 99–100; Hueckel, 'Agriculture during industrialisation', 193–5.

81 Collins and Jones, 'Sectoral advance in English agriculture, 1850–80', 60–71; Hueckel, 'Agriculture during industrialisation', 194; O'Grada, 'Agricultural decline 1860–1914', 188.

82 B. A. Holderness, 'The Victorian farmer', in Mingay, *The Victorian Countryside*, vol. 1, 233.

83 Perren, 'The landlord and agricultural transformation, 1870–1900', 43; T. W. Fletcher, 'Lancashire livestock farming during the Great Depression', *AHR*, 9 (1961), 34–5; S. Farrant, 'The management of four estates in the lower Ouse valley (Sussex) and agricultural change, 1840–1920', *Southern History*, 1 (1979), 165–9.

84 Thompson, *English Landed Society*, 250–6; P. J. Perry, 'High farming in Victorian Britain: the financial foundations', *AH*, 52 (1978), 363–6; B. English, 'On the eve of the Great Depression: the economy of the Sledmere estate 1869–78', *Business History*, 24 (1982), 34; Beckett, *The Aristocracy in England*, 182.

85 D. B. Grigg, 'The development of tenant right in south Lincolnshire', *Lincolnshire Historian*, 2 (1962), 41–3; J. R. McQuiston, 'Tenant right: farmer against landlord in Victorian England, 1847–1883', *AH* 47 (1983), 107, J. H. Brown, 'Agriculture in Lincolnshire during the Great Depression, 1873–96', unpublished University of Manchester Ph.D. thesis (1978), 67–8; J. R. Fisher, 'Landowners and English tenant right, 1845–1852', *AHR*, 31 (1983), 16; Adkin, *Land Drainage in Britain*, 228; P. J. Perry, 'High farming in Victorian Britain: prospect and retrospect', *AH*, 55 (1981), 163.

86 Holderness, 'Capital formation in agriculture', 176–7.
87 Thompson, 'English great estates in the nineteenth century, 1790–1914', 394;
    Orwin and Whetham, *History of British Agriculture*, 196–7; Mingay, *The Agricultural Revolution*, 55.
88 Thompson, *English Landed Society*, 255.
89 Spring, *The English Landed Estate*, 143–58; A. D. M. Phillips, 'Underdraining and agricultural investment in the midlands in the mid-nineteenth century', in A. D. M. Phillips and B. J. Turton (eds.), *Environment, Man and Economic Change* (London, 1975), 254–6; Perry, 'High farming in Victorian Britain: the financial foundations', 367–9.
90 McGregor, 'Introduction: after 1815', cxiii; Spring, *The English Landed Estate*, 141–2; Orwin and Whetham, *History of British Agriculture*, 57–66; Moore, 'The Corn Laws and high farming', 554–5; B. English and J. Saville, *Strict Settlement. A Guide for Historians* (Hull, 1983), 30, 42.
91 B.P.P, 1900, XVII, 'Board of Agriculture. Annual report of proceedings, 1899'; Orwin and Whetham, *History of British Agriculture*, 196; Jones, *Development of English Agriculture*, 22.
92 Ernle, *English Farming Past and Present*, 366; Hoelscher, 'Improvements in fencing and drainage in mid-nineteenth-century England', 76–7; Grigg, *The Agricultural Revolution in South Lincolnshire*, 143; Holderness, 'Capital formation in agriculture', 175–6.
93 Collins and Jones, 'Sectoral advance in English agriculture, 1850–80', 70, but see also A. D. Hall, *A Pilgrimage of British Farming 1910–1912* (London, 1913), 418; Fussell, 'The dawn of high farming in England', 85; Gaut, *A History of Worcestershire Agriculture*, 316–18.
94 Spring, *The English Landed Estate*, 173–4; Darby, 'The draining of the English clay-lands', 200; Trafford, 'Field drainage', 135; Perry, *British Farming in the Great Depression 1870–1914*, 19, 56.
95 David, *Technical Choice*, 283.
96 Trafford, 'Field drainage', 141.
97 Orwin and Whetham, *History of British Agriculture*, 5.
98 R. M. Garnier, *History of the English Landed Interest* (London, 2nd edn, 1908), vol. 2, 440; Fussell, 'The dawn of high farming in England', 84–5; Orwin and Whetham, *History of British Agriculture*, 101; Chambers and Mingay, *The Agricultural Revolution*, 175.
99 T. L. Crosby, *Sir Robert Peel's Administration 1841–1846* (Newton Abbot, 1976), 149; Moore, 'The Corn Laws and high farming', 550.
100 Thompson, *English Landed Society*, 255–6; Mingay, *The Gentry*, 169.
101 Sturgess, 'The agricultural revolution on the English clays: a rejoinder', 87. A similar relationship between underdraining and conversion of arable to pasture but in the 1880s has been suggested by Perren, 'The landlord and agricultural transformation, 1870–1900', 48.
102 Collins and Jones, 'Sectoral advance in English agriculture, 1850–80', 72–8, 80–1. See also R. Morgan, 'The root-crop in English agriculture, 1650–1870', unpublished University of Reading Ph.D. thesis (1978), 596–605, where it is argued that underdraining did not permit a general expansion of turnips on claylands.

103 Fussell, 'Home counties farming, 1840–80', 344; Phillips, 'Underdraining and the English claylands, 1850–80: a review', 55. The three county studies have made partial use of the loan data: Porter, 'Agricultural change in Cheshire during the nineteenth century', 42–6; Walton, 'Aspects of agrarian change in Oxfordshire, 1750–1880', 84–9; Holt, 'Upland farming in northern England, *circa* 1840 to *circa* 1880: some evidence from Cumbria and Northumberland', 213–20.

104 Thompson, *English Landed Society*, 182–3; Spring, *The English Landed Estate*, 53–4.

105 A. D. M. Phillips, 'The landlord and agricultural improvements: underdraining on the Lincolnshire estate of the earls of Scarbrough in the first half of the nineteenth century', *East Midland Geographer*, 7 (1979), 168–77, and 'Agricultural improvement on a Durham estate in the nineteenth century: the Lumley estate of the earls of Scarbrough', *Durham University Journal*, new series, 42 (1981), 161–8.

106 Bateman, *The Great Landowners*, 501–11; Thompson, *English Landed Society*, 32, 113–17.

107 Spring, *The English Landed Estate*, 3–19.

2 The need for underdraining in the nineteenth century

1 W. Marshall, *A Review of the Reports to the Board of Agriculture from the Western Department of England* (London, 1810), 260.

2 Lavergne, *The Rural Economy of England*, 182.

3 BPP (Lords Sessional Papers), 1845, XVIII, 'Select Committee of the House of Lords...to enable possessors of entailed estates to charge such estates...for the purpose of draining...', q. 138.

4 P. Pusey, 'Editorial note', *JRASE*, 2 (1841), 103, and 'On the progress of agricultural knowledge during the last four years', *JRASE*, 3 (1842), 169–70.

5 *Farmer's Magazine*, second series, 8 (1843), 9, 346; H. Hutchinson, *A Treatise on the Practical Drainage of Land* (London, 1844), 1–3; 'Select Committee...on the improvement of land', q. 4126.

6 J. B. Denton, 'General drainage and distribution of water', *Farmer's Magazine*, second series, 6 (1842), 64.

7 Trimmer, *On the Improvement of Land*, 1–3.

8 J. B. Denton, *The Under-Drainage of Land: its Progress and Results* (London, 1855), 3–5, and *Agricultural Drainage*, 33–6.

9 Denton, *Agricultural Drainage*, 33.

10 BPP, 1870, LXVIII; 1871, LXIX; 1872, LXIII; 1873, LXIX; 1874, LXIX; 1875, LXXIX; 1876, LXXVIII; 1877, LXXXV; 1878, LXXVII; 1878–9, LXXV, 'Agricultural returns, 1870–9'.

11 An Act for the Incorporation of the General Land Drainage and Improvement Company and for Facilitating the Execution of Land Drainage and Other Improvements, 12 & 13 Vict., c. xci, preamble; 'Royal Commission on the depressed condition of the agricultural interest', q. 4690.

12 Nicholson, *The Principles of Field Drainage*, 147–51.

13 Darby, 'The draining of the English clay-lands', 191.

14 Professor Darby kindly provided a copy of the original map. The areas of

clayland and of other soil types discussed in this chapter have been computed from the respective original maps by means of a Calcomp Digitizer.

15 Darby, 'The draining of the English clay-lands', 192.

16 Livesley, *Field Drainage*, chap. 5; A. J. Thomasson, 'Soil properties affecting drainage design', in Thomasson, *Soils and Field Drainage*, 18–29.

17 W. F. Karkeek, 'On the farming of Cornwall', *JRASE*, 6 (1845), 421–3; H. Tanner, 'The farming of Devonshire', *JRASE*, 9 (1848), 470–1; T. Rowlandson, 'Farming of Herefordshire', *JRASE*, 14 (1853), 453–4.

18 A less rigorous survey of drainage needs in the country was conducted by the National Farmers' Union in the middle of the 1930s by circularizing a simple questionnaire to its county branches. The results produced an estimate of 7.5 million acres of land capable of improvement by underdraining, land already drained being excluded. This figure compares favourably to that identified in 1968–9: H. H. Nicholson, 'Field drainage and increased production', *JRASE*, 109 (1948), 214.

19 Trafford, 'Farm drainage', 135–6.

20 Avery *et al.*, *Soil Map of England and Wales, 1:1,000,000,* introduction and classification.

21 D. Mackney, 'Soil maps and classification', in Thomasson, *Soils and Field Drainage*, 35–48.

22 Soil Survey of England and Wales, *Soil Map of England and Wales, 1:250,000,* sheets 1–6.

23 Soil Survey of England and Wales, *Legend for the 1:250,000 Soil Map of England and Wales* (Harpenden, 1983), 6–21.

24 McGregor, 'Introduction: after 1815', xcviii–c; H. C. Prince, 'England *circa* 1800', in Darby, *A New Historical Geography of England*, 400–2.

25 *Annals of Agriculture*, 1–45 (1784–1815); *Farmer's Magazine*, 1–15 (1800–16); *Communications to the Board of Agriculture*, 1–5 (1802–6); Board of Agriculture, *The Agricultural State of the Kingdom in ... 1816* (London, 1816); J. Johnstone, *An Account of the Mode of Draining Land According to the System Practised by Mr Joseph Elkington* (London, 2nd edn, 1801); J. Anderson, *A Practical Treatise on Draining Bogs and Swampy Grounds* (London, 1797); W. Smith, *Observations on the Utility, Form and Management of Water Meadows and the Draining and Irrigating of Peat Bogs* (Norwich, 1806); W. Marshall, *The Rural Economy of Norfolk* (2 vols., London, 1787), *The Rural Economy of Yorkshire* (2 vols., London, 1788), *The Rural Economy of Gloucestershire* (2 vols., Gloucester, 1789), *The Rural Economy of the Midland Counties* (2 vols., London, 1790), *The Rural Economy of the West of England* (2 vols., London, 1796), *The Rural Economy of the Southern Counties* (2 vols., London, 1798).

26 McGregor, 'Introduction: after 1815', xcix–c.

27 H. Holland, *General View of the Agriculture of Cheshire* (London, 1808), 210–15 (hereafter *General View ... County*); J. Farey, *General View ... Derbyshire* (3 vols., London, 1811–17), vol. 2, 360–95.

28 H. E. Strickland, *General View ... East Riding of Yorkshire* (York, 1812), 194–201.

29 J. Bailey and G. Culley, *General View ... Cumberland* (London, 1805), 238, and *General View ... Northumberland* (London, 1805), 128; J. Bailey, *General View ... Durham* (London, 1810), 202.

30 A. Pringle, *General View... Westmorland* (London, 1805), 322; J. Tuke, *General View... North Riding of Yorkshire* (London, 1800), 221; Strickland, *General View... East Riding of Yorkshire*, 306.

31 J. Holt, *General View... Lancashire* (London, 1794), 57–60; R. W. Dickson, *General View... Lancashire* (London, 1815), 456–62.

32 W. Pitt, *General View... Staffordshire* (London, 1813), 148–51, 254; Marshall, *Rural Economy of Gloucestershire*, vol. 1, 190–1.

33 A. Young, *General View... Lincolnshire* (London, 1799), 242.

34 R. Lowe, *General View... Nottinghamshire* (London, 1798), 98; R. Parkinson, *General View... Huntingdonshire* (London, 1813), 292.

35 Farey, *General View... Derbyshire*, vol. 1, 453.

36 C. Vancouver, *General View... Devon* (London, 1813), 309; J. Billingsley, *General View... Somerset* (London, 1974), 21.

37 Marshall, *Rural Economy of Gloucestershire*, vol. 2, 148; T. Davis, *General View... Wiltshire* (London, 1813), 191–3.

38 Rev. A. Young, *General View... Sussex* (London, 1808), 191–7.

39 C. Vancouver, *General View... Cambridgeshire* (London, 1794), 13–142 *passim*; W. Gooch, *General View... Cambridgeshire* (London, 1813), 239–45.

40 A. Young, *General View... Norfolk* (London, 1804), 389–95, and *General View... Suffolk* (London, 1794), 27.

41 A. Young, *General View... Essex* (2 vols., London, 1807), vol. 2, 166, and *General View... Hertfordshire* (London, 1804), 155.

42 G. E. Mingay, *Arthur Young and his Times* (London, 1975), 15–19.

43 J. Sinclair, 'Report to the committee of the Board of Agriculture respecting Mr Elkington's mode of draining', *Annals of Agriculture*, 24 (1795), 525–9.

44 Johnstone, *An Account of the Mode of Draining Land*, 105–7.

45 J. Sinclair, *The Code of Agriculture* (London, 1817), 178–80.

46 Bailey and Culley, *General View... Cumberland*, 238; Pringle, *General View... Westmorland*, 322.

47 Bailey, *General View... Durham*, 201–2; Dickson, *General View... Lancashire*, 159–60, 476–80.

48 T. Batchelor, *General View... Bedfordshire* (London, 1808), 174; T. Rudge, *General View... Gloucestershire* (London, 1807), 258–61; J. Duncumb, *General View... Herefordshire* (London, 1805), 98–100; Parkinson, *General View... Huntingdonshire*, 204.

49 Johnstone, *An Account of the Mode of Draining Land*, plan 1; Smith, *Observations on the Utility, Form and Management of Water Meadows*, 94–111; *Farmer's Magazine*, 3 (1802), 471; *Staffordshire Advertiser*, 8 November 1806.

50 G. B. Worgan, *General View... Cornwall* (London, 1811), 108–10; C. Vancouver, *General View... Hampshire* (London, 1813), 95–6, 327–31; J. Boys, *General View... Kent* (London, 1813), 149–55; W. Mavor, *General View... Berkshire* (London, 1813), 124–5; W. Stevenson, *General View... Surrey* (London, 1813), 132; Young, *General View... Sussex*, 59; Davis, *General View... Wiltshire*, 193.

51 Vancouver, *General View... Devon*, 310.

52 Lord Petre, 'Concerning land-ditching', *Annals of Agriculture*, 4 (1785), 294–6.

53 Young, *General View ... Essex*, vol. 2, 166–9; *General View ... Hertfordshire*, 154, and *General View ... Norfolk*, 389–93; Gooch, *General View ... Cambridgeshire*, 239–45.

54 A. Young, 'Mole plough drawn by the force of women applied mechanically', *Annals of Agriculture*, 42 (1804), 413–22; J. Middleton, *General View ... Middlesex* (London, 2nd edn, 1807), 361–4; Sinclair, *The Code of Agriculture*, 186–7; Gooch, *General View ... Cambridgeshire*, 245.

55 Johnstone, *An Account of the Mode of Draining Land*, 92–9; Sinclair, 'Report to the committee of the Board of Agriculture respecting Mr Elkington's mode of draining', 525–9.

56 J. Thurlow, 'On the mole-plough', *Annals of Agriculture*, 43 (1805), 486–8; C. Vancouver, *General View ... Essex* (London, 1795), 139–40; *Farmer's Magazine*, 4 (1803), 278–80; Young, *General View ... Essex*, vol. 2, 168, and *General View ... Norfolk*, 390–2; Gooch, *General View ... Cambridgeshire*, 239–45.

57 Young, *General View ... Suffolk*, 27; Middleton, *General View ... Middlesex*, 356; Davis, *General View ... Wiltshire*, 193.

58 W., J. and J. Malcolm, *General View ... Buckinghamshire* (London, 1794), 59–60; Mavor, *General View ... Berkshire*, 109–14; Middleton, *General View ... Middlesex*, 651.

59 J. Plymley, *General View ... Shropshire* (London, 1803), 347; Duncumb, *General View ... Herefordshire*, 41.

60 Stevenson, *General View ... Surrey*, 485.

61 J. Lawrence, *The Modern Land Steward* (London, 1801), 225–6.

62 Rudge, *General View ... Gloucestershire*, 19; Young, *General View ... Essex*, vol. 1, 23; Young, *General View ... Sussex*, 191.

63 Johnstone, *An Account of the Mode of Draining Land*, 132–7; Lawrence, *The Modern Land Steward*, 204–10; W. Marshall, *On the Landed Property of England* (London, 1804), 59–60, 98–100; Sinclair, *The Code of Agriculture*, 180–1; E. Kerridge, 'Ridge-and-furrow and agrarian history', *EcHR*, second series, 4 (1951), 14–36.

64 Bailey, *General View ... Durham*, 103–4; Holt, *General View ... Lancashire*, 19–20; Bailey and Culley, *General View ... Northumberland*, 66–7; Tuke, *General View ... North Riding of Yorkshire*, 102–3, 224–5.

65 Rudge, *General View ... Gloucestershire*, 258–9; Batchelor, *General View ... Bedfordshire*, 279–80; W. Pitt, *General View ... Leicestershire* (London, 1809), 89.

66 A. and W. Driver, *General View ... Hampshire* (London, 1794), 28.

67 Mavor, *General View ... Berkshire*, 159; Boys, *General View ... Kent*, 149–52; Middleton, *General View ... Middlesex*, 158–60, 355–8.

68 Young, *General View ... Essex*, vol. 1, 16; *General View ... Hertfordshire*, 12, and *General View ... Norfolk*, 190.

69 W. R. Mead, 'Ridge and furrow in Buckinghamshire', *Geographical Journal*, 120 (1954), 34–42; M. J. Harrison, W. R. Mead and D. J. Pannett, 'A midland ridge-and-furrow map', *Geographical Journal*, 131 (1965), 366–9; W. R. Mead and R. J. P. Kain, 'Ridge and furrow in Kent', *Archaeologia Cantiana*, 92 (1976–7), 165–71; R. Kain and W. R. Mead, 'Ridge and furrow in Cambridgeshire', *Proceedings of the Cambridgeshire Antiquarian Society*, 67

(1977), 131–7; W. R. Mead, 'A right-and-furrow map of Leicestershire and Northamptonshire', *East Midland Geographer*, 6 (1977), 382–5. See also J. E. G. Sutton, 'Ridge and furrow in Berkshire and Oxfordshire', *Oxoniensia*, 29–30 (1964–5), 99–115; Royal Commission on Historical Monuments, *Northamptonshire: An Archaeological Atlas* (London, 1980), 9 and map 18.

3 The intensity and location of underdraining, 1845–1899

1 'Board of Agriculture. Annual report of proceedings, 1899'.

2 F. M. L. Thompson, 'The economic and social background of the English landed interest, 1840–70, with particular reference to the estates of the duke of Northumberland', unpublished University of Oxford D.Phil. thesis (1956), 42–7, 132–45; Spring, *The English Landed Estate*, 135–77.

3 'Select Committee...for the purpose of draining', i–iii; Caird, *The Landed Interest*, 79–80; Spring, *The English Landed Estate*, 142–3; E. Spring, 'Landowners, lawyers and land law reform in nineteenth-century England', *American Journal of Legal History*, 21 (1977), 40–2

4 Thompson, *English Landed Society*, 66–8; English and Saville, *Strict Settlement*, 30–1, 114–15.

5 An Act to Enable the Owners of Settled Estates to Defray the Expense of Draining the Same by Way of Mortgage, 3 & 4 Vict., c. 55.

6 An Act...to Defray the Expense of Draining...by Way of Mortgage, ss. 2–4; 'Select Committee...for the purpose of draining', qq. 1–2, 1430–7.

7 'Select Committee...for the purpose of draining', i–iii, q. 1–2, 15–19, 1432.

8 'Select Committee...for the purpose of draining', qq. 1430–7; An Act...to Defray the Expense of Draining...by Way of Mortgage, s. 2.

9 An Act...to Defray the Expense of Draining...by Way of Mortgage, s. 4. Two landowners who made use of this act, J. Mills of Hampshire and J. Bowes with estates in Durham and Yorkshire, charged their tenants 7 per cent on the outlay; 'Select Committee...for the purpose of draining', qq. 14–19, 1447–8.

10 BPP (Lords Sessional Papers), 1845, VIII, 'Return of the number of applications to the court of Chancery or Exchequer in England for leave to make permanent improvement in land under 3 & 4 Vict., c. 55', 287. All applications were made to the court of Chancery and four landowners can be traced proceeding under the legislation: Sir Edward Dering on his estate in Kent between 1842 and 1879; George Wyndham on the Egremont estates in Sussex and Yorkshire from 1846 to 1852; John Mills on his Hampshire property in 1843 and 1844; and John Bowes for estates in Durham and Yorkshire in 1843: PRO, Reports and certificates to the court of Chancery, C. 38/1808, 1838, 1905, 1943, 2101, 2141, 2319, 2448, 2490, 2533, 2615, 2657, 2741, 2787, 2843, 2896, 2942, 3001, 3006, 3213, 3326, 3369, 3601; C. 38/2007, 2047, 2089, 2165, 2167, 2204; C. 38/1849; C. 38/1832.

11 An Act to Alter and Amend the Act...to Enable the Owners of Settled Estates to Defray the Expenses of Draining..., 8 & 9 Vict., c. 56; 'Select Committee...on the improvement of land', qq. 4–17; J. B. Denton, 'On land drainage and improvement by loans from government or public companies', *JRASE*, second series, 4 (1868), 124–5.

12 An Act to Authorize the Advance of Public Money to a Limited Amount to Promote the Improvement of Land in Great Britain and Ireland by Works of Drainage, 9 & 10 Vict., c. 101; An Act to Authorize Further Advances of Money for Drainage and the Improvement of Landed Property in the United Kingdom..., 13 & 14 Vict., c. 31.

13 An Act to Authorize the Advance of Public Money...to Promote...Drainage, ss. 4, 36; Denton, 'On land drainage and improvement by loans from government or public companies', 127.

14 An Act to Authorize the Advance of Public Money...to Promote...Drainage, ss. 16–17; Spring, *The English Landed Estate*, 161–5; Phillips, 'Underdraining and agricultural investment in the midlands in the mid-nineteenth century', 254–6.

15 An Act to Promote the Advance of Private Money for Drainage of Lands in Great Britain and Ireland, 12 & 13 Vict., c. 100; Denton, 'On land drainage and improvement by loans from government or public companies', 127.

16 Denton, 'On land drainage and improvement by loans from government or public companies', 127–8; B. Supple, *The Royal Exchange Assurance* (Cambridge 1970), 319.

17 The Improvement of Land Act, 1864, 27 & 28 Vict., c. 114; Denton, 'On land drainage and improvement by loans from government or public companies', 142.

18 J. H. Charnock, *Suggestions for the More General Extension of Land Draining* (London, 1843), 8–18; Denton, 'On land drainage and improvement by loans from government or public companies', 124.

19 An Act for Incorporating the Landowners Drainage and Inclosure Company and for Enabling the Owners of Settled Estates Drained...by the Said Company to Charge the Same..., 10 & 11 Vict., c. ccxii; Cheshire Record Office, Baker-Wilbraham MSS: DBW/G/B/9, Notice of Landowners' Drainage and Inclosure Company, 11 August 1847; An Act for Incorporating the West of England and South Wales Land Draining Company; and for Enabling Owners of Limited Interests in Land to Charge the Same for the Purpose of Drainage..., 11 & 12 Vict., c. cxlii. Although not receiving its act until 1848, the West of England Company was in existence in 1845 carrying out improvement: 'Select Committee...for the purpose of draining', qq. 420–7; *Farmer's Herald*, 3 (1846), 3, 15.

20 An Act for Incorporating the Landowners Drainage and Inclosure Company, ss. 5–8, 20; An Act for Incorporating the West of England and South Wales Land Draining Company, ss. 5–9, 34; BPP, 1854–5, VII, 'Select Committee of the House of Lords...[on]...the powers now vested in the companies for the improvement of land', qq. 4, 60, 136, 220, 376.

21 An Act for the Incorporation of the General Land Drainage and Improvement Company; 'Select Committee...[on]...the companies for the improvement of land', qq. 5–6; 'Select Committee...on the improvement of land', qq. 399–403; Denton, 'On land drainage and improvement by loans from government or public companies', 133–7.

22 An Act for Incorporating the Lands Improvement Company..., 16 & 17 Vict., c. cliv; An Act for Incorporating and Granting Other Powers to the Land Loan

and Enfranchisement Company, 23 & 24 Vict., c. clxix; 'Select Committee ... [on] ... the companies for the improvement of land', qq. 308–73; 'Select Committee ... on the improvement of land', qq. 893–905; Denton, 'On land drainage and improvement by loans from government or public companies', 138–42.

23 'Select Committee ... [on] ... the companies for the improvement of land', q. 321; 'Select Committee ... on the improvement of land', q. 33; DRO, Drake MSS: 346M/T1332–41, Draining reports, 1854–8; Clinton MSS: 96M/Box 13, Letter from A. S. Parker to H. Drew, 3 May 1863, Letter from J. C. Knollys to H. Drew, 9 December 1865, Letter from W. B. Johnson to H. Drew, 17 December 1865; Sidmouth MSS: 152M/Drainage memoranda, 1850–67; Fortescue MSS: 1262M/E1/102, Letter from H. C. Mules to Lord Fortescue, 17 December 1847; 1262M/E1/103, Letter from W. Blamire to Lord Fortescue, 24 January 1851; Spring, *The English Landed Estate*, 161–4.

24 K. U., Sneyd MSS: Reports of Andrew Thompson to the Inclosure Commissioners, vols. 4–8, 1857–68; Phillips, 'Underdraining and agricultural investment in the midlands in the mid-nineteenth century', 256–74.

25 Three main sources of record of draining loans exist, deriving from the Treasury, the Inclosure Commissioners and the improvement companies. (A) PRO, 1R3/6–38, Registers of certificates of drainage advances from the Treasury under the Public Money Draining Acts, 1846 and 1850. These 33 volumes contain the 4,964 certificates issued for draining loans under these acts in England and Wales and sanctioned by the Inclosure Commissioners. A second summary set of registers of certificates is found in PRO, 1R3/2–4. (B) The Inclosure Commissioners' record of loans under the land-improvement legislation is incomplete, particularly for the Lands Improvement Company: PRO, MAF 66/1–2, Registers of applications and absolute charges under the General Land Drainage Company, 1851–96; MAF 66/3, Register of applications and absolute charges under the Lands Improvement Company, 1861–9; MAF 66/4–6, Registers of applications and absolute charges under the Land Loan Company, 1861–1912; MAF 66/8–9, Registers of work executed under loans from the Land Loan Company, 1861–96; MAF 66/11, Register of applications and absolute charges under the Improvement of Land Act 1864, 1864–1940. (C) Records of loans were kept by the various improvement companies and survived in the hands of the Lands Improvement Company. This company is still (1988) in existence. It absorbed the Land Loan Company in 1896–8 and amalgamated with the General Land Drainage Company between 1909 and 1911, taking possession of those companies' loan registers and depositing them with its own in its muniment room in St James's Square, London, where they were examined and where data on the history of the three companies were obtained: material on the liquidation of the Land Loan Company 1897–1924; notes and letters on the amalgamation of the Lands Improvement and General Land Drainage Companies, 1909–11. For present purposes, the material comprised a complete series of 15 volumes of registers of absolute orders and charges of loans issued by the Lands Improvement Company from 1853 to 1899; a complete series of 5 volumes of registers of absolute orders and charges of loans issued by the General Land Drainage Company from 1852 to 1911; and a complete series of

10 volumes of registers of absolute orders and charges of loans issued by the Land Loan Company from 1862 to 1903. These volumes of registers have subsequently been deposited in the PRO as respectively MAF 66/25–39; MAF 66/43–7; and MAF 66/13–22 but some have been mislaid in the transfer.

26 'Select Committee…[on]…the companies for the improvement of land', qq. 128–31; Wilbraham MSS, Rode Hall, Cheshire: Report of the Landowners' Drainage and Inclosure Company, 28 June 1848; *Farmer's Herald*, 6 (1848), 27, and 9 (1851), 33.

27 'Select Committee…[on]…the companies for the improvement of land', qq. 375–80; 'Select Committee…on the improvement of land', q. 399; BPP, 1884, XXII – 1900, XVII, *passim*, 'Reports of the Land Commissioners, 1883–88, and of the Board of Agriculture, 1889–99, on the improvement of land'.

28 'Select Committee…on the improvement of land', qq. 1139–40.

29 BPP, 1893–4, XXIII, part 2, 'Board of Agriculture. Annual report of proceedings', 59.

30 Denton, 'On land drainage and improvement by loans from government or public companies', 127–8.

31 'Select Committee…on the improvement of land', qq. 500–9.

32 'Select Committee…[on]…the companies for the improvement of land', qq. 413–31; 'Select Committee…on the improvement of land', qq. 143–58, 473–87; BPP, 1865, XLVII, 'Rules and practices of the Inclosure Commissioners', 429; KU, Sneyd MSS: Thompson's reports, vol. 6, 61; DRO, Troyte MSS: 2547/E7, Report of J. M. Martin, 18 March 1880, and letter of Inclosure Commissioners to J. C. Knollys, 19 April 1880.

33 'Select Committee…on the improvement of land', qq. 133–4, 612, 4141–5.

34 'Select Committee…on the improvement of land', qq. 401–3, 886–92; Denton, 'On land drainage and improvement by loans from government or public companies', 138–9.

35 PRO, MAF 48/189, Improvement of Land Bill, 1899; Improvement of Land Act, 1899, 62 & 63 Vict., c. 46; s. 1.

36 PRO, MAF 66/1–3, 11.

37 Spring, *The English Landed Estate*, 153.

38 *London Gazette*, 1846–78, *passim*.

39 An Act to Explain and Amend the Act Authorizing the Advance of Money for the Improvement of Land by Draining in Great Britain, 10 Vict., c. 11, s. 7; An Act to Authorize Further Advances of Money for Drainage and the Improvement of Landed Property in the United Kingdom…, 13 & 14 Vict., c. 31, s. 5; BPP (Lords Sessional Papers), 1849, XXX, 'Select Committee of the House of Lords…to enable possessors of entailed estates to charge such estates…for the purpose of draining…', qq. 2–7.

40 *London Gazette*, 1846–9, *passim*; PRO, 1R3/6–38. The five landowners were the earls of Derby and of Ellesmere with estates in Lancashire, the earl of Lovelace with estates in Surrey, and J. A. Gordon and the Dean of Windsor with land in Somerset.

41 Sturgess, 'The agricultural revolution on the English clays: a rejoinder', 84–5.

42 'Select Committee…[on]…the companies for the improvement of land', qq. 353–68; 'Select Committee…on the improvement of land', qq. 893–905;

Denton, 'On land drainage and improvement by loans from government or public companies', 135; PRO, MAF 66/25–39, 43–7, Registers of absolute orders and charges issued by the Lands Improvement Company and the General Land Drainage Company which list assignments of rentcharges.

43 PRO, MAF 66/1–2, 4–6, 11, 13–22, 25–39, 43–7, Registers of absolute orders and charges, 1851–99; S. Homer, *A History of Interest Rates* (New Brunswick, New Jersey, 1963), 195–7.

44 Supple, *The Royal Exchange Assurance*, 319–20, 331.

45 'Select Committee... on the improvement of land', qq. 291, 819–27.

46 'Select Committee... on the improvement of land', iii–v, qq. 1207, 2018–21, 3025–7; 'Select Committee... [on]... the companies for the improvement of land', q. 411; Denton, 'On land drainage and improvement by loans from government or public companies', 129.

47 Ministry of Town and Country Planning, *Rainfall: Annual Average, 1881–1915, 1:625,000* (Southampton, 1949), sheets 1 and 2.

48 Robinson, 'The extent of farm underdrainage in England and Wales, prior to 1939', 82–5.

49 H. C. Darby, *The Changing Fenland* (Cambridge, 1983), 148–93; J. A. Sheppard, *The Draining of the Hull Valley* (East Yorkshire Local History Society, 1958), 14–22, and *The Draining of the Marshlands of South Holderness and the Vale of York* (East Yorkshire Local History Society, 1966), 9–11, 20–5; M. Williams, *The Draining of the Somerset Levels* (Cambridge, 1970), 169–229.

50 S. Siegel, *Nonparametric Statistics for the Behavioural Sciences* (New York, 1956), 202–13.

51 'Agricultural returns, 1870–9'.

52 Siegel, *Nonparametric Statistics*, 202–13.

53 *London Gazette*, 1849–64, *passim*.

54 'Select Committee [1849]... for the purpose of draining', qq. 35–7; 'Select Committee... [on]... the companies for the improvement of land', qq. 96–7; 'Select Committee... on the improvement of land', v. qq. 4125–30; 'Royal Commission on the depressed condition of the agricultural interest', qq. 4703, 4719–24.

55 KU, Sneyd MSS: Thompson's reports, vols. 4–8.

56 Phillips, 'Underdraining and agricultural investment in the midlands in the mid-nineteenth century', 262.

57 'Select Committee... on the improvement of land', qq. 586–9, 4125–30; 'Royal Commission on the depressed condition of the agricultural interest', qq. 4705–24; 'Agricultural depression', 29.

58 For a general discussion of the difficulties of distinguishing between new and repair work in agricultural improvement, see Holderness, 'Capital formation in agriculture', 170–9 and 'Landlord's capital formation in East Anglia 1750–1870', 434–7, 445–7.

59 'Return of owners of land in England and Wales, 1873'; Bateman, *The Great Landowners*.

60 Northumberland MSS, Alnwick Castle, Northumberland: Business minutes, vol. 119, 21 November 1906; Survey and terrier of the duke of Northumberland's estates, 1850, by T. Bell and sons; Thompson, 'The economic and social background of the English landed interest, 1840–70...', 157.

61 Grey MSS, Howick Estate Office, Northumberland: Maps of Ancroft, Chevington and Howick estates, 1848–98; DPD, Grey MSS: Estate cropping book, 1845–78; Bateman, *The Great Landowners*, 195; M. Hughes, 'Lead, land and coal as sources of landlord income in Northumberland between 1700 and 1850', unpublished University of Durham Ph.D. thesis (2 vols., 1963), vol. 2. 60.

62 NdRO, Sample of Bothal MSS: ZSA/18/2/1 and 2, A schedule of the area of the Bothal estate belonging to the duke of Portland, 1861; DPD, Howard of Naworth MSS: N99/2, Survey of the Northumberland estates, 1886. This analysis deals only with the Carlisle estates around Morpeth. The earls of Carlisle owned other land in the west of the county, but which was administered with their Cumberland estates.

63 Newcastle Central Library, L.622.33, J. T. W. Bell, Map of the Blyth and Warkworth Coal District, 1851, $2\frac{1}{2}$ in to 1 mile.

64 NdRO, Sample MSS: ZSA/8/4–5, Estate accounts and rentals, 1883–1903.

65 DPD, Howard MSS: Box 121/23, Notes on sales of earl of Carlisle's estates; Uncatalogued Morpeth rentals, 2 vols., 1879–1915; Thompson, *English Landed Society*, 319.

66 NdRO, Ridley MSS: ZR1/44/4, Miscellaneous farm accounts, 1847–85; ZR1/49/11 and 12, Sir Matthew Ridley's fieldbook, 1889; DPD, Howard MSS: 109/1, Letter from R. du Cann to R. Turnbull, 1 May 1889; Newcastle Central Library, L.622.33, J. T. W. Bell, Plan of the Newcastle Coal District, 1847, $2\frac{1}{2}$ in to 1 mile.

67 DPD, Baker-Baker MSS: 119/19, Plan of the estate...belonging to H. J. Baker-Baker, 1847; 23/100–1, Stanton rent sheet, 20 November 1872; 'Return of owners of land in England and Wales, 1873', part 2.

68 NdRO, Blackett MSS: ZBL/4/10, Indenture with Lands Improvement Company, 1861; ZBL/54/1, Indenture with Land Loan and Enfranchisement Company, 1867; ZBL/275/6, Estate rentals, 1875–99; Hughes, 'Lead, land and coal as sources of landlord income in Northumberland between 1700 and 1850', vol. 2, 100–1.

69 NdRO, Belsay (Middleton) MSS: Part III (Supplemental), S. 18, Plans of estate, 1847–62; Box 6/VI/20, Valuation of freehold estate...belonging to Sir Charles Monck and showing rental in 1853; B. 42/2, General cultivation book, 1868–72; PRO, MAF 66/36, Register of absolute orders and charges of the Lands Improvement Company, 1879–83.

70 Northumberland MSS: Draining vols., 1–3, 1844–1903.

71 *Ibid.*

72 Northumberland MSS: Business minutes, vol. 17, 29 October 1855, and vol. 37, 16 February 1866.

73 DPD, Grey MSS: Uncatalogued ledger books, 1840–92; Building and improvement expenditure volume 1841–58; Draining vols., 1841–86.

74 DPD, Grey MSS: Boxes 550–1, Draining papers, 1841–69, and draining reports for 1847.

75 DPD, Howard MSS: N. 101–6, Rental accounts, 1801–55; N. 73/2, Annual accounts, 1865–74; Shelf 55, Estate accounts, 14 vols., 1875–1900; Uncatalogued, Estate draining vols., 1–3, 1856–1901; 80/11, Draining expenditure on the Morpeth estate, 1850–70.

76 DPD, Howard MSS: Uncatalogued, Estate draining vols., 1–3, 1856–1901.

77 NdRO, Sample MSS: ZSA/8/1–5, Estate accounts and rentals, 1828–1903. On this estate, annual draining expenditure is recorded from 1856. Before that date the amount spent on the improvement has to be calculated from the interest paid by tenants on different types of draining introduced from 1829.

78 NdRO, Ridley MSS: ZR1/44/4, Miscellaneous farm accounts, 1847–85.

79 NdRO, Belsay MSS: Box 12/1X, Drainage accounts, 1869–84, and statement of loans received from the Lands Improvement Company and the government, June 1891.

80 NdRO, Blackett MSS: ZBL/54/2 and 3, Draining schedules, 1857–75 and 1884–6; ZBL/277/6, Estate journal, 1826–1901; ZBL/282/2, Draining accounts, 1840–1908.

81 NdRO, Blackett MSS: ZBL/54/2 and 3, Draining schedules, 1857–75 and 1884–6; ZBL/282/2, Draining accounts, 1840–1908.

82 DPD, Baker-Baker MSS: 23/36–7, Abstract of drainage on the Stanton estate, 1851–4; 23/98, 102, 111, 118, Letters from H. R. Goddard to Mrs Baker-Baker, 8 and 23 November 1872, 24 January and 15 March 1873; 24/57, Letter from R. Clark to Mrs Baker-Baker, 12 June 1876.

83 DPD, Grey MSS: Boxes 550–1, Draining papers, 1841-69, and draining reports for 1847.

84 NdRO, Ridley MSS: ZR1/39/19, D. Turner's valuation of the Hawkhope estate, Falstone, 9 February 1837.

85 NdRO, Blackett MSS: ZBL/4/10, Indenture with Lands Improvement Company, 1861; ZBL/87, Letter from Sir Edward Blackett to W. Sample, 17 May 1854; North East Development Association, *A Physical Land Classification of Northumberland, Durham and part of the North Riding of Yorkshire* (Newcastle upon Tyne, 1950), 36–9.

86 T. L. Colbeck, 'On the agriculture of Northumberland', *JRASE*, 8 (1847), map at 436–7.

87 DRO, Fortescue MSS: 1262M/E1/103, Draining sheets and schedules, 1847–64; Sidmouth MSS: 152M/Memoranda, Draining schedules 1851–66; NRO, Dryden MSS: D (CA) 450, Draining schedules, 1858–66; Cartwright MSS: C (A) 5242, 5943, Draining sheets, 1856; Raby MSS, Staindrop, Durham: Draining bundle: Sudborough draining vol., 1848–53; Draining abstracts 1848–53; Draining vols., 1861–72.

88 Buccleuch MSS, Boughton, Northamptonshire: Particular and rental of the estates within the Boughton collection, 1834; Numerical reference, 1895, of the Boughton estate; NRO, Buccleuch MSS: Miscellaneous ledgers, vol. 140, Particular, rental and valuation of the estates … 1813; vol. 144, Barnwell estate, 1823; vol. 137, A particular of the Barnwell estate, 1860; vol. 136, Collected reference of the Barnwell estate, 1903.

89 Spencer MSS, Althorp, Northamptonshire: A survey of the parishes of Strixton and Bozeat in Northamptonshire, 1827; Survey of Harlestone, Northamptonshire, 1831, by John Beasley; Reference to the estate of earl Spencer, 1859; Agricultural holdings on earl Spencer's estate, 1879; An abstract of rents and acreage of Lord Spencer's tenants on agricultural holdings, 1882.

90 NRO, Overstone MSS: O.906–38, Estate accounts, 1832–58; Ov. map 342, Map of the Northamptonshire estate of Lewis Loyd, 1850; Ov. vol. 4, Terrier of Lord

Overstone's estates, 1877; Thompson, *English Landed Society*, 38–40. For a general discussion of the development of Overstone's interest in land, see R. C. Michie, 'Income, expenditure and investment of a Victorian millionaire: Lord Overstone, 1823–1883', *Bulletin of the Institute of Historical Research*, 68 (1985), 59–72.

91 NRO, Grafton MSS: G 1558, List of farms, tenants and acreages, 1822; Bateman, *The Great Landowners*, 190.

92 NRO, Fitzwilliam MSS: Miscellaneous volumes, no. 646, Valuation of estates in Great and Little Harrowden, Orlingbury, Finedon, Mears Ashby and Wellingborough, the property of earl Fitzwilliam, 1857; no. 738, Survey of estate about High Ferrers (*c.* 1860).

93 NRO, Fitzwilliam MSS: Miscellaneous volumes, no. 763, Reference to a map of the Milton estate, 1841 and corrected in 1855; no. 450, A valuation of the parish of Castor, 1835; Estate accounts, 1800–99.

94 Raby MSS: Particulars of Brigstock and Sudborough rents for 1835, 1846, 1849, 1855; Fieldbook, 1871–8; Northamptonshire estate reference book, 1901–2. For further detail on the estate, see A. D. M. Phillips, 'Agricultural land use on a Northamptonshire estate (1849–1899), as revealed by cropping books', *East Midland Geographer*, 8 (1983), 70–8.

95 Bedford MSS, Bedford Estate Office, London: Annual reports for 1857 and 1895.

96 NRO, Cartwright MSS: C (A) 4740–57, Absolute orders under the Lands Improvement Company loans, 1854–65; C (A) 4914, Report of the Cartwright estates, 19 May 1893.

97 NRO, Ellesmere MSS: X.3695, Rental of Brackley and Halse estate, *c.* 1837; X.471–2, Maps of Halse and Brackley, n.d.; Bateman, *The Great Landowners*, 150; J. R. Lowerson, 'Enclosure and farm buildings in Brackley, 1829–51', *Northamptonshire Past and Present*, 6 (1978), 33–48.

98 NRO, Dryden MSS: D (CA) 445, Acreage description and value of estate in 1837 and rates, rateable values and rents in 1844; D (CA) 446, Valuation of estate, *c.* 1847; 'Return of owners of land in England and Wales, 1873', part 2.

99 NRO, Fisher-Sanders MSS: FS1/34, Terrier of the Naseby estate, 1860; FS1/38, Terrier of the Naseby estate, 1896; FS1/14, Sale catalogue Naseby estate, 1903; FS1/13–14, Plans of Naseby estate, 1887.

100 Raby MSS: Brigstock and Sudborough rentals, 1845–99; Sudborough draining vol., 1848–53; Draining abstracts 1848–53; Draining vols., 1861–72; Plans of the Brigstock and Sudborough estates by W. H. Boynes, Staindrop, 1855.

101 Spencer MSS: Map of the Brampton Ash estate, n.d.; Map of the Brington estate, 1851; Map and survey of the Chapel Brampton estate, 1842; Map of the Elkington estate, 1838; Plan of the Harlestone estate, 1879; Map of an estate in Heyford, Flore and Great Brington, 1856; Map of the Silsworth estate, n.d.; Plan of the Strixton estate, n.d.; Estate cropping books for Brington, Nobottle, Flore and Heyford, 1857–95, Strixton, 1879–99, Murcott and Muscott, 1865–99, Harlestone, 1877–99, Brampton Ash, 1864–99 and Chapel and Church Brampton, 1887–99.

102 NRO, Overstone MSS: Ov. maps: 184, Map of Pitsford and Brixworth estate, 1854; 185, Map of Broughton estate, 1853; 187, Map of Sibbertoft and Clipstone estate, 1858; 188, Map of Orlingbury estate, 1853; 190, Map of Stanwick and Chelveston estate, 1862; 191, Map of Braybrook estate, 1864; 192, Map of Holcot and Walgrave estate, 1856; 327, Map of Hackleton and Piddington estate, 1854; 330, Map of Moulton estate, n.d.; 331, Map of Pytchley estate, 1849; 354, Map of Sywell estate, 1851.

103 W. Bearn, 'On the farming of Northamptonshire, *JRASE*, 13 (1852), 48–51, map at 45–6.

104 PRO, IR 29 and 30/9/280, 429, Tithe maps and apportionments for Membury and Upottery, 1840–2; DRO, Sidmouth MSS: 152M/Draft terrier, 1894; 'Return of owners of land in England and Wales, 1873', part 1.

105 DRO, Seymour MSS: 1392M/Estate rentals, bundles 1 and 2.

106 'Return of owners of land in England and Wales, 1873', part 1; Duchy of Cornwall MSS, Buckingham Gate, London: Valuation of the manor of Bradninch by R. Watt, 1855; Valuation of rents of manor of Bradford, c. 1856; Valuation of Marsh estate, 1868; Valuation of Bradford farm, 1869; Valuations of Horslett tenement 1856 and 1861; A survey of the honor of Bradninch, 1788 by W. Simpson; Report on the manor of Bradford by E. C. Marriott, 1862.

107 Bedford MSS: Annual reports for 1857 and 1895.

108 DRO, Courtenay MSS: 1508M/Estate papers/14/A/III, Shelf III, Particulars and valuation of the estates of the earl of Devon by J. Hooper, 1862; 1508M/ Estate vols./14/B/III/2–39, Rentals and accounts, 1848–99.

109 DRO, Fortescue MSS: 1262M/E29/58, List of estates in north Devon showing acreages, 1864; 1262M/E20/73, Rental for south Devon estate, 1860. About 700 acres of the south Devon estate lay in the Cornish parishes of Michaelstow, Jacobstow, Warbstow and St Endellion but have been included in the present analysis: 1262M/E22/45, Map of lands in Lamerton...by T. H. Lakeman, 1825.

110 DRO, Fortescue MSS: 1262M/E20/58–111, Rentals, 1850–80.

111 Bateman, *The Great Landowners*, 503.

112 DRO, Sidmouth MSS: 152M/Memoranda, draining schedules under the Public Money Draining Acts and General Land Drainage Company loans, 1851–66; 152M/Accounts, 1850–99.

113 DRO, Sidmouth MSS: 152M/A general report of the Upottery estate...by F. Thynne, 1850; 152M/Memoranda, report on drainage on the Upottery estate by T. Webber, 1861.

114 DRO, Sidmouth MSS: 152M/Memoranda, draining schedules, 1851–66.

115 Cornwall MSS: Farm bundles, manor of Bradninch, Billingsmoor and Northdown farm, nos. 44 and 63, 1853–66 bundle: valuation by R. Watt, 1853; letter to R. Watt, 29 January 1856; letters between R. Watt and J. W. Bateman, 30 January and 9 May 1864; recommendations of the land steward as to works to be executed in 1865, 25 February 1865; vouchers to R. Watt, 14 July 1856, 9 June and 26 November 1858; Farm bundle for Bathaies tenement, no. 73, 1862–77: valuation, 1874; Inrolment books of patents and warrants, XLVIII, 1862–74, LXX, 1875–90 and XC, 1890–1902.

116 Cornwall MSS: Farm bundles, manor of Bradford, Horslett farm, 1855–63: letter from G. Richardson to J. W. Bateman, 6 November 1861; Horslett farm, 1864–78: letter from G. Richardson to J. W. Bateman, 4 May 1864; letter from J. M. Martin to J. W. Bateman, 9 September 1865; letter from G. Herriot to J. W. Bateman, 1 June 1869; Horslett farm, 1882–1915: letter from G. Herriot to G. W. Wilmshurst, 28 December 1885; Bundle of terriers, Bradford manor: tracing of Horslett draining, 1865.

117 DRO, Seymour MSS: 1392M/Estate/Accounts/Bundle 4, Estate accounts, 1868–75; Michelmore, Loveys and Carter MSS: 867/B, Berry Pomeroy estate books, 1853–69 and day books, 1862–75.

118 DRO, Courtenay MSS: 1508M/Estate vols./14/B/III/2–39, Rentals and accounts, 1848–99.

119 DRO, Fortescue MSS: 1262M/E20/37–42, 59–80, Rentals of south Devon estate, 1845–80.

120 DRO, Fortescue MSS: 1262M/E20/53–70, 93–144, Rentals of north Devon estate, 1845–99; 1262M/E1/103, Draining schedules, 1847–64.

121 DRO, Fortescue MSS: 1262M/E1/102, List of lands required to be drained, 1847.

122 DRO, Fortescue MSS: 1262M/E1/103, Report on the Castle Hill and other estates...by Josiah Parkes, 10 January, 1848.

123 DRO, Fortescue MSS: 1262M/E1/103, Draining schedules, 1847–64.

124 Bedford MSS: Annual reports, 1845–95.

125 Tanner, 'The farming of Devonshire', 458–61; BPP, 1882, XV, 'Report by W. C. Little on Devon, Cornwall, Dorset and Somerset to the Royal Commission on the depressed condition of the agricultural interest', 22, 25.

126 Tanner, 'The farming of Devonshire', 470–1; 'Report by W. C. Little on Devon, Cornwall, Dorset and Somerset...', 23.

127 Bedford MSS: Annual report, 1869: report on the mid-Devon estate; Cornwall MSS: Farm bundles, manor of Bradford, Horslett farm, 1855–63: report on the manor of Bradford by E. C. Marriott, 11 September 1862; Horslett farm, 1882–1915: letter from G. Herriot to M. Holzman, 31 October 1891; Pinkworthy farm, 1870–1928: letter from A. M. Webster to W. Peacock, 11 February 1910.

128 J. Caird, *English Agriculture in 1850–51* (London, 2nd edn, 1852), 145, 152; Holderness, 'Landlord's capital formation in East Anglia 1750–1870', 439–43; Wade Martins, *A Great Estate*, 96–7.

129 Robinson, 'The extent of farm underdrainage in England and Wales, prior to 1939', 83.

130 Thompson, 'Free trade and the land', 105.

131 Sturgess, 'The agricultural revolution on the English clays: a rejoinder', 84–5.

132 Green, 'Field under-drainage before and after 1940', 120–2; Robinson, 'The extent of farm underdrainage in England and Wales, prior to 1939', 80–2; Trafford, 'Field drainage', 131.

133 Collins and Jones, 'Sectoral advance in English agriculture, 1850–80', 69.

134 Holderness, 'Capital formation in agriculture', 165–7; M. Turner, *Enclosures in Britain 1750–1830* (London, 1984), 59–60.

4 The temporal pattern of underdraining in the nineteenth century

1 For an introduction to the diffusion of agricultural innovations over time, see Grigg, *The Dynamics of Agricultural Change*, 153–63.
2 Spring, *The English Landed Estate*, 176.
3 Settled Land Act, 1882, 45 & 46 Vict., c. 38, ss. 21–8.
4 BPP, 1903, XVII, 'Board of Agriculture. Annual report of proceedings, 1902'.
5 BPP, 1894, XVI, parts 1–3, 'Royal Commission on agricultural depression', qq. 7293–4, 7327–9, 42731–74, 42907–36.
6 NRO, Brudenell MSS: ASR/95 and 96, A particular and valuation of sundry estates... in the county of Northampton, 1812.
7 Northumberland MSS: Annual returns of state of farms, 1834–9; Draining vol. 1, 1844–54; NdRO, Bell Collection: ZHE 34/8, Reasons for deep draining, 1848.
8 DRO, Sidmouth MSS: 152M/A general report of the Upottery estate... by F. Thynne, 1850.
9 Cornwall MSS: Valuation of the manor of Bradninch by R. Watt, 1855; Report on the manor of Bradford by E. C. Marriott, 1862; DRO, Forstescue MSS: 1262M/E1/103, Report on the Castle Hill and other estates... by Josiah Parkes, 10 January 1848.
10 NRO, Ellesmere MSS: X.3695, Letter from J. Loch to Lord Francis Egerton, 1 July 1837; Bedfordshire Record Office, Woburn MSS: Solicitor's and steward's papers, Box 353, Correspondence 1844–56: letter from T. Wing to C. Haedy, 1 June 1850.
11 BPP, 1833, V, 'Select Comittee on agriculture', qq. 380, 5826–34, 8520, 10414, 11112–24, 12805–6; BPP, 1836, VIII, 'Select Committee on the state of agriculture', qq. 2861–4, 3478–81, 9341–6, 12410–12, 13142–7; BPP, 1837, V, 'Select Committee of the House of Lords on the state of agriculture', qq. 1809–21, 4172.
12 'Select Committee on agriculture', iv; C. S. Lefevre, *Remarks on the Present state of Agriculture* (London, 1836), 13–14.
13 J. F. Burke, 'On the drainage of land', *JRASE*, 2 (1841), 273–6; Pusey, 'On the progress of agricultural knowledge during the last four years', 169–71; E. Little, 'Farming in Wiltshire', *JRASE*, 5 (1844), 177; G. Buckland, 'On the farming of Kent', *JRASE*, 6 (1845), 293–4; R. W. Corringham, 'Agriculture of Nottinghamshire', *JRASE*, 6 (1845), 30–1, 43; Karkeek, 'On the farming of Cornwall', 421–2.
14 R. J. P. Kain, *An Atlas and Index of the Title Files of mid-Nineteenth-Century England and Wales* (Cambridge, 1986), 562–631; R. J. P. Kain and H. C. Prince, *The Tithe Surveys of England and Wales* (Cambridge, 1985), 103–13.
15 L. Kennedy and T. B. Grainger, *The Present State of the Tenancy of Land in Great Britain* (London, 1828), 289, 317; 'Select Committee of the House of Lords on the state of agriculture', qq. 2728–34; P. Pusey, 'Evidence on the antiquity, cheapness and efficacy of thorough-draining or land-ditching, as practised throughout the counties of Suffolk, Hertford, Essex and Norfolk',

JRASE, 4 (1843), 23–9, and 'On the progress of agricultural knowledge during the last four years', 169–217; H. Raynbird, 'On the farming of Suffolk', *JRASE*, 8 (1847), 278–9, 308–10.

16 Pusey, 'Evidence on the antiquity, cheapness and efficacy of thorough-draining...', 23–5; Wade Martins, *A Great Estate*, 96.

17 Caird, *English Agriculture in 1850–51*, 133, 145; BPP, 1847–8, VII, 'Select Committee on agricultural customs', qq. 1386–90, 3136–46, 3290–301.

18 J. Parkes, 'On the influence of water on the temperature of soils. On the quantity of rain-water and its discharge by drains', *JRASE*, 5 (1844), 146–58.

19 J. B. Denton, *Land Drainage: Arterial Channels and Outfalls* (London, 2nd edn, 1861), 3–4.

20 BPP, 1884, XXII, 'Report of the Land Commissioners', 513.

21 'Royal Commission on the depressed condition of the agricultural interest', qq. 4747–8, 4787, 50465–71, 50495; BPP, 1881, XVI, and 1882, XV, 'Reports of the assistant commissioners', 156, 280, 417 and 52, 317, 322, 349; J. H. Tiffen, 'Prize essay on the agriculture of East and North Ridings of Yorkshire', *North British Agriculturalist*, 29 October 1884, 734. The last reference was kindly brought to my attention by Dr A. Harris.

22 Royal Meteorological Society, *Rainfall Atlas of the British Isles* (London, 1926), map page 6.

23 Siegel, *Nonparametric Statistics*, 202–13.

24 Spencer MSS: An abstract of rents and acreage of Lord Spencer's tenants on agricultural holdings, Michaelmas, 1879; NdRO, Blackett MSS: ZBL/88, Estate correspondence, 1871–80, letters from Sir E. Blackett to T. Sample, 31 May 1879 and 31 July 1880, from J. Lishman and J. Richardson to T. Sample, 22 November 1877 and 16 October 1880.

25 Cornwall MSS: Farm bundles, manor of Bradninch, Park farm, 1852–1902: letter from G. Herriot to G. Wilmshurst, 21 June 1884; Fordishaies farm, 1865–1907: letter from G. Herriot to G. Wilmshurst, 17 August 1881.

26 J. H. Charnock, *On Thorough-Draining; and its Immediate Results to the Agricultural Interest* (London, 2nd edn, 1844), 20; 'Royal Commission on the depressed condition of the agricultural interest', qq. 2438–42; 'Royal Commission on agricultural depression', qq. 29–49; Jones, *Seasons and Prices*, 126–7.

27 'Select Committee [1845]...for the purpose of draining', qq. 148–50.

28 'Select Committee on agriculture', qq. 6982–3; 'Select Committee on the state of agriculture', qq. 1073–7, 8857–60, 15816; 'Royal Commission on the depressed condition of the agricultural interest', q. 2476; 'Royal Commission on agricultural depression', qq. 7293–4; BPP, 1897, XV, 'Final report and report by F. A. Chaning to the Royal Commission on agricultural depression', 23, 287, 292.

29 Denton, *The Under-Drainage of Land*, 1–2.

30 Siegel, *Nonparametric Statistics*, 202–13.

31 J. C. Stamp, *British Incomes and Property* (London, 1916), 36, 54–5; C. H. Feinstein, *National Income, Expenditure and Output of the United Kingdom 1855–1965* (Cambridge, 1972), table 23; A. Offer, 'Ricardo's paradox and the movement of rents in England, c. 1870–1910', *EcHR*, second series, 33 (1980), 240, 250.

32 Buccleuch MSS: Barnwell estate account, 1822.

33 DPD, Howard MSS: N. 122, Letter from G. A. Grey to R. Du Cann, 4 February 1880.

34 Northumberland MSS: Business minutes, vol. 63, J. Snowball to duke of Northumberland, 14 February 1897; NdRO, Blackett MSS: ZBL/89, Estate correspondence, 1881–92, letter from Sir E. Blackett to T. Sample, 10 November 1881.

35 NdRO, Blackett MSS: ZBL/89, Estate correspondence, 1881–92, letter from T. Sample to C. H. Sample, 31 January 1888.

36 J. Loch, *An Account of the Improvements on the Estates of the Marquess of Stafford in the Counties of Stafford and Salop and on the Estate of Sutherland* (London, 1820), 189–200; J. C. Loudon, *An Encyclopaedia of Agriculture* (London, 1825), 625–45; J. Yule, 'An account of the mode of draining by means of tiles as practised on the estate of Netherby in Cumberland, the property of Sir James Graham', *Prize Essays and Transactions of the Highland Society of Scotland*, new series, 1 (1829), 388–400; G. Stephens, *The Practical Irrigator and Drainer* (Edinburgh, 1834), 87–114; Anon., *On Land–Drainage, Subsoil-Ploughing and Irrigation* (London, 1841), 3–7, 27–39; D. Low, *On Landed Property and the Economy of Estates* (London, 1844) 254–310.

37 J. Smith, *Remarks on Thorough Draining and Deep Ploughing* (Stirling, 1831), 4–15; 'Select Committee on the state of agriculture', qq. 14976–15223; Trimmer, *On the Improvement of Land*, 1–11; T. Gisborne, *Agricultural Drainage* (London, 2nd edn, 1852), 11–17.

38 J. Parkes, 'Report on drain-tiles and drainage', *JRASE*, 4 (1843), 378; 'On draining', *JRASE*, 7 (1846), 249–72; *Essays on the Philosophy and Art of Land Drainage* (London, 1848), and *Fallacies on Land Drainage Exposed* (London, 1851); H. Davis, *Farming Essays* (London, 1848), 74–9; G. Monckton, *A Treatise on Deep Draining* (Wolverhampton, 1847); 'Select Committee [1845]...for the purpose of draining', qq. 54–169, 477–82.

39 See, for example, C. Newman, *Practical Hints on Land Draining* (London, 1845), 8; C. Arbuthnot, 'On deep draining', *JRASE*, 6 (1845), 129–31, 573–4; J. C. Clutterbuck, 'On the theory of deep draining', *JRASE*, 6 (1845), 489–93; W. B. Webster, 'On the failure of deep draining on certain strong clay subsoils with a few remarks on the injurious effect of sinking the water too far below the roots of plants in very porous, alluvial and peaty soils', *JRASE*, 9 (1848), 237–48.

40 BPP, 1821, IX, 'Select Committee...[on]...the depressed state of agriculture', 106–10.

41 'Select Committee on the state of agriculture', qq. 10730–5, 11047–66, 13830; Newman, *Practical Hints*, 9–14; Pusey, 'Evidence on the antiquity, cheapness and efficacy of thorough-draining...', 23–49.

42 Smith, *Remarks on Thorough Draining*, 16; R. White, 'Report of several operations in thorough-draining and subsoil-ploughing at Oakley Park', *JRASE*, 1 (1840), 33–7; T. Arkell, 'On the drainage of land', *JRASE*, 4 (1843), 331–4; 'Select Committee [1845]...for the purpose of draining', qq. 626–40, 1378–82.

43 An Act...to Exempt Tiles Made for the Purpose of Draining Lands from the Duties of Excise, 46 Geo. III, c. 138; An Act for Allowing Certain Tiles to be Made Duty Free for Draining, 55 Geo. III, c. 176; An Act...for Exempting Tiles

Made for Draining Lands from Duty, 1 & 2 Geo. IV, c. 102; An Act...to Amend Certain Laws of Excise Relating...to the Duty on Draining Tiles, 5 Geo. IV, c. 75.

44 Yule, 'An account of the mode of draining by means of tiles as practised on the estate of Netherby...', 388–400; J. Wiggins, 'On the mode of making and using tiles for underdraining practised on the Stow Hall estate in Norfolk', *JRASE*, 1 (1840), 350–6; R. Beart, 'On the economical manufacture of draining-tiles and soles', *JRASE*, 2 (1841), 102–4.

45 'Select Committee...[on]...the companies for the improvement on land', table at 54.

46 Beart, 'On the economical manufacture of draining-tiles and soles', 93–9; J. Hunt, 'On the marquis of Tweeddale's tile making machine', *JRASE*, 2 (1841), 148–50; Parkes, 'Report on drain–tiles and drainage', 369–71.

47 Parkes, 'Report on drain-tiles and drainage', 369–79; P. Pusey, 'Note to T. L. Hodges, On the cheapest method of making and burning draining tiles', *JRASE*, 5 (1844), 556–9.

48 Parkes, 'Report on drain–tiles and drainage', 370–2.

49 J. Read, 'On pipe-tiles', *JRASE*, 4 (1843), 273–4; 'Select Committee [1845]...for the purpose of draining', qq. 726–38.

50 Parkes, 'Report on drain-tiles and drainage', 372–9.

51 J. Parkes, 'Report on the exhibition of implements at the Newcastle upon Tyne meeting, 1846', *JRASE*, 7 (1846), 681–96 and 'Report on the exhibition of implements at the Northampton meeting, 1847', *JRASE*, 8 (1847), 330–61.

52 J. Parkes, 'Report on the exhibition of implements at the Shrewsbury meeting in 1845', *JRASE*, 6 (1845), 303–23; H. S. Thompson, 'Report on the exhibition and trial of implements at the York meeting, 1848', *JRASE*, 9 (1848), 377–422; R. Boyle, 'On drain-tile and pipe machines', *Transactions of the Highland and Agricultural Society of Scotland*, new series, 14 (1853–5), 40–54, 75–90.

53 J. Parkes, 'Report on the exhibition of implements at the Southampton meeting in 1844', *JRASE*, 5 (1844), 361–91; T. L. Hodges, 'On the cheapest method of making and burning draining tiles', *JRASE*, 5 (1844), 553–6.

54 'Select Committee [1845]...for the purpose of draining', qq. 54–60, 1384–90; J. Parkes, 'On reducing the cost of permanent drainage', *JRASE*, 6 (1845), 125–9; P. Pusey, 'On cheapness of draining', *JRASE*, 7 (1846), 520–4 and 'On the progress of agricultural knowledge during the last eight years', *JRASE*, 11 (1850), 381–443; Denton, *Agricultural Drainage*, 3–4.

55 Arbuthnot, 'On deep draining', 130; 'Select Committee [1845]...for the purpose of draining', qq. 1387–94.

56 'Select Committee on the state of agriculture', q. 12522; Baron Western, *Practical Remarks on the Improvement of Grass Land by Means of Irrigation, Winter Flooding and Drainage* (London, 1838), 21; T. L. Hodges, *The Use and Advantages of Pearson's Draining Plough* (London, 1839), 7–9; H. H. Brown, 'On the drainage of land', *JRASE*, 3 (1842), 165–8; Pusey, 'Evidence on the antiquity, cheapness and efficacy of thorough-draining...', 23–49.

57 'Select Committee [1845]...for the purpose of draining', qq. 92–5.

58 'Select Committee [1845]...for the purpose of draining', qq. 561–3, 762; Pusey, 'On the progress of agricultural knowledge during the last four years', 169–217.

59 J. B. Denton, *What Can Now Be Done for British Agriculture?* (London, 1842), 11; Pusey, 'On the progress of agricultural knowledge during the last four years', 169–217; Charnock, *On Thorough-Draining*, 17; Newman, *Practical Hints*, 27; 'Select Committee [1845]…for the purpose of draining', qq. 342–5, 827–33.

60 BPP, 1852, XIX, 'General Board of Health: minutes of information collected in respect to the drainage of land', appendix 7, 69; 'Select Committee [1845]…for the purpose of draining', i, qq. 590, 657; 'Select Committee on agricultural customs', qq. 1450–1, 4119–22; 'Select Committee [1849]…for the purpose of draining', qq. 4–7.

61 Charnock, *On Thorough-Draining*, 6–7; 'Select Committee [1845]…for the purpose of draining', i.

62 N. Goddard, 'The development and influence of agricultural periodicals and newspapers, 1780–1880' *AHR*, 31 (1983), 128.

63 NdRO, Sample MSS: ZSA/1/2, Wm Sample's letters to the duke of Portland, 1827–57, letter dated 6 September 1830.

64 NdRO, Sample MSS: ZSA/1/2, Letters dated 4 and 16 December 1829, 5 February, 12 March, 9 April and 11 October 1830.

65 NdRO, Sample MSS: ZSA/1/2, Letters dated 13 August and 11 September 1833.

66 NdRO, Sample MSS: ZSA/1/2, Letters dated 25 October and 12 November 1833. For a discussion of the use of agricultural labourers in the adoption of new agricultural improvements, see S. Macdonald, 'Agricultural improvement and the neglected labourer', *AHR*, 31 (1983), 81–90.

67 NdRO, Sample MSS: ZSA/1/2, Letters dated 20 February, 21 March and 21 August 1834.

68 NdRO, Sample MSS: ZSA/1/2, Letters dated 7 and 19 January 1836.

69 DPD, Howard MSS: N. 101–6, Rental accounts, 1837–42.

70 DPD, Grey MSS: Letters from earl Grey to Sir T. Lander, 7 December 1837 and 9 January 1838; Draining vol. 1841–54; Boxes 550 and 551, Draining papers 1841–69, letters from J. Hunt of the Tweeddale Company to R. Robson, estate agent, 15 and 29 March 1844.

71 NdRO, Blackett MSS: ZBL/281/1, Account of the establishment of the East Matfen tilery; ZBL/282/2, Drainage accounts, 1840–1908; ZBL/86, Estate correspondence, 1835–44.

72 Northumberland MSS: Annual returns of state of farms, Lucker bailiwick, 1840; NdRO, Bell Collection: ZHE 34/8, T. Bell's proposed history of agriculture in Northumberland: underdraining, circular letter by J. C. Blackden and W. Laws, commissioners to the duke of Northumberland, to tenants, July 1844.

73 Northumberland MSS: Tile works account, 1844–6; F. W. Etheredge, 'On the cheapest and best method of establishing a tile yard', *JRASE*, 6 (1845), 463–77.

74 Northumberland MSS: Tile works account, 1844–6.

75 Northumberland MSS: Business minutes, vol. 32, 21 April 1963, and vol. 37, 16 February 1866; NdRO, Bell Collection: ZHE 34/8, Circular letter by J. C. Blackden and W. Laws, July 1844.

76 NdRO, Sample MSS: ZSA/1/2, Letters dated 16 September and 31 December 1846, 8 and 23 January and 27 March 1847; Parkes, 'On draining', 249–72.

77 NdRO, Blackett MSS: ZBL/281/1, Account of the establishment of the East Matfen tilery.
78 Northumberland MSS: Business minutes, vol. 1, 27 December 1847, and vol. 5, 28 January 1850.
79 Northumberland MSS: Business minutes, vol. 1, 6 March 1848, and vol. 2, 29 May 1848.
80 Northumberland MSS: Business minutes, vol. 1, 27 December 1847; vol. 2, 5 and 26 June 1848; vol. 4, 2 April 1849; NdRO, Bell Collection: ZHE 34/3, Report by R. Hall on Lord Lonsdale's tileries at Whitehaven, 21 July 1848; ZHE 34/8, Instructions to draining superintendents on the duke of Northumberland's estate, 1848.
81 Northumberland MSS: Business minutes, vol. 17, 29 October and 31 December 1855.
82 DPD, Howard MSS: N. 101–6, Rental accounts, 1847–50.
83 NdRO, Belsay (Middleton) MSS: Box 6/VI/26, Valuation of Sir Charles Monck's estate for mortgage and comment upon that valuation, 1849.
84 NdRO, Belsay MSS: Box 12/IX, Statement of draining and building expenditure on each farm, 1868–79; Macdonald, 'The development of agriculture and the diffusion of agricultural innovation in Northumberland, 1750–1850', 96–7.
85 Trafford, 'Field drainage', 131; Robinson, 'The extent of farm underdrainage in England and Wales, prior to 1939', 79.
86 Thompson, 'English great estates in the nineteenth century, 1790–1914', 394, and *English Landed Society*, 231–6
87 This trend was also identified on the Holkham estate in Norfolk by Wade Martins, *A Great Estate*, 98.
88 Moore, 'The Corn Laws and high farming', 546–8.
89 Ernle, *English Farming Past and Present*, 349–50, 365–8; Thompson, 'The second agricultural revolution, 1815–1880', 63–5; Perry, 'High farming in Victorian Britain, prospect and retrospect', 159–63.

5 Capital provision and the management of the improvement

1 Thompson, 'The second agricultural revolution, 1815–1880', 73; Holderness, 'Capital formation in agriculture', 170–9; T. W. Beastall, 'Landlords and tenants', in Mingay, *The Victorian Countryside*, vol. 2, 428–30.
2 Charnock, *On Thorough-Draining*, 6–7; H. Stephens, *The Book of the Farm* (3 vols., Edinburgh, 1844), vol. 1, 609–11.
3 Buccleuch MSS: Barnwell and Boughton estate accounts, 1800–99; Bearn, 'On the farming of Northamptonshire', 96,
4 NRO, Overstone MSS: Ov. 45, Box 816, General letter from J. Beasley, 20 July 1840; O. 906, 912–30, Estate accounts, 1832–58.
5 Spencer MSS: Estate accounts of J. Harrison, 1800–26; Estate accounts and rentals, 1827–99; NRO, Overstone MSS: Ov. 45, Box 816, General letter from J. Beasley, 20 July 1840; J. Beasley, *The Duties and Privileges of the Landowners, Occupiers and Cultivators of the Soil* (London, 1860), 9, 45.
6 Caird, *English Agriculture in 1850–51*, 431–3; Bearn, 'On the farming of Northamptonshire', 96.

7 NRO, Ellesmere MSS: Box X.461, Rentals, 1837–74; X.3695, Letter from J. Loch to Lord Francis Egerton, 18 October 1841; memorandum about draining by J. Loch, 10 August 1841; X.3696, Letters from J. Loch to Lord Francis Egerton, 19 January, 11 and 19 December 1846.

8 PRO, MAF 66/31, Lands Improvement Company's register of absolute charges and orders, 1866–8; NRO, Fitzwilliam MSS: Box X.970, Letter from Lands Improvement Company to Messrs Sharp, 11 March 1864; letter from Messrs Sharp to J. Yeoman, 16 October 1866; Box X.971, Letter from Smith and Gore to J. Yeoman, 24 January 1876; letter from F. G. Butler to T. W. Fitzwilliam, 4 January 1879; letter from B. Measures to T.W. Fitzwilliam, 25 September 1885.

9 Bedfordshire Record Office, Woburn MSS: Solicitor's and steward's papers, Box 353, Correspondence: letter from T. Wing to C. Haedy, 1 June 1850; letter from R. Mein to C. Haedy, 28 June 1852.

10 Bedfordshire Record Office, Woburn MSS: Solicitor's and steward's papers, Box 353, Correspondence: letter from R. Mein to C. Haedy, 28 June 1852.

11 Raby MSS: Rentals, 1806–99; Applications to the Inclosure Commissioners for a draining loan, 9 February 1847; NRO, Cartwright MSS: C (A) 3502–3, 3574–81, Estate accounts, 1799–1846; C (A) 3784–5, 3787–9, Rental accounts, 1783–1809, 1830–45; PRO, 1R3/6 and 18, Registers of draining advances 1847–8 and 1853.

12 DRO, Fortescue MSS: 1262M/E20/16–144, Rentals, 1830–99; 1262M/E29/42, Contracts and conditions for letting, 1799–1851.

13 Bedford MSS: Annual reports for 1848, 1850, 1851 and 1855.

14 DRO, Sidmouth MSS: 152M/Memoranda, leases dated 11 October 1834, 28 February 1841, 2 June 1843, 27 May 1845, 23 January 1847; 152M/ Miscellaneous, agent's book, 1842; PRO, 1R3/12, Register of draining advances, 1851–2.

15 NdRO, Blackett MSS: ZBL/274/6–10, Estate ledgers, 1802–45; ZBL/277/5, Estate journal, 1804–20; Sample MSS: ZSA/8/1–5, Estate accounts and rentals, 1828–1903; ZSA/1/2. Correspondence, letters from W. Sample to the duke of Portland, 4 and 16 December 1828, 5 February 1830.

16 DPD, Howard MSS: N. 101–6, Rental accounts, 1801–55.

17 DPD, Howard MSS: N. 122, Letter from G. A. Grey to R. du Cann, 11 March 1873.

18 NdRO, Bell Collection: ZHE 34/8, Circular letter by J. C. Blackden and W. Laws, July 1844.

19 Northumberland MSS: Business minutes, vol. 4, 30 April 1849; vol. 37, 16 February 1866.

20 NdRO, Bell Collection: ZHE 34/8, Form for the return of the progress of draining on the Northumberland estate; ZAN 77/36, Mr Tate's report on the draining done on the duke of Northumberland's north estate, June 1847.

21 Northumberland MSS: Draining vols., 1844–1903.

22 Thompson, 'English great estates in the nineteenth century, 1790–1914', 393; Spring, 'English landed society in the eighteenth and nineteenth centuries', 149.

23 NdRO, Bell Collection: ZHE 34/8, Circular letter by J. C. Blackden and W. Laws, July 1844.

24 Northumberland MSS: Business minutes, vol. 5, 15 October 1849.

25 Northumberland MSS: Business minutes, vol. 20, 14 September 1857.

26 Northumberland MSS: Business minutes, vol. 37, 16 February 1866.

27 Northumberland MSS: Draining vols., 1844–1903.

28 NdRO, Sample MSS: ZSA/1/2, Correspondence, letter from W. Sample to the duke of Portland, 29 January 1836.

29 NdRO, Sample MSS: ZSA/1/2, Correspondence, letters 3 March and 5 December 1845, 15 April 1854.

30 NdRO, Sample MSS: ZSA/3/37, Letters from T. Sample to E. Bailey, 24 and 25 March 1885; ZSA/8/3, Estate accounts, 1870–82.

31 DRO, Fortescue MSS: 1262M/E1/102, List of lands required to be drained; Sigmouth MSS: 152M/A general report of the Upottery estate... by F. Thynne, 1850; DPD, Grey MSS: Boxes 550–1, Draining papers, 1841–69: draining reports for 1847; Howard MSS: N. 80/11, G. A. Grey's statement of proposed drainage expenditure, 1862–5; Mr Grey's report of the drainage required to be done and estimate of cost, 10 April 1868.

32 Bedford MSS: Annual report, 1857.

33 NdRO, Sample MSS: ZSA/1/2, Correspondence, letter from W. Sample to the duke of Portland, 15 April 1854.

34 NdRO, Sample MSS: ZSA/8/4–5, Estate accounts, 1883–1903; ZSA/3/37, Bothal improvement loan; PRO, MAF 66/17–18, Land Loan Company's registers of absolute charges and orders, 1881–5.

35 NdRO, Blackett MSS: ZBL/274/6–10, Estate ledgers, 1802–45; ZBL/277/5 and 6, Estate journals, 1804–1901; ZBL/282/2, Drainage accounts, 1840–1908; ZBL/87, Correspondence, letter from Sir E. Blackett to W. Sample, 17 May 1854.

36 NdRO, Blackett MSS: ZBL/87, Correspondence, letter from G. A. Grey to T. Sample, 29 May and 1 October 1856; ZBL/54/1, Sir E. Blackett's application for a Public Money Draining Acts loan, 3 June 1856.

37 NdRO, Blackett MSS: ZBL/87, Correspondence, letter from Sir E. Blackett to T. Sample, 31 May 1859; letter from G. A. Grey to T. Sample, 3 June 1859; ZBL/54/1, Lands Improvement Company loan, 1862–6.

38 NdRO, Blackett MSS: ZBL/54/1, Land Loan and Enfranchisement Company loans, 1867–72 and 1885–6; ZBL/87, Correspondence, letters from Lands Improvement, General Land Drainage and Land Loan Companies to T. Sample, 20 and 29 June and 12 November 1866.

39 DPD, Baker-Baker MSS 23/36 and 37, Abstract of draining on the Stanton estate, 1851–4; 23/100 and 101, Stanton rent sheet, 1872; PRO, 1R3/12, Register of draining advances, 1851–2.

40 NRO, Dryden MSS: D. (CA) 450, Draining schedules, 1858–66; D (CA) 445, Acreage description and value of estate in 1837 and rates, rateable values and rents in 1844; PRO, 1R3/31, Register of draining advances, 1857–8.

41 DRO, Sidmouth MSS: 152M/A general report of the Upottery estate... by F. Thynne, 1850; 152M/Estate accounts, 1850–99.

42 PRO, 1R3/12, Register of draining advances 1851–2; MAF 66/1 and 44, General Land Drainage Company's registers of absolute charges and orders, 1851–72.

43 NdRO, Belsay (Middleton) MSS: Box 12/IX/Drainage accounts: statement of loans received from the Lands Improvement Company and government, June 1891; PRO, 1R3/38, Register of draining advances, 1866–84; MAF 66/3, 32–7, Lands Improvement Company's registers of absolute charges and orders, 1868–88.

44 PRO, MAF 66/2 and 47, General Land Drainage Company's registers of absolute charges and orders, 1868–1911.

45 Bedford MSS: Annual report, 1867: report on Devon estates by G. Martin.

46 Lavergne, *The Rural Economy of England*, 98–103. F. A. Channing in his 1897 report to the Royal Commission on agricultural depression considered that outlay on small estates from the early 1880s had been much less than on other sizes of estate, while in 1873, in evidence to the Select Committee of the House of Lords on the improvement of land, G. Hope, speaking from Scottish experience, discerned little difference in landlord agricultural investment between estates of 1,000 acres and those of 5,000 acres: 'Select Committee...on the improvement of land', qq. 2814–16; 'Final report and report by F. A. Channing to the Royal Commission on agricultural depression', 287.

47 Bateman, *The Great Landowners*; Thompson, *English Landed Society*, 32, 113–17.

48 Thompson, *English Landed Society*, 32, 113–17.

49 'Select Committee [1845]...for the purpose of draining', i, qq. 590, 657; 'Select Committee on agricultural customs', qq. 1450–1, 4119–22; 'Select Committee [1849]...for the purpose of draining', qq. 4–7; 'General Board of Health...in respect to the drainage of land', appendix 7, 69.

50 BPP, 1863, VII, 'Select Committee of the House of Lords on the charging of entailed estates for railways', q. 306.

51 'Select Committee...on the improvement of land', iii–iv, qq. 976–80, 1207, 2018–21, 3025–7.

52 Where the full cost of draining was provided on these estates before 1880, interest of 5 per cent on outlay was charged. Buccleuch MSS: Boughton estate accounts, 1855 and 1874; NRO, Fitzwilliam MSS: Estate accounts, 1863–7; Box X.971, Letter from Smith and Gore to J. Yeoman, 24 January 1876.

53 DRO, Sidmouth MSS: 152M/Memoranda, agreeements with Upottery tenants as to drainage interest, 4 March 1861.

54 See, for example, DRO, Sidmouth MSS: 152M/A general report of the Upottery estate...by F. Thynne, 1850; 152M/Letterbooks, letter from F. Thynne to J. Jennings, 10 October 1867; 152M/Memoranda, agreements with Upottery tenants as to drainage interest, 4 March 1861; NRO, Cartwright MSS: C (A) 3882, Letterbook, letter from T. R. Cartwright to J. Dodwell, 15 May 1851; Dryden MSS: D (CA) 448, Tenancy agreements, 30 November 1874, 19 July and 25 September 1875.

55 Northumberland MSS: Business minutes, vol. 2, 24 January 1848; NdRO, Bell Collection: ZHE 34/8, Circular letter by J. C. Blackden and W. Laws, July 1844; Blackett MSS: ZBL/274/6–10, Estate ledgers, 1802–45; ZBL/277/5, Estate journal, 1804–20; Sample MSS: ZSA/1/2, Letters from W. Sample to the duke of Portland, 4 and 15 April 1834.

56 KU, Sneyd MSS: Reports of Andrew Thompson, vol. 5, 176.

57 Bedford MSS: Annual report, 1867: report on Devon estates by G. Martin. For a general discussion of the effect of life leases in Devon, R. Stanes, 'Landlord and tenant and husbandry covenants in eighteenth-century Devon', in W. Minchinton (ed.), *Agricultural Improvement: Medieval and Modern* (University of Exeter, 1981), 41–64.

58 Northumberland MSS: Draining vols. 1844–1903; NdRO, Bell Collection: ZHE 34/8, Circular letter by J. C. Blackden and W. Laws, July 1844; Sample MSS: ZSA/1/2, Letters from W. Sample to the duke of Portland, 4 and 15 April 1834; ZSA/3/37, Letter from T. Sample to E. Bailey, 24 March 1885; ZSA/8/1–5, Estate accounts, 1828–1903; DPD, Grey MSS: Draining vols., 1841–86; Howard MSS: N. 101–6, Rental accounts, 1801–55; N. 122, Letter from G. A. Grey to R. du Cann, 11 March 1873; Uncatalogued, Description of Northumberland estates, 1855–85: letters from W. Laurie and J. Shotton, 6 January 1860.

59 NdRO, Blackett MSS: ZBL/54/1, Details of draining loans; ZBL/78, Leases, valuations and correspondence: farm agreements, Steel farm, 1 May 1816, Willimontswyke farm, 22 December 1823, Leazes farm, 13 November 1840, Haltonshields West farm, 7 December 1840, Clarewood North and South farm, 22 December 1847, Mill Hill, Aydon Castle and Partridge Nest farms, 13 May 1864, Blackfell farm, 13 May 1872, letter from J. Armstrong to T. Sample, 26 December 1885; ZBL/274/6–10, Estate ledgers, 1802–45; ZBL/277/5 and 6, Estate journals, 1804–1901; ZBL/282/2, Draining accounts, 1840–1908.

60 Raby MSS: Sudborough draining vol., 1848–53; Rentals, 1848–54; NRO, Cartwright MSS: C (A) 3882, Letterbook, letter from T. R. Cartwright to J. Dodwell, 15 May 1851; C (A) 4014–17, Memoranda of evidence at Cartwright v. Falkner trial, 1856; Dryden MSS: D (CA) 448, Tenancy agreements, 30 November 1874, 19 July and 25 September 1875; NdRO, Belsay (Middleton) MSS: Box 12/IX/Statement of draining and building expenditure on each farm 1869–84; DPD, Baker-Baker MSS: 23/99, Letter from H. R. Goddard to Mrs Baker-Baker, 16 November 1872; 23/100–1, Stanton rent sheet, 20 November 1872.

61 DRO, Sidmouth MSS: 152M/Memoranda, account of interest on draining to Lady Day, 1856; agreements with Upottery tenants as to drainage interest, 4 March 1861.

62 See, for example, DRO, Fortescue MSS: 1262M/E20/16–144, Rentals, 1830–99; NRO, Cartwright MSS: C (A) 3844–71, Rentals, 1851–71; NdRO, Sample MSS: ZSA/8/2–5, Estate accounts, 1855–1903.

63 Northumberland MSS: Draining vols., 1844–1903; DPD, Grey MSS: Draining vols., 1841–86; Howard MSS: Draining vols., 1856–1901; NdRO, Blackett MSS: ZBL/78 Leases, valuations and correspondence, letter from J. Armstrong to T. Sample 26 December 1885; ZBL/282/2, Drainage accounts, 1855–1903; ZSA/3/37, Letter from T. Sample to E. Bailey, 24 March 1885.

64 Buccleuch MSS: Barnwell letting book, 1886–1900; Spencer MSS: Estate accounts and rentals, 1874–99; Cornwall MSS: Farm bundles, manor of Bradninch; DRO, Fortescue MSS: 1262M/E20/16–144, Rentals, 1830–99; 'Royal Commission on agricultural depression', q. 5838.

65 Bedford MSS: Annual report, 1849, dated 15 June 1850.

66 Bedford MSS: Annual reports, 1865–73; Annual report, 1867: report on Devon estates by G. Martin.

67 DPD, Howard MSS: Uncatalogued, Description of Northumberland estates, 1855–85: advertisements for Ulgham Fence and Dovecott farms, 1858, Tranwell farm, 1862, Duddo farm, 1864 and Dovecott farm, 1874

68 NdRO, Blackett MSS: ZBL/89, Correspondence, letter from T. Sample to C. H. Sample, 3 January 1888.

69 Raby MSS: Northamptonshire estate letterbooks, letter from W. T. Scarth to Messrs Parkin and Woodhouse, 5 March 1886; Rentals, 1848–70; NRO, Cartwright MSS: C (A) 4740–57, Absolute orders and notices of applications under the draining loans; C (A) 3844–71, Rentals, 1851–71.

70 NRO, Fisher-Sanders Collection: FS1/7, Farm agreements, 1887–1904; FS1/14, Sale catalogue Naseby estate, 1903; FS1/20, Provisional contracts with the Lands Improvement Company; FS1/31, Draft valuation, Naseby, 1877, FS1/45, Summary of G. A. Ashby's bankruptcy, 1888; FS1/46, Report of proceedings of G. A. Ashby's bankruptcy, 1887–8; PRO, MAF 66/39, Lands Improvement Company's register of absolute charges and orders, 1883–8.

71 For example, draining outlay formed only 27 per cent of that on farm buildings on the dukes of Bedford's Devon estate in the period 1870–95; 4 per cent on the Northamptonshire estate of the earls Spencer from 1840 to 1899; and 44 per cent on the dukes of Northumberland estate between 1844 and 1899. Bedford MSS: Annual reports, 1870–99; Spencer MSS: Estate accounts and rentals, 1840–99; Northumberland MSS: Business minutes, vol. 119, 21 November 1906.

72 In the analysis for Devon those estates with the greatest intensity of draining outlay have been selected, while for Northamptonshire for purposes of comparison those which relied mainly on providing allowances or materials have been employed.

73 Perren, 'The landlord and agricultural transformation, 1870–1900', 43, 50.

74 Northumberland MSS: Return of state of farms on Warkworth bailiwick, 1845; Returns of state of farms on bailiwicks, 1827–45.

75 NdRO, Sample MSS: ZSA/1/2, Letters from W. Sample to the duke of Portland, 13 August and 11 September 1833, 26 October 1846.

76 Bedford MSS: Annual report, 1867: report on Devon estates by G. Martin; Cornwall MSS: Valuation of the manor of Bradninch, 1831, by T. Davis; Farm bundles, manor of Bradninch: Gingerland farm, 1851–1915; Wishay farm, 1860–66; Downhead farm, 1862–72.

77 Orwin and Whetham, *History of British Agriculture*, 170–3, 298.

78 NdRO, Sample MSS: ZSA/1/2, Letters from W. Sample to the duke of Portland, 11 September 1833, 3 March and 5 December 1845; ZSA/3/37, Letter from J. S. Clark to T. Sample, 24 January 1885.

79 Northumberland MSS: Business minutes, vol. 5, 15 October 1849; vol. 7, 14 October 1850; vol. 9, 6 October 1851; vol. 14, 25 December 1854; NdRO, Bell Collection: ZHE 34/8, Circular letter by J. C. Blackden and W. Laws, July 1844; Tenant application form for draining on the Northumberland estate, 1844.

80 Northumberland MSS: Business minutes, vol. 54, 18 September 1874.

81 NRO, Overstone MSS: Ov. 45, Box 816, General letter from J. Beasley, 20 July

1840; Ellesmere MSS: X.3695, Memorandum about draining by J. Loch, 10 August 1841; Caird, *English Agriculture in 1850–51*, 431–3.

82 NdRO, Sample MSS: ZSA/1/2, Letters from W. Sample to the duke of Portland, 7 and 19 January 1836.

83 NdRO, Bell Collection: ZHE 34/8, Lectures addressed at his Grace's request to the tenants...on the estate of the duke of Northumberland by Professor Johnston, 1843. Johnston had produced a pamphlet in 1842 on agricultural improvement: J. F. W. Johnston, *What Can Be Done for English Agriculture?* (Durham, 1842), 26–7.

84 For example, Bedfordshire Record Office, Woburn MSS: Solicitor's and steward's papers, Box 353, Correspondence: letter from R. Mein to C. Haedy, 30 June 1852, referring to the duke of Bedford's Northamptonshire estate; NdRO, Sample MSS: ZSA/1/2, Letter from W. Sample to the duke of Portland, 5 December 1845; Bell Collection: ZHE 34/8, Circular letter by J. C. Blackden and W. Laws, July 1844; Northumberland MSS: Business minutes, vol. 1, 24 January 1848.

85 Northumberland MSS: Business minutes, vol. 5, 15 October 1849; NdRO, Sample MSS: ZSA/1/2, Letters from W. Sample to the duke of Portland, 3 March and 5 December 1845.

86 Northumberland MSS: Returns of state of farms on Newburn bailiwick, 1850 and 1853.

87 NdRO, Bell Collection: ZHE 34/8, Concluding remarks on J. D. Ferguson's lecture on draining to the Newcastle upon Tyne Farmers' Club, 5 February 1859.

88 Cornwall MSS: Farm bundles, manor of Bradford: Bradford and Marsh farms, 1862–1914, letter from G. Herriot to J. W. Bateman, 21 December 1867.

89 Denton, *What Can Now Be Done for British Agriculture?*, 11; Pusey, 'On the progress of agricultural knowledge during the last four years', 169–217; 'Select Committee [1845]...for the purpose of draining', qq. 827–33.

90 Caird, *English Agriculture in 1850–51*, 474; B. R. Mitchell and P. Deane, *Abstract of British Historical Statistics* (Cambridge, 1962), 488–9.

91 'Select Committee...on the improvement of land', qq. 4219–28; Mitchell and Deane, *British Historical Statistics*, 488–9; Kain, *An Atlas and Index of the Tithe Files*, 460.

92 Caird, *English Agriculture in 1850–51*, 474; Kain, *An Atlas and Index of the Tithe Files*, 213, 319.

93 'Select Committee...on the improvement of land', qq. 872–6, 4125–30; E. P. Squarey, 'Farm capital', *JRASE*, second series, 14 (1878), 433.

94 Northumberland MSS: Draining vols., 1844–1903.

95 DPD, Howard MSS: Draining vols., 1856–1901; NdRO, Blackett MSS: ZBL/282/2, Draining accounts, 1840–1908.

96 Bedford MSS: Annual reports, 1880–9; DRO, Sidmouth MSS: 152M/Memoranda, draining schedules, 1851–66.

97 Mitchell and Deane, *British Historical Statistics*, 488–9.

98 NdRO, Bell Collection: ZAN 65/18, Statistical report on Tindale bailiwick for 1850.

99 Cornwall MSS: Report on the manor of Bradford by E. C. Marriott, 11

September 1862; Farm bundles, manor of Bradford: Bradford and Marsh farms, 1862–1914, letter from G. Herriot to J. W. Bateman, 21 December 1867; Farm bundles, manor of Bradford: Horslett farm, 1882–1915, letter from G. Herriot to M. Holzman, 31 October 1891; 'Report by W. C. Little on Devon, Cornwall, Dorset and Somerset...', 23.

100 Northumberland MSS: Business minutes, vol. 20, 14 September 1857.

101 Bedford MSS: Annual report, 1864: report by M. Benson on the north Devon estate; Annual report, 1867: report on Devon estates by G. Martin.

102 Bearn, 'On the farming of Northamptonshire', 96.

103 Phillips 'Underdraining and agricultural investment in the midlands in the mid-nineteenth century', 267–71.

104 Collins and Jones, 'Sectoral advance in English agriculture, 1850–80', 69–71; Hueckel, 'Agriculture during industrialisation', 194–5.

105 O'Grada, 'Agricultural decline 1860–1914', 188.

106 This view was expressed clearly in the separate report made by F. A. Channing to the Royal Commission on agricultural depression: 'Final report and report by F. A. Channing to the Royal Commission on agricultural depression', 292–4.

6. The success of underdraining as an agricultural improvement

1 'Select Committee...on the improvement of land', iv, qq. 188–90, 523–30, 4212–17.

2 'Select Committee...[on]...the companies for the improvement of land', qq. 72–85, 418–30; 'General Board of Health...in respect to the drainage of land', appendix 7, question 6; 'Rules and practices of the Inclosure Commissioners', 429; Denton, *The Under-Drainage of Land*, 19, and *Agricultural Drainage*, 3.

3 KU, Sneyd MSS: Reports of Andrew Thompson, vol. 6, 61.

4 Phillips, 'Underdraining and agricultural investment in the midlands in the mid-nineteenth century', 265–6.

5 Northumberland MSS: Business minutes, vol. 2, 5 June 1848; vol. 20, 14 September, 1857; NdRO, Bell Collection: ZHE 34/8, Instructions to draining superintendents on the duke of Northumberland's estate, 1848.

6 NRO, Overstone MSS: Ov. 45, Box 816, General letter from J. Beasley, 20 July 1840

7 NRO, Ellesmere MSS: X.3696, Letter from J. Loch to Lord Francis Egerton, 19 January 1846.

8 NRO, Cartwright MSS: C (A) 5242, Draining schedules, 1856; Dryden MSS: D (CA) 450, Draining schedules 1858–66.

9 Caird, *English Agriculture in 1850–51*, 2, 75, 89, 187, 217, 256–7

10 Bearn, 'On the farming of Northamptonshire', 96.

11 NRO, Ellesmere MSS: X.461, Rentals, 1836–74; Fitzwilliam MSS: Estate accounts, 1840–59; Spencer MSS: Estate accounts and rentals, 1840–59.

12 Raby MSS: Rentals, 1840–9.

13 Northumberland MSS: Business minutes, vol. 1, 10 January 1848.

14 Northumberland MSS: Business minutes, vol. 6, 29 April 1850; NdRO, Bell Collection: ZHE 34/3. Circular letter by H. Taylor to bailiffs, 11 November

1847; Bailiffs' replies to Taylor's letter; Projected brick manufactory at Percy Main near Tynemouth, 27 December 1852; Walbottle Tile works: cost of making tiles, 29 October 1848.

15 Several Inclosure Commissioners commented on a general growth in the use of pipes with larger diameter from the late 1840s: 'Select Committee...on the improvement of land', qq. 511–18, 4184–6.

16 NdRO, Sample MSS: ZSA/1/2, Letters from W. Sample to the duke of Portland, 23 January and 27 March 1847; ZSA/8/2–5, Estate accounts and rentals, 1855–1903.

17 NRO, Ellesmere MSS: X.3696, Letter from J. Loch to Lord Francis Egerton, 19 January 1846. John Beasley, agent to the Overstone and Spencer estates in Northamptonshire, also expressed a dislike of 1 in pipes: *The Duties and Privileges of the Landowners*, 9.

18 Northumberland MSS: Business minutes, vol. 1, 27 December 1847; vol. 2, 5 June 1848; vol. 6, 29 April 1850; vol. 20, 14 September 1857; NdRO, Bell Collection: ZHE 34/8, Draining notes by J. Loraine, 27 April 1850.

19 Raby MSS: Sudborough draining vol., 1848–52; Draining abstracts, 1848–53.

20 NdRO, Blackett MSS: ZBL/281/1, 4 and 5, Accounts of the Matfen and Melkridge tileries, 1839–1908.

21 P. Pusey, 'Report to H.R.H. the President of the commission for the exhibition of the works of industry of all nations', *JRASE*, 12 (1851), 587–649; M. R. Lane, 'John Fowler and the company he founded', *Steaming*, 23 (1980), 77–8; W. Vamplew, 'The cost of best practice in the mid-nineteenth century', *Tools and Tillage*, 3 (1980), 206–7; E. J. T. Collins, 'The age of machinery', in Mingay, *The Victorian Countryside*, vol. 1, 209.

22 Spencer MSS: Estate accounts and rentals, 1885–6.

23 Buccleuch MSS: Barnwell estate accounts, 1884 and 1893.

24 Raby MSS: Northamptonshire estate letterbooks, letter from W. T. Scarth to T. Tilley, 4 November 1881.

25 NRO., Overstone MSS: Ov. maps, 184–5, 187–8, 190–2, 327, 330–1, 334.

26 Buccleuch MSS: Boughton drainage sheets, 1888; NRO, Fisher-Sanders Collection: FS1/20, Provisional contracts with the Lands Improvement Company; 'Royal Commission on the depressed condition of the agricultural interest', qq. 2088, 2379–82, 4740.

27 H. F. French, *Farm Drainage* (New York, 1879), 165–7; A. Roland, *Farming for Pleasure and Profit* (London, 1880), 34–7; W. Fream, *Elements of Agriculture* (London, 1892), 37–9; G. S. Mitchell, *A Handbook of Land Drainage* (London, 1898), 67–70; H. H. Smith, *The Principles of Landed Estate Management* (London, 1898), 220–2.

28 'Report by W. C. Little on Devon, Cornwall, Dorset and Somerset...', 24. This example was used by E. J. T. Collins and E. L. Jones to illustrate ineffective draining: 'Sectoral advance in English agriculture, 1850–80', 72.

29 DRO, Fortescue MSS: 1262M/E1/102, Letter from H. C. Mules to earl Fortescue, 17 December 184; letter from N. J. Graban to D. T. Brewer, agent at Castle Hill, 23 May 1850; 1262M/E1/103, Report on the Castle Hill and other estates...by J. Parkes, 10 January 1848; letter from W. Blamire to earl Fortescue, 14 January 1851; H. Davis' report on Lord Fortescue's drainage,

February 1851; letter from Lord Wharncliffe to Lord Fortescue, 5 February 1853; letter from Lord Fortescue to the Inclosure Commissioners, 16 February 1854; draining sheets and schedules, 1847–64; Lord Wharncliffe, 'On draining, under certain conditions of soil and climate', *JRASE*, 12 (1851), 41–62.

30 NRO, Dryden MSS: D (CA) 448, Agreements and inventories of tenant right, 1874–89.

31 Bedford MSS: Annual report, 1869.

32 Cornwall MSS: Farm bundles, manor of Bradford: Horslett farm, letter from J. Toy to J. W. Bateman, 26 March 1867; letter from J. M. Martin to R. Watt, 5 April 1867; letter from G. Herriot to G. Wilmshurst, 22 November 1883; DRO, Sidmouth MSS: 152M/Memoranda, draining schedules under the General Land Drainage Company's loan, 1864–6.

33 Bedford MSS: Annual report, 1885; Raby MSS: Northamptonshire estate letterbooks, letters from W. T. Scarth to A. Mace 11 November 1859, to W. Mace, 8 December 1860, 4 and 11 November 1861, to J. Perkins, 28 November 1867.

34 BPP, 1895, XVII, 'Report by R. Hunter Pringle to the Royal Commission on agricultural depression on the counties of Bedford, Huntingdon and Northampton', 5, 7.

35 Spencer MSS: Agricultural holdings on earl Spencer's estate, 1879.

36 Denton, *What Can Now Be Done for British Agriculture?*, 11; Johnston, *What Can Be Done for English Agriculture?* 26–7.

37 *Hansard*, 3rd series, vol. 83, 27 January 1846, 270; vol. 84, 2 March 1846, 456. Caird, *The Landed Interest*, 80–1.

38 'Select Committee [1845]...for the purpose of draining', q. 599; Davis, *Farming Essays*, 85; J. C. Morton, 'On the management of grass lands', *Journal of the Bath and West of England Society for the Encouragement of Agriculture*, 13 (1865), 61–73; Taylor, 'The English dairy industry, 1860–1930', 598.

39 NdRO, Sample MSS: ZSA/1/2, Letters from W. Sample to the duke of Portland, 6 September 1830, 15 September 1832, 5 December 1845.

40 NdRO, Blackett MSS: ZBL/87, Letter from J. Smith to W. Sample, 14 December 1853; letter from J. R. Spraggon to T. Sample, 8 November 1858.

41 NRO, Overstone MSS: Ov. 45, Box 816, General letter from J. Beasley, 20 July 1840.

42 DRO, Fortescue MSS: 1262M/E1/103, Report on the Castle Hill and other estates...by J. Parkes, 10 January 1848.

43 Raby MSS: Draining correspondence, letters from T. F. Scarth to J. Parkes, 17 June 1851 and to W. Blamire, 5 July 1851.

44 'Royal Commission on agricultural depression', q. 4607.

45 Stamp, *The Land of Britain*, 362–81.

46 Cornwall MSS: Valuation of the manor of Bradninch, 1855, by R. Watt; Farm bundles, manor of Bradninch: Chapelhaies tenement, letter from J. M. Martin to T. Hyatt, 16 July 1868; letter from T. Hyatt to J. W. Bateman, 24 December 1869; map of draining on Chapelhaies farm, 1868–9.

47 Raby MSS: Sudborough draining vol., 1848–52; Draining abstracts, 1848–53; Particulars of rent, 1846 and 1855.

48 Northumberland MSS: Business minutes, vol. 37, 16 February 1866.

49 Pusey, 'On the progress of agricultural knowledge during the last four years', 167–72; Newman, *Practical Hints*, 16.

50 Parkes, 'On draining', 258.

51 Northumberland MSS: Business minutes, vol. 4, 3 September 1849; vol. 5, 28 January 1850; NdRO, Bell Collection: ZHE 34/8, Instructions to draining superintendents on the duke of Northumberland's estate, 1848.

52 NdRO, Blackett MSS: ZBL/54/2 and 3, Draining schedules, 1857–86; letter from J. Rutherford to T. Sample, 28 March 1885; ZBL/89, Letters to T. Sample from W. Woodman, 5 March 1884 and from T. Weddle, 15 October 1885.

53 Spencer MSS: Maps of Althorp Park, 1858; Brampton Ash; Chapel and Church Brampton, 1842; Elkington, 1838; Heyford, Flore and Great Brington, 1856; Harlestone, 1879; NRO, Overstone MSS: Ov. maps, 184–5, 187–8, 190–2, 327, 330–1, 334.

54 Charnock, *On Thorough-Draining*, 20; Davis, *Farming Essays*, 79; 'Select Committee [1845]...for the purpose of draining', qq. 510–16.

55 Bedfordshire Record Office, Woburn MSS: Solicitor's and steward's papers, Box 353, Correspondence: letter from R. Mein to C. Haedy, 30 June 1852; Bedford MSS: Annual report, 1864.

56 DRO, Fortescue MSS: 1262M/E1/103, Report on the Castle Hill and other estates...by J. Parkes, 10 January 1848; H. Davis' report on Lord Fortescue's drainage, February 1851.

57 Bedford MSS: Annual report, 1869.

58 The range in these may be seen in 'General Board of Health...in respect to the drainage of land', appendix 7, question 9.

59 As examples, Burke, 'On the drainage of land', 273–96; R. Peel, 'Account of a field thorough-drained at Drayton in Staffordshire', *JRASE*, 3 (1842), 18–21; 'Select Committee [1845]...for the purpose of draining', qq. 182–3, 460, 477–82. Throughout 1848, the *Gardeners' Chronicle and Agricultural Gazette* published draining reports, detailing successful and high-yielding projects: 15 January, 20 and 27 May, 26 August and 7 October 1848.

60 The series of reports by R. White on draining at Oakley Park published in the *JRASE* formed one of the few exceptions: 'Report of several operations in thorough-draining and subsoil-ploughing at Oakley Park', 33–7 and continued in the *Journal* in 1 (1840), 248–52; 2 (1841), 346–53; 4 (1843), 172–6; and 6 (1845), 229–36.

61 NdRO, Sample MSS: ZSA/1/2, Letter from W. Sample to the duke of Portland, 5 December 1845.

62 NdRO, Sample MSS: ZSA/1/2, Letters from W. Sample to the duke of Portland, 11 October 1838 and 8 October 1848.

63 NdRO, Bell Collection: ZHE 34/2, Circular letter to bailiffs from H. Taylor, August 1850.

64 Siegel, *Nonparametric Statistics*, 75–83.

65 Northumberland MSS: Statistical report on farms on Barrasford bailiwick, 1850.

66 NdRO, Sample MSS: ZSA/1/2, Letter from W. Sample to the duke of Portland, 9 October 1848.

67 Smith, *Remarks on Thorough Draining*, 20.

68 P. Pusey, 'Some introductory remarks on the present state of agriculture as a science in England', *JRASE*, 1 (1840), 6; Charnock, *On Thorough-Draining*, 20; Trimmer, *On the Improvement of Land*, 11; J. L. Morton, *The Resources of Estates* (London, 1858), 381–5; 'Select Committee [1845]...for the purpose of draining', qq. 182–3, 337–9, 512–16, 574–83, 1064–8.

69 J. Grey, 'A view of the past and present state of agriculture in Northumberland', *JRASE*, 2 (1841), 159.

70 'General Board of Health...in respect to the drainage of land', appendix 7, question 10; Colbeck, 'On the agriculture of Northumberland', 437.

71 Caird, *English Agriculture in 1850–51*, 485–6.

72 DRO, Fortescue MSS: 1262M/E1/103, Report on the Castle Hill and other estates...by J. Parkes, 10 January 1848; Sidmouth MSS: 152M/A general report of the Upottery estate...by F. Thynne, 1850.

73 For a discussion of the value of cropping books in determining change in farming systems, see A. D. M. Phillips, 'Agricultural land use and cropping in Cheshire around 1840: some evidence from cropping books', *Transactions of the Lancashire and Cheshire Antiquarian Society*, 84 (1987), 46–63, and 'Agricultural land use on a Northamptonshire estate (1849–99) as revealed by cropping books', 70–8.

74 NdRO, Sample MSS: ZSA/8/2, Estate accounts and rentals, 1855–69; Colbeck, 'On the agriculture of Northumberland', map at 436–7.

75 NdRO, Sample MSS: ZSA/1/2, Letters from W. Sample to the duke of Portland, 18 November 1837, 26 April 1843, 17 December 1844, 13 October 1852.

76 NdRO, Sample MSS: ZSA/3/31, Account by T. Sample of the corn scale regulating rents, 3 February 1885.

77 NdRO, Sample MSS: ZSA/18/5, Tillage account and valuation for Pegswood Middle Moor farm, 1864.

78 NdRO, Sample MSS: ZSA/1/2, Letters from W. Sample to the duke of Portland, 18 September 1847 and 31 July 1849.

79 Sturgess, 'The agricultural revolution on the English clays: a rejoinder', 87.

80 Northumberland MSS: Draining vols., 1–3: T. Bell and sons, Survey and terrier...1850; Annual returns of state of farms for Newburn and Shilbottle, 1850–70.

81 NdRO, Sample MSS: ZSA/3/31, Account by T. Sample of the corn scale regulating rents, 3 February 1885.

82 Collins and Jones, 'Sectoral advance in English agriculture, 1850–80', 80–1.

83 Sturgess, 'The agricultural revolution on the English clays: a rejoinder', 87.

84 Collins and Jones, 'Sectoral advance in English agriculture, 1850–80', 75.

85 Trafford, 'Field drainage', 141.

86 M. J. R. Healy and E. L. Jones, 'Wheat yields in England, 1815–59', *Journal of the Royal Statistical Society*, series A, 125 (1962), 576; J. A. Venn, *The Foundations of Agricultural Economics* (Cambridge, 2nd edn, 1933), 555.

87 Thompson, *English Landed Society*, 255.

7 Findings about underdraining

1 Jones, *Agriculture and Economic Growth*, 4; Mingay, *The Agricultural Revolution*, 5; M. Overton, 'Agricultural revolution? Development of the agrarian economy in early modern England', in A. R. H. Baker and D. Gregory (eds.), *Explorations in Historical Geography: Interpretative Essays* (Cambridge, 1984), 127–9.

2 Caird, *The Landed Interest*, 98–9; D. B. Grigg, 'An index of regional change in English farming', *Transactions of the Institute of British Geographers*, 36 (1965), 62–3; Thompson, 'Free trade and the land', 106–7.

3 See, for example, N. Riches, *The Agricultural Revolution in Norfolk* (London, 2nd edn, 1967); Overton, 'Agricultural revolution? Development of the agrarian economy in early modern England', 128–30.

4 E. H. Hunt, 'Labour productivity in English agriculture, 1850–1914', *EcHR*, second series, 20 (1967), 284; Holderness, 'Landlord's capital formation in East Anglia 1750–1870', 439–41.

5 C. W. Hoskyns, *Talpa: Or the Chronicles of a Clay Farm* (London, 1852), 9–26.

6 Hueckel, 'Agriculture during industrialisation', 194–5.

7 For a discussion of the development of tenants' commercial attitudes to farming in the nineteenth century, see J. Obelkevich, *Religion and Rural Society: South Lindsey, 1825–1875* (Oxford, 1976), 46–52.

8 Sturgess, 'The agricultural revolution on the English clays: a rejoinder', 82–7.

9 E. L. Jones, 'The changing basis of English agricultural prosperity, 1853–73', *AHR*, 10 (1962), 104.

10 Collins and Jones, 'Sectoral advance in English agriculture, 1850–80', 69, 73, 80.

11 J. Thirsk, *England's Agricultural Regions and Agrarian History, 1500–1750* (Basingstoke, 1987), 56–61.

12 Whetham, 'Sectoral advance in English agriculture 1850–80: a summary', 47.

13 Collins and Jones, 'Sectoral advance in English agriculture, 1850–80', 75–8.

14 Thompson, 'The second agricultural revolution, 1815–1880', 68–71, 75–77.

15 Jones, *Development of English Agriculture*, 24–5, and 'The changing basis of English agricultural prosperity, 1853–73, 117–18; J. R. Fisher, 'The economic effects of cattle disease in Britain and its containment, 1850–1900', *AH*, 54 (1980), 286.

16 Given the range of productive improvements in the third quarter of the nineteenth century, the view of an overall decline in domestic cereal output, argued most forcibly by S. Fairlie for wheat, would benefit from thorough re-examination: S. Fairlie, 'The Corn Laws and British wheat production, 1829–76', *EcHR*, second series, 22 (1969), 96–101.

17 Thompson, 'The second agricultural revolution, 1815–1880', 63–4.

18 T. W. Fletcher, 'The great depression of English agriculture 1873–1896', *EcHR*, second series, 13 (1960), 430–1; P. J. Perry, 'Where was the Great Agricultural Depression? A geography of agricultural bankruptcy in late Victorian England and Wales', *AHR*, 20 (1972), 30–45.

Bibliography

This bibliography is not exhaustive, but it includes the main MS collections and printed works used in the preparation of this study. Of the nineteenth-century agricultural periodicals consulted, only the articles in the *Journal of the Royal Agricultural Society of England* have been listed, as most of the major developments in draining were first and most fully reported within its pages.

MANUSCRIPT COLLECTIONS

Alnwick Castle, Northumberland: Northumberland MSS.

Althorp, Northamptonshire: Spencer MSS.

Bedford Estates Office, London: Bedford MSS.

Bedfordshire Record Office: Woburn MSS.

Boughton House, Northamptonshire: Buccleuch MSS.

Buckingham Gate, London: Duchy of Cornwall MSS.

Cheshire Record Office: Baker-Wilbraham MSS.

Department of Palaeography and Diplomatic, Durham University: Baker-Baker MSS; Grey MSS; Howard of Naworth MSS.

Devon Record Office: Courtenay of Powderham MSS; Fortescue MSS; Michelmore, Lovey and Carter MSS; Seymour MSS; Sidmouth MSS.

Howick Estate Office, Northumberland: Grey MSS.

Keele University: Sneyd MSS.

Newcastle Central Library: J. T. W. Bell's plans of the Durham and Northumberland coal districts.

Northamptonshire Record Office: Brudenell MSS; Cartwright MSS; Dryden MSS; Ellesemere MSS; Fisher-Sanders Collection; Fitzwilliam MSS; Grafton MSS; Overstone MSS.

Northumberland Record Office: Bell Collection; Belsay (Middleton) MSS; Blackett of Matfen MSS; Ridley MSS; Sample MSS.

Public Record Office: C36–38, Reports and Certificates to the Court of Chancery; 1R3, Registers of draining advances under the Public Money Draining Acts; 1R29 and 30, Tithe maps and apportionments; MAF 66, Registers of loans under the land-improvement legislation.

Rode Hall, Cheshire: Wilbraham MSS.

St James's Square, London: Lands Improvement Company MSS.
Staindrop Estate Office, Durham: Raby MSS.

PARLIAMENTARY PAPERS

1821, IX, 'Select Committee...[on]...the depressed state of agriculture'.
1833, V, 'Select Committee on agriculture'.
1836, VIII, 'Select Committee appointed to enquire into the state of agriculture'.
1837, V, 'Select Committee of the House of Lords appointed to enquire into the state of agriculture'.
1845, VIII, 'Return of the number of applications to the Court of Chancery or Exchequer...under 3 & 4 Vict., c. 55'.
1845, XVIII, 'Select Committee of the House of Lords...to enable possessors of entailed estates to charge such estates...for the purpose of draining'.
1847–8, VII, 'Select Committee on agricultural customs'.
1849, XXX, 'Select Committee of the House of Lords...to enable possessors of entailed estates to charge such estates...for the purpose of draining'.
1852, XIX, 'General Board of Health: minutes of information collected in respect to the drainage of land'.
1854–5, VII, 'Select Committee of the House of Lords...[on]...the powers now vested in the companies for the improvement of land'.
1863, VII, 'Select Committee of the House of Lords on charging of entailed estates for railways'.
1865, XLVII, 'Rules and practices of the Inclosure Commissioners'.
1870, LXVIII–1878–9, LXXV, *passim*, 'Agricultural returns, 1870–9'.
1873, XVI, 'Select Committee of the House of Lords on the improvement of land'.
1874, LXXII, 'Return of owners of land in England and Wales, 1873'.
1878–9, LXV, 'Return of the average prices of butchers' meat at the metropolitan cattle market, 1828–78'.
1881, XV–XVII; 1882, XIV, XV, 'Royal Commission on the depressed condition of the agricultural interest'.
1884, XXII–1903, XVII, *passim*, 'Annual reports of the Land Commissioners and the Board of Agriculture, 1883–1902'.
1894, XVI; 1895, XVI, XVII; 1896, XVI, XVII; 1897, XV, Royal Commission on agricultural depression'.
Hansard's Parliamentary Debates, 3rd series.
Statutes at large.

NEWSPAPERS, PERIODICALS AND DIRECTORIES

Annals of Agriculture
Communications to the Board of Agriculture
Edinburgh Review
Farmer's Herald
Farmer's Magazine
Gardener's Chronicle and Agricultural Gazette

Journal of the Newcastle Farmers' Club
London Gazette
Quarterly Journal of Agriculture
Post Office and Kelly's Directories
Transactions of the Institution of Surveyors

CONTEMPORARY BOOKS, PAMPHLETS AND ARTICLES

Acland, T. D. 'On the farming of Somersetshire', *JRASE*, 11 (1850), 666–764.
Almack, B. 'On the agriculture of Norfolk', *JRASE*, 5 (1846), 307–57.
Anderson, J. *A Practical Treatise on Draining Bogs and Swampy Grounds*, London, 1797.
Andrews, G. H. *Modern Husbandry*, London, 1853.
Anon, *On Land-Drainage, Subsoil-Ploughing and Irrigation*, London, 1841.
Arbuthnot, C. 'Letter on deep draining', *JRASE*, 6 (1845), 129–31.
  'On deep draining', *JRASE*, 6 (1845), 573–4.
  'On the advantage of deep drainage', *JRASE*, 10 (1849), 496–506.
Arkell, T. 'On the drainage of land', JRASE, 4 (1843), 318–40.
Bailey, J. *General View of the Agriculture of the County of Durham*, London, 1810.
Bailey, J., and Culley, G. *General View of the Agriculture of the County of Cumberland*, London, 1974 and 1805.
  General View of the Agriculture of the County of Northumberland, London, 1794 and 1805.
Baker, J. L. *An Essay on the Farming of Northamptonshire*, London, 1852.
Baker, R. 'On the farming of Essex', *JRASE*, 5 (1844), 1–43.
Batchelor, T. *General View of the Agriculture of the County of Bedford*, London, 1808.
Bateman, J. *The Great Landowners of Great Britain and Ireland*, London, 4th edn, 1883.
Bearn, W. 'On the farming of Northamptonshire', *JRASE*, 13 (1852), 44–113.
Beart, R. 'On the economical manufacture of draining-tiles and soles', *JRASE*, 2 (1841), 93–104.
  'On the proper materials for filling up drains and the mode in which water enters them', *JRASE*, 4 (1843), 411–30.
Beasley, J. *The Duties and Privileges of the Landowners, Occupiers and Cultivators of the Soil*, London, 1860.
Bedford, Duke of *A Great Agricultural Estate*, London, 1897.
Bell, T. G. 'A report upon the agriculture of the county of Durham', *JRASE*, 17 (1856), 86–123.
Bennett, W. 'The farming of Bedfordshire', *JRASE*, 18 (1857), 1–29.
Billingsley, J. *General View of the Agriculture in the County of Somerset*, London, 1794, and Bath, 1798.
Bishton, J. *General View of the Agriculture of the County of Salop*, Brentford, 1794.
Board of Agriculture *The Agricultural State of the Kingdom in February, March and April 1816*, London, 1816.
Boyle, R. 'On drain-tile and pipe machines', *Transactions of the Highland and Agricultural Society of Scotland*, new series, 14 (1853–5), 40–54, 75–90.

Boys, J. *General View of the Agriculture of the County of Kent*, Brentford, 1794, and London, 1813.

Bravendar, J. 'Farming of Gloucestershire', *JRASE*, 11 (1850), 116–77.

Brodrick, G. C. *English Land and English Landlords*, London, 1881.

Brown, H. H. 'On the drainage of land', *JRASE*, 3 (1842), 165–8.

Brown, R. *General View of the Agriculture of the West Riding of Yorkshire*, Edinburgh, 1799.

Brown, T. *General View of the Agriculture of the County of Derby*, London, 1794.

Buckland, G. 'On the farming of Kent', *JRASE*, 6 (1845), 251–302.

Burke, J. F. 'On the drainage of land', *JRASE*, 2 (1841), 273–96.

Cadle, C. 'The agriculture of Worcestershire', *JRASE*, second series, 3 (1867), 439–66.

'The improvement of grass lands', *JRASE*, second series, 5 (1869), 317–36.

Caird, J. *English Agriculture in 1850–51*, London, 2nd edn, 1852.

The Landed Interest and the Supply of Food, London, 4th edn, 1880.

Callander, R. W. *An Essay on Irrigating and Draining Different Sorts of Land*, Glasgow, 1856.

Challoner, Col. 'Report on the exhibitions and trials of implements at the Exeter meeting, 1850', *JRASE*, 11 (1850), 452–94.

Charnock, J. H. *Suggestions for the More General Extension of Land-Draining*, London, 1843.

On Thorough-Draining; and its Immediate Results to the Agricultural Interest, London, 2nd edn, 1844.

On Land Drainage, London, 1848.

'On the farming of the West Riding of Yorkshire', *JRASE*, 9 (1848), 284–311.

'On suiting the depth of drainage to the circumstances of the soil', *JRASE*, 10 (1849), 507–19.

Claridge, J. *General View of the Agriculture in the County of Dorset*, London, 1793.

Clark, J. *General View of the Agriculture of the County of Hereford*, London, 1794.

Clarke, J. A. 'Farming of Lincolnshire', *JRASE*, 12 (1851), 259–414.

Clutterbuck, J. C. 'On the theory of deep draining', *JRASE*, 6 (1845), 489–93.

'The Farming of Middlesex', *JRASE*, second series, 5 (1869), 3–27.

Colbeck, T. L. 'On the agriculture of Northumberland', *JRASE*, 8 (1847), 422–37.

Colman, H. *European Agriculture and Rural Economy from Personal Observation*, Boston, 1846.

Corringham, R. W. 'Agriculture of Nottinghamshire', *JRASE*, 6 (1845), 1–43.

Crutchley, J. *General View of the Agriculture in the County of Rutland*, London, 1794.

Davis, H. *Farming Essays*, London, 1848.

Davis, R. *General View of the Agriculture of the County of Oxford*, London, 1794.

Davis, T. *General View of the Agriculture of the County of Wiltshire*, London, 1794.

Davis, T. jun. *General View of the Agriculture of Wiltshire*, London, 1813.

Dean, G. A. *The Land Steward*, London, 1851.

The Culture, Management and Improvement of Landed Estates, London, 1872.

Dempsey, G. D. *On the Drainage of Lands, Towns and Buildings*, London, 1887.

Denton, J. B. 'General drainage and distribution of water', *Farmer's Magazine*, second series, 6 (1842), 64.

What Can Now Be Done for British Agriculture?, London, 1842.

Land Drainage and Drainage Systems, London, 1855.

The Under-Drainage of Land: Its Progress and Results, London, 1855.

A Letter to the Lord Berners on the Modes of Draining Practised at Keythorpe, London, 1857.

'Hinxworth drainage...', *JRASE*, 20 (1859), 273–94.

Land Drainage: Arterial Channels and Outfalls, London, 2nd edn, 1861.

'On the discharge from under-drainage and its effect on the arterial channels and outfalls of the country', *Minutes of Proceedings of the Institution of Civil Engineers*, 21 (1861–2), 48–131.

'The effect of underdrainage on our rivers and arterial channels', *JRASE*, 24 (1863), 573–89.

'On land drainage and improvement by loans from government or public companies', *JRASE*, second series, 4 (1868), 123–43.

Agricultural Drainage, London, 1883.

Dickinson, W. 'On the farming of Cumberland', *JRASE*, 13 (1852), 207–300.

Dickson, R. W. *General View of the Agriculture of Lancashire*, London, 1815.

Dobson, E. *A Rudimentary Treatise on the Manufacture of Bricks and Tiles*, London, 1850.

Donald, J. *Land Drainage, Embankment and Irrigation*, London, 1851.

Donaldson, J. *General View of the Agriculture of the County of Northampton*, Edinburgh, 1794.

Driver, A., and W. *General View of the Agriculture of Hampshire*, London, 1794.

Duncumb, J. *General View of the Agriculture of the County of Hereford*, London, 1805.

Etheredge, F. W. 'On the cheapest and best method of establishing a tile-yard', *JRASE*, 6 (1845), 463–77.

Evershed, H. 'On the farming of Surrey', *JRASE*, 14 (1853), 395–424.

'Farming of Warwickshire', *JRASE*, 17 (1856), 475–93.

'Agriculture of Herefordshire', *JRASE*, 25 (1864), 269–302.

'The agriculture of Staffordshire', *JRASE*, second series, 5 (1869), 263–317.

Farey, J. *General View of the Agriculture and Minerals of Derbyshire*, 3 vols., London, 1811–17.

Farncombe, J. 'On the farming of Sussex', *JRASE*, 11 (1850), 75–88.

Farrall, T. 'A report on the agriculture of Cumberland chiefly with regard to the production of meat', *JRASE*, second series, 10 (1874), 402–29.

Foot, P. *General View of the Agriculture of the County of Middlesex*, London, 1794.

Fraser, R. *General View of the County of Cornwall*, London, 1794.

General View of the County of Devon, London, 1794.

Fream, W. *Elements of Agriculture*, London, 1892.

French, H. F. *Farm Drainage*, New York, 1879.

Garnett, W. J. 'Farming of Lancashire', *JRASE*, 10 (1849), 1–51.

Gisborne, T. *Agricultural Drainage*, London, 2nd edn, 1852.

Glynn, J. *Draining Land by Steam Power*, London, 1838.

Gooch, W. *General View of the Agriculture of the County of Cambridge*, London, 1813.

Graham, Sir James 'On Deanstonizing, as distinguished from and compared with the furrow-draining and deep-ploughing of the midland counties of England', *JRASE*, 1 (1840), 29–33.

Granger, J. *General View of the Agriculture of the County of Durham*, London, 1794.

Grantham, R. B. *The Land Drainage Act, 1861; Its Provisions, Working and Results*, London, 1865.

Green, R. *An Address to the Nobility ... on the Subject of Under-Draining Wet and Cold Land by a Plough*, London, 1832.

Grey, J. 'A view of the past and present state of agriculture in Northumberland', *JRASE*, 2 (1841), 151–92.

Griggs, Messrs *General View of the Agriculture of the County of Essex*, London, 1794.

Hamond, A. 'Report on the exhibition of implements at the Lincoln meeting of the society', *JRASE*, 15 (1854), 363–79.

Hawes, S. 'Notes on the Wealden clay of Sussex and on its cultivation', *JRASE*, 19 (1858), 182–98.

Henriques, A. G. *On Some Legal and Economic Questions Connected with Land Credit and Mortgage Companies*, London, 1864.

Hillyard, C. *Practical Farming and Grazing*, Northampton, 3rd edn, 1840.

Hodges, T. L. *The Use and Advantages of Pearson's Draining Plough*, London, 1839.

'On the cheapest method of making and burning draining tiles', *JRASE*, 5 (1844), 551–9.

'On temporary tile kilns', *JRASE*, 9 (1848), 193–9.

Holland, H. *General View of the Agriculture of Cheshire*, London, 1808.

Holt, J. *General View of the Agriculture of the County of Lancaster*, London, 1794.

Hoskyns, C. W. *Talpa: Or the Chronicles of a Clay Farm*, London, 1852.

'On ridge and furrow pasture land and a method of levelling it', *JRASE*, 17 (1856), 327–31.

Hozier, W. W. *Practical Remarks on Agricultural Drainage*, Edinburgh, 1870.

Hunt, J. 'On the marquis of Tweeddale's tile making machine', *JRASE*, 2 (1841), 148–50.

Hutchinson, H. *A Treatise on the Practical Drainage of Land*, London, 1844.

Hutchinson, S. *Practical Instructions on the Drainage of Land*, Grantham, 1851.

Jonas, S. 'On the farming of Cambridgeshire', *JRASE*, 7 (1846), 35–72.

Johnston, J. F. W. *What Can Be Done for English Agriculture?*, Durham, 1842.

Johnstone, J. *An Account of the Mode of Draining Land According to the System Practised by Mr Joseph Elkington*, London, 2nd edn, 1801.

Karkeek, W. F. 'On the farming of Cornwall', *JRASE*, 6 (1845), 400–62.

Kennedy, L., and Grainger, T. B. *The Present State of the Tenancy of Land in Great Britain*, London, 1828.

Kent, N. *General View of the Agriculture of the County of Norfolk*, London, 1794.

Lavergne, L. de *The Rural Economy of England, Scotland and Ireland*, Edinburgh and London, 1855.

Lawrence, J. *The Modern Land Steward*, London, 1801.

Laws, P. *The Prize Essay of the Newcastle upon Tyne Farmers' Club on Draining Strong Clays*, Newcastle upon Tyne, 1850.

Leatham, I. *General View of the Agriculture in the East Riding of Yorkshire*, London, 1794.

Lefevre, C. S. *Remarks on the Present State of Agriculture*, London, 1836.

Legard, G. 'Farming of the East Riding of Yorkshire', *JRASE*, 9 (1848), 85–136.

Little, E. 'Farming of Wiltshire', *JRASE*, 5 (1844), 161–80.

Loch, J. *An Account of the Improvements on the Estates of the Marquess of Stafford in the Counties of Stafford and Salop and on the Estate of Sutherland*, London, 1820.

Loudon, J. C. *An Encyclopaedia of Agriculture*, London, 1825.

Low, D. *On Landed Property and the Economy of Estates*, London, 1844.

Lowe, R. *General View of the Agriculture of the County of Nottingham*, London, 1794 and 1798.

Macdonald, D. G. F. *Hints on Farming and Estate Management*, London, 5th edn, 1866.

Malcolm, W., J., and J. *General View of the Agriculture of the County of Buckingham*, London, 1794.

General View of Agriculture of the County of Surrey, London, 1794.

Marshall, W. *The Rural Economy of Norfolk*, 2 vols., London, 1787.

The Rural Economy of Yorkshire, 2 vols., London, 1788.

The Rural Economy of Gloucestershire, 2 vols., Gloucester, 1789.

The Rural Economy of Midland Counties, 2 vols., London, 1790.

The Rural Economy of the West of England, 2 vols., London, 1796.

The Rural Economy of the Southern Counties, 2 vols., London, 1798.

On the Landed Property of England, London, 1804.

The Review and Abstract of the County Reports to the Board of Agriculture, 5 vols., London and York, 1808–18.

Mavor, W. *General View of the Agriculture of Berkshire*, London, 1813.

Maxwell, G. *General View of the Agriculture of the County of Huntingdon*, London, 1793.

Mechi, J. J. *Mechi's Experience in Drainage*, London, 1847.

Middleton, J. *View of the Agriculture of Middlesex*, London, 2nd edn, 1807.

Milburn, M. M. 'On the farming of the North Riding of Yorkshire', *JRASE*, 9 (1848), 496–521.

Milward, R. 'Experiment on drainage at different depths', *JRASE*, 14 (1853), 210–11.

'On improving grass land', *JRASE*, 14 (1853), 430–1.

Mitchell, G. S. *A Handbook of Land Drainage*, London, 1898.

Monckton, G. *A Treatise on Deep Draining*, Wolverhampton, 1847.

Monk, J. *General View of the Agriculture of the County of Leicester*, London, 1794.

Moore, E. W. 'On the blocking up of drains by the roots of mangold', *JRASE*, 10 (1849), 622–3.

Morton, J. C. *The Cyclopedia of Agriculture*, Edinburgh and London, 1856.

'On the management of grass lands', *Journal of the Bath and West of England Society for the Encouragement of Agriculture*, 13 (1865), 61–73.

'Some of the agricultural lessons of 1868', *JRASE*, second series, 5 (1869), 27–73.

Morton, J. L. *The Resources of Estates*, London, 1858.

Moscrop, W. J. 'A report on the farming of Leicestershire', *JRASE*, second series, 2 (1866), 289–337.

Murray, A. *General View of the Agriculture of the County of Warwickshire*, London, 1815.

Murray, G. 'On the farming of Huntingdon' *JRASE*, second series, 4 (1868), 251–77.

Newman, C. *Practical Hints on Land Draining*, London 1845.

Palin, W. 'The farming of Cheshire', *JRASE*, 5 (1844), 57–111.

Parkes, J. 'Report on drain-tiles and drainage', *JRASE*, 4 (1843), 369–79.

'On the influence of water on the temperature of soils. On the quantity of rain-water and its discharge by drains', *JRASE*, 5 (1844), 119–58.

'Report on the exhibition of implements at the Southampton meeting in 1844', *JRASE*, 5 (1844), 361–91.

'On reducing the cost of permanent drainage', *JRASE*, 6 (1845), 125–9.

'Report on the exhibition of implements at the Shrewsbury meeting in 1845', *JRASE*, 6 (1845), 303–23.

'On draining', *JRASE*, 7 (1846), 249–72.

'Report on the exhibition of implements at the Newcastle upon Tyne meeting, 1846', *JRASE*, 7 (1846), 681–96.

'Report on the exhibition of implements at the Northampton meeting, 1847', *JRASE*, 8 (1847), 330–61.

Work on Draining, Worksop, 1847.

Essays on the Philosophy and Art of Land Drainage, London, 1848.

Fallacies on Land Drainage Exposed, London, 1851.

Parkinson, J. 'On improvements in agriculture in the county of Nottinghamshire since the year 1800' *JRASE*, 22 (1861), 159–66.

Parkinson, R. *General View of the Agriculture of the County of Rutland*, London, 1808.

General View of the Agriculture of the County of Huntingdon, London, 1813.

Pearce, W. *General View of the Agriculture in Berkshire*, London, 1794.

Peel, Sir Robert 'Account of a field thorough-drained at Drayton in Staffordshire', *JRASE*, 3 (1842), 18–21.

Pell, A. 'The making of the land in England: a retrospect', *JRASE*, second series, 23 (1887), 355–74.

Petre, Lord 'Concerning land-ditching', *Annals of Agriculture*, 4 (1785), 294–8.

Pidgeon, D. 'The evolution of agricultural implements', *JRASE*, third series, 3 (1892), 238–58.

Pitt, W. *General View of the Agriculture of the County of Stafford*, London, 1794 and 1813.

General View of the Agriculture of the County of Leicester, London, 1809.

General View of the Agriculture of the County of Northampton, London, 1813.

General View of the Agriculture of the County of Worcester, London, 1813.

Plymley, J. *General View of the Agriculture of Shropshire*, London, 1803.

Pomeroy, W. T. *General View of the Agriculture of the County of Worcester*, London, 1794.

Porter, G. R. *Progress of the Nation*, London, 3rd edn, 1851.

Portman, Lord 'On the stoppage of drains by a strong deposit', *JRASE*, 10 (1849), 119–21.

Priest, St John *General View of the Agriculture of Buckinghamshire*, London, 1813.

Pringle, A. *General View of the Agriculture of the County of Westmorland*, Edinburgh, 1794, and London, 1805.

Pusey, P. 'Some introductory remarks on the present state of agriculture as a science in England', *JRASE*, 1 (1840), 1–22.

'On the progress of agricultural knowledge during the last four years', *JRASE*, 3 (1842), 169–217.

'Evidence on the antiquity, cheapness and efficacy of thorough-draining or land-ditching, as practised throughout the counties of Suffolk, Hertford, Essex and Norfolk. With some notice of improved machines for tile-making', *JRASE*, 4 (1843), 23–49.

'On the agricultural improvements of Lincolnshire', *JRASE*, 4 (1843), 287–316.

'On cheapness of draining', *JRASE*, 7 (1846), 520–4.

'On the progress of agricultural knowledge during the last eight years', *JRASE*, 11 (1850), 381–443.

'Report to H.R.H. the President of the commission for the exhibition of the works of industry of all nations', *JRASE*, 12 (1851), 587–648.

Raynbird, H. 'On the farming of Suffolk', *JRASE*, 8 (1847), 261–329.

Read, C. S. 'On the farming of Oxfordshire', *JRASE*, 15 (1854), 189–276.

'Report on the farming of Buckinghamshire', *JRASE*, 16 (1855), 269–322.

'Recent improvements in Norfolk farming', *JRASE*, 19 (1858), 265–311.

Read, J. 'On pipe-tiles', *JRASE*, 4 (1843), 273–4.

Rennie, G., Brown, R., and Shirreff, J. *General View of the Agriculture of the West Riding of Yorkshire*, London, 1794.

Roland, A. *Farming for Pleasure and Profit*, London, 1880.

Rowlandson, T. 'Farming of Herefordshire', *JRASE*, 14 (1853), 433–56.

Rowley, J. J. 'The farming of Derbyshire', *JRASE*, 14 (1853), 17–66.

Rudge, T. *General View of the Agriculture of the County of Gloucester*, London, 1807.

Ruegg, L. H. 'Farming of Dorsetshire', *JRASE*, 15 (1854), 389–454.

Sinclair, Sir John 'Report of the committee of the Board of Agriculture respecting Mr Elkington's mode of draining', *Annals of Agriculture*, 24 (1795), 525–9.

The Code of Agriculture, London, 1817.

Smith, H. H. *The Principles of Landed Estate Management*, London, 1898.

Smith, J. *Remarks on Thorough Draining and Deep Ploughing*, Stirling, 1831.

Smith, R. 'The management of grass land', *JRASE*, 9 (1848), 1–22.

Smith, W. *Observations on the Utility, Form and Management of Water Meadows and the Draining and Irrigating of Peat Bogs*, Norwich, 1806.

Spearing, J. B. 'On the agriculture of Berkshire', *JRASE*, 21 (1860), 1–46.

Squarey, E. P. 'Farm capital', *JRASE*, second series, 14 (1878), 425–44.

Stephens, G. *The Practical Irrigator and Drainer*, Edinburgh, 1834.

Stephens, H. *The Book of the Farm*, 3 vols., Edinburgh, 1844.

A Manual of Practical Draining, London, 3rd edn, 1848.

Stephenson, C. *Essay on the Farming of Tyneside*, Hexham, 1861.

Stevenson, W. *General View of the Agriculture of the County of Surrey*, London, 1813.

General View of the Agriculture of the County of Dorset, London, 1815.

Stone, T. *General View of the Agriculture of the County of Bedford*, London, 1794.

General View of the Agriculture of the County of Lincoln, London, 1794.

Strickland, H. E. *General View of the Agriculture of the East Riding of Yorkshire*, York, 1812.

Tanner, H. 'The farming of Devonshire', *JRASE*, 9 (1848), 454–95.

'The agriculture of Shropshire', *JRASE*, 19 (1858), 1–64.

Thompson, H. S. 'On subsoil ploughing', *JRASE*, 2 (1841), 26–37.

'Report on the exhibition and trial of implements at the York meeting, 1848', *JRASE*, 9 (1848), 377–422.

'Report on the exhibition and trial of implements at the Norwich meeting, 1849', *JRASE*, 10 (1849), 526–70.

'Report on the exhibition and trial of implements at the Lewes meeting, 1852', *JRASE*, 13 (1852), 301–46.

Thurlow, J. 'On the mole-plough', *Annals of Agriculture*, 43 (1805), 486–8.

Tiffen, J. H. 'Prize essay on the agriculture of East and North Ridings of Yorkshire', *North British Agriculturalist*, 29 October 1884.

Tooke, T., and Newmarch, W. *A History of Prices*, vols. 5 and 6, London, 1857.

Trimmer, J. *On the Improvement of Land as an Investment for Capital*, London, 1847.

'Notes on the geology of the Keythorpe estate and its relations to the Keythorpe system of draining', *JRASE*, 14 (1853), 96–105.

Tuke, J. *General View of the Agriculture of the North Riding of Yorkshire*, London, 1794 and 1800.

Turner, G. *General View of the Agriculture of the County of Gloucester*, London, 1794.

Turner, N. *An Essay on Draining and Improving Peat Bogs*, London, 1784.

Vancouver, C. *General View of the Agriculture in the County of Cambridge*, London, 1794.

General View of the Agriculture in the County of Essex, London, 1795.

General View of the Agriculture of the County of Devon, London, 1813.

General View of the Agriculture of Hampshire, London, 1813.

Walker, D. *General View of the Agriculture of the County of Hertford*, London, 1795.

Webster, C. 'On the farming of Westmorland', *JRASE*, second series, 4 (1868), 1–37.

Webster, W. B. 'On the failure of deep draining on certain strong clay subsoils...', *JRASE*, 9 (1848), 237–48.

'On the mischief arising from draining certain clay soils too deeply', *JRASE*, 11 (1850), 311–13.

Wedge, J. *General View of the Agriculture of the County of Warwick*, London, 1794.

Wedge, T. *General View of the Agriculture of the County Palatine of Chester*, London, 1794.

Western, Baron *Practical Remarks on the Improvement of Grass Land by Means of Irrigation, Winter Flooding and Drainage*, London, 1838.

Wharncliffe, Lord 'On draining under certain conditions of soil and climate', *JRASE*, 12 (1851), 41–62.

White, R. 'Report of several operations in thorough-draining and subsoil-ploughing at Oakley Park', *JRASE*, 1 (1840), 33–7.

'Second report of several operations in thorough-draining and subsoil-ploughing at Oakley Park', *JRASE*, 1 (1840), 248–52.

'Report of results obtained in thorough-draining and subsoil-ploughing in the years 1840 and 1841', *JRASE*, 2 (1841), 346–53.

'Report to the Honourable Robert Clive M.P. on his improvements by draining and subsoil-ploughing', *JRASE*, 4 (1843), 172–6.

'Report to the Honourable Robert Henry Clive of his Poles farm improvements, effected by thorough-draining', *JRASE*, 6 (1845), 229–36.

Wiggins, J. 'On the mode of making and using tiles for underdraining practised on the Stow Hall estate in Norfolk', *JRASE*, 1 (1840), 350–6.

Wilkinson, J. 'The farming of Hampshire', *JRASE*, 22 (1861), 239–371.

Williams, G. M. 'On the tenant's right to unexhausted improvements, according to the custom of north Lincolnshire', *JRASE*, 6 (1845), 44–6.

Worgan, G. B. *General View of the Agriculture of the County of Cornwall*, London, 1811.

Wright, W. 'On the improvements in the farming of Yorkshire since the date of the last reports in the journal', *JRASE*, 22 (1861), 87–131.

Young, A. *General View of the Agriculture of the County of Suffolk*, London, 1794 and 1804.

General View of the Agriculture of the County of Lincoln, London, 1799.

General View of the Agriculture of Hertfordshire, London, 1804.

General View of the Agriculture of the County of Norfolk, London, 1804.

'Mole plough drawn by the force of women applied mechanically', *Annals of Agriculture*, 42 (1804), 413–22.

General View of the Agriculture of the County of Essex, 2 vols., London, 1807.

General View of the Agriculture of Oxfordshire, London, 1809.

Young, Rev. A. *General View of the Agriculture of the County of Sussex*, London, 1793 and 1808.

Yule, J. 'An account of the mode of draining by means of tiles as practised on the estate of Netherby in Cumberland, the property of Sir James Graham', *Prize Essays and Transactions of the Highland Society of Scotland*, new series, 1 (1829), 388–400.

TWENTIETH-CENTURY BOOKS, ARTICLES AND THESES

Adkin, B. W. *Land Drainage in Britain*, London, 1933.

Ayres, Q. C., and Scoates, D. *Land Drainage and Reclamation*, New York, 2nd edn, 1939.

Avery, B. W., Findlay, D. C., and Mackney, D. *Soil Map of England and Wales, 1:1,000,000,* Southampton, 1975.

Beastall, T. W. *A North Country Estate*, Chichester, 1975.

The Agricultural Revolution in Lincolnshire, Lincoln, 1978.

'Landlords and tenants', in G. E. Mingay (ed.), *The Victorian Countryside*, London, 1981, vol. 2, 428–38.

Beckett, J. V., *The Aristocracy in England 1660–1914*, Oxford, 1986.

Brown, J. H. 'Agriculture in Lincolnshire during the great depression', unpublished University of Manchester Ph.D. thesis (1978).

Burke's Landed Gentry, London, various editions.

Burke's Peerage, Baronetage and Knightage, London, various editions.

Butlin, R. A. *The Transformation of Rural England c. 1580–1800*, Oxford, 1982.

Cannadine, D. 'Aristocratic indebtedness in the nineteenth century: the case reopened', *EcHR*, second series, 30 (1977), 624–50.

Chambers, J. D., and Mingay, G. E. *The Agricultural Revolution 1750–1880*, London, 1966.

Clapham, J. H. *An Economic History of Modern Britain: 1, the Early Railway Age, 1820–50*, Cambridge, 1926.

An Economic History of Modern Britain: 2, Free Trade and Steel, 1850–1886, Cambridge, 1932.

Clay, C. 'Lifeleasehold in the western counties of England 1650–1750', *AHR*, 29 (1981), 83–96.

Clayton, C. H. J. *Land Drainage from Field to Sea*, London, 1919.

Clemenson, H. A. *English Country Houses and Landed Estates*, London, 1982.

Cokayne, G. E. *The Complete Peerage*, 13 vols., London, 1910–59.

Cole, G. 'Land drainage in England and Wales', *Journal of the Institute of Water Engineers and Scientists*, 30 (1976), 345–61.

Collins, E. J. T. 'The age of machinery', in G. E. Mingay (ed.), *The Victorian Countryside*, London, 1981, vol. 1, 200–13.

Collins, E. J. T., and Jones, E. L. 'Sectoral advance in English agriculture, 1850–80', *AHR*, 15 (1967), 65–81.

Coppock, J. T. 'The changing face of England: 1850–*circa* 1900', in H. C. Darby (ed.), *A New Historical Geography of England*, Cambridge, 1973, 595–673.

Crosby, T. L. *Sir Robert Peel's Administration 1841–1846*, Newton Abbot, 1976.

English Farmers and the Politics of Protection 1815–1852, Hassocks, 1977.

Darby, H. C. 'The draining of the English clay-lands', *Geographische Zeitschrift*, 52 (1964), 190–201.

The Changing Fenland, Cambridge, 1983.

Daubney, R. 'The influence of good drainage in relation to certain parasitic diseases of stock', *Journal of the Ministry of Agriculture*, 31 (1924), 616–21.

David, P. A. *Technical Choice, Innovation and Economic Growth*, Cambridge, 1975.

Davies, C. S. *The Agricultural History of Cheshire 1750–1850*, Manchester, 1960.

Dittmer, B. R. 'An agricultural geography of northwest Wiltshire, 1773–1840', unpublished University of London M.A. thesis (1963).

Dodd, J. P. 'High farming in Shropshire 1845–1870', *Midland History*, 8 (1983), 148–68.

Edwards, R., and Perren, R. 'A note on regional differences in British meat prices, 1828–1865', *Economy and History*, 22 (1979), 123–34.

English, B. 'On the eve of the Great Depression: the economy of the Sledmere estate 1869–1878', *Business History*, 24 (1982), 24–47.

'Patterns of estate management in east Yorkshire c. 1840–c. 1880', *AHR*, 32 (1984), 29–48.

English, B., and Saville, J. *Strict Settlement. A Guide for Historians*, Hull, 1983.

Erickson, A. B. 'Sir James Graham, agricultural reformer', *AH*, 24 (1950), 170–4.

Ernle, Lord *English Farming Past and Present*, London, 6th edn, 1961.

Fairlie, S. 'The nineteenth-century Corn Law reconsidered', *EcHR*, second series, 18 (1965), 562–75.

'The Corn Laws and British wheat production, 1829–76', *EcHR*, second series, 22 (1969), 88–116.

Farrant, S. 'The management of four estates in the lower Ouse valley (Sussex) and agricultural change, 1840–1920', *Southern History*, 1 (1979), 155–70.

Feinstein, C. H. *National Income, Expenditure and Output of the United Kingdom 1855–1965*, Cambridge, 1972.

'Capital formation in Great Britain', in P. Mathias and M. Postan (eds.), *Cambridge Economic History of Europe*, Cambridge, 1978, vol. 7, 28–96.

Fisher, J. R. 'The economic effects of cattle disease in Britain and its containment, 1850–1900', *AH.*, 54 (1980), 278–93.

'Landowners and English tenant right, 1845–1852', *AHR*, 31 (1983), 15–25.

Fletcher, T. W. 'The great depression of English agriculture 1873–1896', *EcHR*, second series, 13 (1960), 417–32.

'Lancashire livestock farming during the Great Depression', *AHR*, 9 (1961), 17–42.

'The agrarian revolution in arable Lancashire', *Transactions of the Lancashire and Cheshire Antiquarian Society*, 72 (1962), 93–122.

Found, W. C., Hill, A. R., and Spence, E. S. *Economic and Environmental Impacts of Land Drainage in Ontario*, Toronto, 1974.

Fussell, G. E. 'The evolution of field drainage', *Journal of the Bath and West and Southern Counties Society*, sixth series, 4 (1929–30), 59–72.

'English agriculture from Cobbett to Caird (1830–80)', *EcHR*, 15 (1945), 79–85.

'Home counties farming, 1840–80', *Economic Journal*, 57 (1947), 321–45.

'The dawn of high farming in England', *AH*, 22 (1948), 83–95.

'High farming in southwestern England, 1840–1880', *Economic Geography*, 24 (1948), 53–73.

'High farming in the north of England, 1840–1880', *Economic Geography*, 24 (1948), 296–310.

'High farming in the west midland counties, 1840–1880', *Economic Geography*, 25 (1949), 159–79.

'High farming in the east midlands and East Anglia', *Economic Geography*, 27 (1951), 72–89.

The Farmers' Tools, London, 1952.

'The evolution of farm implements: 4, field drainage', *Journal of the Chartered Land Agents' Society*, 58 (1959), 361–72.

Garnier, R. M. *History of the English Landed Interest*, 2 vols., London, 2nd edn, 1908.

Gaut, R. C. *A History of Worcestershire Agriculture and Rural Evolution*, Worcester, 1939.

Glasspoole, J. 'Two centuries of rain', *Meteorological Magazine*, 63 (1928), 1–6.

Goddard, N. 'The development and influence of agricultural periodicals and newspapers, 1780–1880', *AHR*, 31 (1983), 116–31.

Green, F. H. W. 'Aspects of the changing environment: some factors affecting the aquatic environment in recent years', *Journal of Environmental Management*, 1 (1973), 377–91.

'Ridge and furrow, mole and tile', *Geographical Journal*, 141 (1975), 88–93.

'Recent changes in land use and treatment', *Geographical Journal*, 142 (1976), 12–26.

'Current trends in field drainage practice', *Journal of Environmental Management*, 5 (1977), 207–13.

'Field, forest and hill drainage in Scotland', *Scottish Geographical Magazine*, 95 (1979), 159–64.

'Field under-drainage before and after 1940, *AHR*, 28 (1980), 120–3.

'Quantification of areas of agricultural and forestry drainage, as they affect extrapolation of the results of representative and experimental basins', in *The Influence of Man on the Hydrological Regime: Proceedings of the Helsinki Symposium, 1980*, 385–8.

Grigg, D. B. 'The development of tenant right in south Lincolnshire', *Lincolnshire Historian*, 2 (1962), 41–8.

'A note on agricultural rent and expenditure in nineteenth-century England', *AH*, 39 (1965), 147–54.

'An index of regional change in English farming', *Transactions of the Institute of British Geographers*, 36 (1965), 55–67.

The Agricultural Revolution in South Lincolnshire, Cambridge, 1966.

The Dynamics of Agricultural Change, London, 1982.

Haggard, H. R. *Rural England*, 2 vols., London, 1902.

Hall, A. D. *A Pilgrimage of British Farming 1910–1912*, London, 1913.

Harley, J. B. 'England *circa* 1850', in H. C. Darby (ed.), *A New Historical Geography of England*, Cambridge, 1973, 527–94.

Harris, A. *The Rural Landscape of the East Riding of Yorkshire 1700–1850*, Oxford, 1961.

'Changes in the early railway age: 1800–1850', in H. C. Darby (ed.), *A New Historical Geography of England*, Cambridge, 1973, 465–526.

Harrison, M. J., Mead, W. R., and Pannett, D. J. 'A midland ridge-and-furrow map', *Geographical Journal*, 131 (1965), 366–9.

Healy, M. J. R., and Jones, E. L. 'Wheat yields in England, 1815–59', *Journal of the Royal Statistical Society*, series A, 125 (1962), 574–9.

Hoelscher, L. 'Improvements in fencing and drainage in mid-nineteenth-century England', *AH*, 37 (1963), 75–9.

Holderness, B. A. 'Capital formation in agriculture', in J. P. P. Higgins and S. Pollard (eds.), *Aspects of Capital Investment in Great Britain 1750–1850*, London, 1971, 159–83.

'Landlord's capital formation in East Anglia 1750–1870', *EcHR*, second series, 25 (1972), 434–47.

'Agriculture and industrialization in the Victorian economy', in G. E. Mingay (ed.), *The Victorian Countryside*, London, 1981, vol. 1, 179–99.

'The Victorian farmer', in G. E. Mingay (ed.), *The Victorian Countryside*, London, 1981, vol. 1, 277–44.

Holt, H. M. E. 'Upland farming in northern England, *circa* 1840 to *circa* 1880: some evidence from Cumbria and Northumberland', unpublished University of Exeter Ph.D. thesis (1985).

Homer, S. *A History of Interest Rates*, New Brunswick, New Jersey, 1963.

Hooper, S. G. *The Finance of Farming in Great Britain*, London, 1955.

Hoskins, W. G., and Stamp, L. D. *The Common Lands*, London, 1963.

Hueckel, G. 'Agriculture during industrialisation', in R. Floud and D. McCloskey (eds.), *The Economic History of Britain since 1700*, Cambridge, 1981, vol. 1, 182–203.

Hughes, M. 'Lead, land and coal as sources of landlord income in Northumberland between 1700 and 1850', unpublished University of Durham Ph.D. thesis (2 vols., 1963).

Hunt, E. H. 'Labour productivity in English agriculture, 1850–1914', *EcHR*, second series, 20 (1967), 280–92.

Regional Wage Variations in Britain 1850–1914, Oxford, 1973.

Jones, E. L. 'The changing basis of English agricultural prosperity, 1853–73', *AHR*, 10 (1962), 102–19.

'English farming before and during the nineteenth century', *EcHR*, second series, 15 (1962), 145–52.

Seasons and Prices, London, 1964.

(ed.), *Agriculture and Economic Growth in England 1650–1815*, London, 1967.

The Development of English Agriculture 1815–1873, London, 1968.

Agriculture and the Industrial Revolution, Oxford, 1974.

'The environment and the economy', in P. Burke (ed.), *New Cambridge Modern History*, Cambridge, 1978, vol. 13, 15–42.

'Agriculture 1700–80', in R. Floud and D. McCloskey (eds.), *The Economic History of Britain since 1700*, Cambridge, 1981, vol. 1, 66–86.

Jones, G. E. 'The diffusion of agricultural innovations', *Journal of Agricultural Economics*, 15 (1962), 387–409.

Kain, R. J. P. *An Atlas and Index of the Tithe Files of Mid-Nineteenth-Century England and Wales*, Cambridge, 1986.

Kain, R. J. P., and Mead, W. R. 'Ridge-and-furrow in Cambridgeshire', *Proceedings of the Cambridgeshire Antiquarian Society*, 67 (1977), 131–7.

Kain, R.J.P., and Prince, H.C. *The Tithe Surveys of England and Wales*, Cambridge, 1985.

Kendall, R. G. *Land Drainage*, London, 1950.

Kerridge, E. 'Ridge and furrow and agrarian history', *EcHR*, second series, 4 (1951), 14–36.

The Agricultural Revolution, London, 1967.

'The agricultural revolution reconsidered', *AH*, 43 (1969), 463–75.

Land Utilization Survey *Land Classification, 1:625,000*, Southampton, 1945.

Lane, M. R. 'John Fowler and the company he founded', *Steaming*, 23 (1980), 73–87, 145–53.

Livesley, M. C. *Field Drainage*, London, 1960.

Lowerson, J. R. 'Enclosure and farm buildings in Brackley, 1829–51', *Northamptonshire Past and Present*, 6 (1978), 33–48.

Macdonald, S. 'The development of agriculture and the diffusion of agricultural innovation in Northumberland, 1750–1850', unpublished University of Newcastle Ph.D. thesis (1974).

'The diffusion of knowledge among Northumberland farmers, 1780–1815', *AHR*, 27 (1979), 30–9.

'Agricultural response to a changing market during the Napoleonic Wars', *EcHR*, second series, 33 (1980), 59–71.

'Agricultural improvement and the neglected labourer', *AHR*, 31 (1983), 81–90.

McGregor, O. R. 'Introduction: after 1815', in Lord Ernle, *English Farming Past and Present*, London, 6th edn, 1961, lxxix–cxlv.

Mackney, D. 'Soil maps and classification', in A. J. Thomasson (ed.), *Soils and Field Drainage*, Harpenden, 1975, 35–48.

McQuiston, J. R. 'Tenant right: farmers against landlord in Victorian England, 1847–1883', *AH*, 47 (1973), 95–113.

Martins, S. Wade *A Great Estate at Work*, Cambridge, 1980.

Mead, W. R. 'Ridge and furrow in Buckinghamshire', *Geographical Journal*, 120 (1954), 34–42.

'A ridge-and-furrow map of Leicestershire and Northamptonshire', *East Midland Geographer*, 6 (1977), 382–5.

Mead, W. R., and Kain, R. J. P. 'Ridge-and-furrow in Kent', *Archaeologia Cantiana*, 92 (1976–7), 165–71.

Michie, R. 'Income, expenditure and investment of a Victorian millionaire: Lord Overstone, 1823–83', *Bulletin of the Institute of Historical Research*, 68 (1985), 59–77.

Mingay, G. E. 'The agricultural revolution in English history: a reconsideration', *AH*, 37 (1963), 123–33.

Arthur Young and his Times, London, 1975.

The Gentry, London, 1976.

The Agricultural Revolution: Changes in Agriculture 1650–1880, London, 1977.

Ministry of Agriculture, Fisheries and Food, *Old Underdrainage Systems*, Pinner, 1973.

History of Agricultural Drainage, Ministry of Agriculture, Land Drainage Service, 1977.

Underdrainage Information Sheet for Draining Advisers' Reports, Ref. CG4, rev. 1977.

Ministry of Town and Country Planning, *Rainfall: Annual Average 1881–1915, 1:625,000*, Southampton, 1949.

Mitchell, B. R., and Deane, P. *Abstract of British Historical Statistics*, Cambridge, 1962.

Molland, R. 'Agriculture *c.* 1793–*c.* 1870', in *Victoria County History of Wiltshire*, London, 1959, vol. 4, 65–91.

Moore, D. C. 'The Corn Laws and high farming', *EcHR*, second series, 18 (1965), 544–61.

'The landed aristocracy', in G. E. Mingay (ed.), *The Victorian countryside*, London, 1981, vol. 2, 367–82.

Morgan, R. 'The root-crop in English agriculture, 1650–1870', unpublished University of Reading Ph.D. thesis (1978).

Nicholas, J., and Glasspoole, J. 'General monthly rainfall over England and Wales, 1727 to 1931', *British Rainfall*, 71 (1931), 299–306.

Nicholson, H. H. *The Principles of Field Drainage*, Cambridge, 1946.

'Field drainage and increased production', *JRASE*, 109 (1948), 212–21.

North East Development Association *A Physical Land Classification of Northumberland, Durham and part of the North Riding of Yorkshire*, Newcastle upon Tyne, 1950.

Obelkevich, J. *Religion and Rural Society: South Lindsey, 1825–1875*, Oxford, 1976.

Offer, A. 'Ricardo's paradox and the movement of rents in England, *c.* 1870–1910', *EcHR*, second series, 33 (1980), 236–52.

O'Grada, C. 'The landlord and agricultural transformation, 1870–1900: a comment on Richard Perren's hypothesis', *AHR*, 27 (1979), 40–2.

'Agricultural decline 1860–1914', in R. Floud and D. McCloskey (eds.), *The Economic History of Britain since 1700*, Cambridge, 1981, vol. 2, 175–97.

Orwin, C. S., and Whetham, E. H. *History of British Agriculture 1846–1914*, London, 1964.

Overton, M. 'Agricultural revolution? Development of the agrarian economy in early modern England', in A. R. H. Baker and D. Gregory (eds.), *Explorations in Historical Geography: Interpretative Essays*, Cambridge, 1984, 118–39.

Parker, R. A. C. *Coke of Norfolk*, Oxford, 1975.

Parton, A. G. 'Town and country in Surrey *c.* 1800–1870: a study in historical geography', unpublished University of Hull Ph.D. thesis (1973).

Perkins, J. A. 'Tenure, tenant right and agricultural progress in Lindsey, 1750–1850', *AHR*, 23 (1975), 1–22.

'The prosperity of farming on the Lindsey uplands, 1813–37', *AHR*, 24 (1976), 126–43.

Perren, R. 'The landlord and agricultural transformation, 1870–1900', *AHR*, 18 (1970), 36–51.

Perry, P. J. 'Where was the great agricultural depression? A geography of agricultural bankruptcy in late Victorian England and Wales', *AHR*, 29 (1972), 30–45.

British Farming in the Great Depression 1870–1914, Newton Abbot, 1974.

'High farming in Victorian Britain: the financial foundations', *AH*, 52 (1978), 364–79.

'High farming in Victorian Britain: prospect and restrospect', *AH*, 55 (1981), 156–66.

Phillips, A. D. M. 'Underdraining and the English claylands, 1850–80: a review', *AHR*, 17 (1969), 44–55.

'The development of underdraining on a Yorkshire estate during the nineteenth century, *Yorkshire Archaeological Journal*, 44 (1972), 195–206.

'A study of farming practices and soil types in Staffordshire around 1840', *North Staffordshire Journal of Field Studies*, 13 (1973), 27–52.

'Underdraining and agricultural investment in the midlands in the mid-nineteenth century', in A. D. M. Phillips and B. J. Turton (eds.), *Environment, Man and Economic Change*, London, 1975, 253–74.

'Agricultural land use, soils and the Nottinghamshire tithe surveys, *circa* 1840', *East Midland Geographer*, 6 (1976), 284–301.

'The landlord and agricultural improvement: underdraining on the Lincolnshire estate of the earls of Scarbrough in the first half of the nineteenth century', *East Midland Geographer*, 7 (1979), 168–77.

'Agricultural improvement on a Durham estate in the nineteenth century: the Lumley estate of the earls of Scarbrough', *Durham University Journal*, new series, 42 (1981), 161–8.

'Agricultural land use on a Northamptonshire estate (1849–99) as revealed by cropping books', *East Midland Geographer*, 8 (1983), 70–8.

'Agricultural land use and cropping in Cheshire around 1840: some evidence from cropping books', *Transactions of the Lancashire and Cheshire Antiquarian Society*, 84 (1987), 46–63.

Phillips, A. D. M., and Clout, H. D. 'Underdraining in France during the second half of the nineteenth century', *Transactions of the Institute of British Geographers*, 51 (1970), 71–94.

Pickersgill, A. C. 'The agricultural revolution in Bassetlaw, Nottinghamshire, 1750–1873', unpublished University of Nottingham Ph.D. thesis (1979).

Porter, R. E. 'Agricultural change in Cheshire during the nineteenth century', unpublished University of Liverpool Ph.D. thesis (1974).

Prince, H. C. 'England *circa* 1800', in H. C. Darby (ed.), *A New Historical Geography of England*, Cambridge, 1973, 389–464.

'Victorian rural landscapes', in G. E. Mingay (ed.), *The Victorian Countryside*, London, 1981, vol. 1, 17–29.

Richards, E. 'Leviathan of wealth: west midland agriculture, 1800–50', *AHR*, 22 (1974), 97–117.

'The land agent', in G. E. Mingay (ed.), *The Victorian Countryside*, London, 1981, vol. 2, 439–56.

Richardson, H. G., and Fussell, G. E. 'The beginnings of field drainage', *Journal of the Ministry of Agriculture*, 29 (1922), 585–91.

Riches, N. *The Agricultural Revolution in Norfolk*, London, 2nd edn., 1967.

Robinson, M. 'The extent of farm underdrainage in England and Wales, prior to 1939', *AHR*, 34 (1986), 79–85.

Royal Commission on Historical Monuments, *Northamptonshire: An Archaeological Atlas,* London, 1980.

Royal Meterological Society, *Rainfall Atlas of the British Isles*, London, 1926.

Sheppard, J. A. *The Draining of the Hull Valley*, East Yorkshire Local History Society, 1958.

The Draining of the Marshlands of South Holderness and the Vale of York, East Yorkshire Local History Society, 1966.

Short, B. M. 'Agriculture in the High Weald of Kent and Sussex 1850 to 1953', unpublished University of London Ph.D. thesis (1973).

Siegel, S. *Nonparametric Statistics for the Behavioural Sciences*, New York, 1956.

Smith, W. *An Historical Introduction to the Economic Geography of Great Britain*, London, 1968 reprint.

Soil Survey of England and Wales, *Soil Map of England and Wales, 1:250,000,* Southampton, 1983.

Legend for the 1:250,000 Soil Map of England and Wales, Harpenden, 1983.

Spring, D. 'Earl Fitzwilliam and the Corn Laws', *American Historical Review*, 59 (1953), 287–304.

'A great agricultural estate: Netherby under Sir James Graham, 1820–1845', *AH*, 29 (1955), 73–81.

'English landownership in the nineteenth century: a critical note', *EcHR*, second series, 9 (1956), 472–84.

The English Landed Estate in the Nineteenth Century: Its Administration, Baltimore, 1963.

'English landed society in the eighteenth and nineteenth centuries', *EcHR*, second series, 17 (1964), 146–53.

'Lord Chandos and the farmers, 1816–1846', *Huntingdon Library Quarterly*, 33 (1969), 257–81.

'Land and politics in Edwardian England', *AH*, 58 (1984), 17–42.

Spring, E. 'The settlement of land in nineteenth-century England', *American Journal of Legal History*, 8 (1964), 209–23.

'Landowners, lawyers and land law reform in nineteenth-century England', *American Journal of Legal History*, 21 (1977), 40–59.

'The family, strict settlement and historians', *Canadian Journal of History*, 18 (1983), 379–98.

Stamp, J. C. *British Incomes and Property*, London, 1916.

Stamp, L. D. *The Land of Britain: Its Use and Misuse*, London, 3rd edn, 1962.

Stanes, R. 'Landlord and tenant and husbandry covenants in eighteenth-century Devon', in W. Minchinton (ed.), *Agricultural Improvement: Medieval and Modern*, University of Exeter, 1981, 41–64.

Sturgess, R. W. 'The response of agriculture in Staffordshire to the price changes of the nineteenth century', unpublished University of Manchester Ph.D. thesis (1965).

'The agricultural revolution on the English clays', *AHR*, 14 (1966), 104–21.

'The agricultural revolution on the English clays: a rejoinder', *AHR*, 15 (1967), 82–7.

Supple, B. *The Royal Exchange Assurance*, Cambridge, 1970.

Sutton, J. E. G. 'Ridge and furrow in Berkshire and Oxfordshire', *Oxoniensia*, 29–30 (1964–5), 99–115.

Taylor, D. 'The English dairy industry, 1860–1930', *EcHR*, second series, 29 (1976), 585–601.

Thirsk, J. *English Peasant Farming*, London, 1957.

England's Agricultural Regions and Agrarian History, 1500–1750, Basingstoke, 1987.

Thirsk, J., and Imray, J. *Suffolk Farming in the Nineteenth Century*, Suffolk Records Society, 1958.

Thomasson, A. J. 'Other site factors: climate and land use', in A. J. Thomasson (ed.), *Soils and Field Drainage*, Harpenden, 1975, 30–4.

Thomasson, A. J., and Trafford, B. D. 'Introduction', in A. J. Thomasson (ed.), *Soils and Field Drainage*, Harpenden, 1975, 1–4.

Thompson, F. M. L. 'The end of a great estate', *EcHR*, second series, 8 (1955), 36–52.

'The economic and social background of the English landed interest, 1840–70, with particular reference to the estates of the duke of Northumberland', unpublished University of Oxford D. Phil. thesis (1956).

'English landownership: the Ailesbury trust, 1832–56', *EcHR*, second series, 11 (1958), 121–32.

'Agriculture since 1870', in *Victoria County History of Wiltshire*, London, 1959, vol. 4, 92–114.

'English great estates in the nineteenth century, 1790–1914', in *Contributions to the First International Conference of Economic History*, Paris, 1960, 385–97.

English Landed Society in the Nineteenth Century, London, 1963.

'Land and politics in England in the nineteenth century', *Transactions of the Royal Historical Society*, fifth series, 15 (1965), 23–44.

'The social distribution of landed property in England since the sixteenth century', *EcHR*, second series, 19 (1966), 505–17.

'The second agricultural revolution, 1815–1880', *EcHR*, second series, 21 (1968), 62–77.

'Free trade and the land', in G. E. Mingay (ed.), *The Victorian Countryside*, London, 1981, vol. 1, 103–17.

Thompson, R. J. 'An enquiry into the rent of agricultural land in England and Wales during the nineteenth century', *Journal of the Royal Statistical Society*, 70 (1907), reprinted in W. E. Minchinton (ed.), *Essays in Agrarian History*, Newton Abbot, 1968, vol. 2, 55–86.

Trafford, B. D. 'Field drainage', *JRASE*, 131 (1970), 129–52.

'Farm drainage', *Journal of the Royal Society of Arts*, (vol. for 1973), 134–45.

'Drainage design', in A. J. Thomasson (ed.), *Soils and Field Drainage*, Harpenden, 1975, 5–17.

'Recent progress in field drainage', *JRASE*, 138 (1977), 27–42.

Trafford, B. D., and Walpole, R. A. 'Drainage design in relation to soil series', in A. J. Thomasson (ed.), *Soils and Field Drainage*, Harpenden, 1975, 49–61.

Turner, M. *English Parliamentary Enclosure*, Folkestone, 1980.

Enclosures in Britain 1750–1830, London, 1984.

Vamplew, W. 'The cost of best practice in the mid-nineteenth century', *Tools and Tillage*, 3 (1980), 204–14.

Venn, J. A. *The Foundations of Agricultural Economics*, Cambridge, 2nd edn, 1933.

Walton, J. R. 'Aspects of agrarian change in Oxfordshire, 1750–1880', unpublished University of Oxford D. Phil. thesis (1976).

'Agriculture 1730–1900', in R. A. Dodgshon and R. A. Butlin (eds.), *An Historical Geography of England and Wales*, London, 1978, 239–65.

'Mechanization in agriculture: a study of the adoption process', in H. S. A. Fox and R. A. Butlin (eds.), *Change in the Countryside*, London, 1979, 23–42.

Ward, J. T. 'The earls Fitzwilliam and the Wentworth Woodhouse estate in the nineteenth century', *Yorkshire Bulletin of Economic and Social Research*, 12 (1960), 19–27.

'A nineteenth-century Yorkshire estate: Ribston and the Dent family', *Yorkshire Archaeological Journal*, 41 (1963), 43–51.

East Yorkshire Landed Estates in the Nineteenth Century, East Yorkshire Local History Society, 1967.

Sir James Graham, London, 1967.

Wasson, E. A. 'The third earl Spencer and agriculture 1818–1845', *AHR*, 26 (1978), 89–99.

Weaver, M. M. *History of Tile Drainage in America prior to 1900*, Waterloo, New York, 1964.

Whetham, E. H. 'Sectoral advance in English agriculture 1850–80: a summary', *AHR*, 16 (1968), 46–8.

Wilkes, A. R. 'Adjustments in arable farming after the Napoleonic Wars', *AHR*, 28 (1980), 90–113.

Williams, M. *The Draining of the Somerset Levels*, Cambridge, 1970.

Woodward, D. 'Agricultural revolution in England 1500–1900: a survey', *Local Historian*, 9 (1971), 323–33.

Wyndham, H. 'The farming activities of the third earl Spencer', *Northamptonshire Past and Present*, 3 (1961), 40–8.

Index